THERMAL ASPECTS
OF FLUID FILM
TRIBOLOGY

RELATED TITLES FROM ASME PRESS

Handbook of Friction Units of Machines, by I. V. Kragel'skii and N. M. Mikhin, 1988, ISBN 0-7918-0001-6

Failure Atlas for Hertz Contact Machine Elements, by T. E. Tallian, 1991, ISBN 0-7918-0008-3

THERMAL ASPECTS
OF FLUID FILM
TRIBOLOGY

Oscar Pinkus
Sigma Tribology Consultants
Sandia Park, New Mexico

ASME PRESS   NEW YORK   1990

Library of Congress Cataloging-in-Publication Data
Pinkus, Oscar.
    Thermal aspects of fluid film tribology / Oscar Pinkus.
      p. cm.
    Includes index.
    ISBN 0-7918-0011-3
    1. Tribology.   2. Fluid-film bearings—Thermal properties.
I. Title.
TJ1075.P53   1990
621.8′9—dc20   90-661
                                          CIP

TABLE OF CONTENTS

Foreword ix
Nomenclature xi

Part I. THERMOHYDRODYNAMICS 1

1.0 PERVASIVENESS OF THERMAL EFFECTS 3
   1.1 ANALYTICAL RAMIFICATIONS 4
      1.1.1 Impact on Calculated Results 4
      1.1.2 Computational Procedures 12
      1.1.3 Boundary Conditions 13
      1.1.4 Gas Lubrication 15
      1.1.5 Thermal Parameters 18
   1.2 EXPERIMENTAL FIELD 24
      1.2.1 Viscosity 25
      1.2.2 Transverse Measurements 25
      1.2.3 Methodology 27
   1.3 SPECIAL COMPLEXITIES 28
      1.3.1 Non-Isotropic Films 28
      1.3.2 Reverse Flows 30
      1.3.3 Rheodynamic Models 30
      1.3.4 System Interaction 31

2.0 APPROACHES TO THE ADIABATIC PROBLEM 35
   2.1 THE ENERGY EQUATION 35
   2.2 METHOD OF EFFECTIVE VISCOSITY 37
   2.3 THE COUETTE APPROXIMATION 41
   2.4 THE TWO-DIMENSIONAL SOLUTION 48
      2.4.1 Journal Bearings 51
      2.4.2 Thrust Bearings 53
      2.4.3 Pumping Rings 60

3.0 TRANSVERSE TEMPERATURE VARIATIONS 79
   3.1 THE BASIC DIFFERENTIAL EQUATIONS 80
      3.1.1 The Energy Equation 80
      3.1.2 Compressibility Effects 85
      3.1.3 The Reynolds Equation 86
   3.2 INFINITE PARALLEL SURFACES 87
   3.3 HYDROSTATIC FILMS 92

3.4 HYDRODYNAMIC SLIDERS 96
   3.4.1 Parallel Films 96
   3.4.2 Tapered Films 101

4.0 HEAT TRANSFER SOLUTIONS 121
4.1 THE HEAT TRANSFER EQUATIONS 121
4.2 SLIDERS 126
   4.2.1 Infinitely Long Sliders 126
   4.2.2 Finite Sliders 131
4.3 JOURNAL BEARINGS 147
   4.3.1 The Couette Approximation 149
   4.3.2 The Integral Method 155
   4.3.3 The (x, y) Solution 157
   4.3.4 Three-Dimensional Solutions 161
4.4 THRUST BEARINGS 172
   4.4.1 The Governing Equations 173
   4.4.2 Tapered Land Bearings 175
   4.4.3 Crowned Pads 176
   4.4.4 Pivoted Pads 180

5.0 BOUNDARY CONDITIONS AND SYSTEM INTERACTION 187
5.1 INLET CONDITIONS 188
   5.1.1 The Nature of Inlet Mixing 188
   5.1.2 The Mixing Coefficient 194
   5.1.3 Reverse Flow 197
   5.1.4 Groove Parameters 203
   5.1.5 Starvation 203
5.2 RUNNER TEMPERATURE 207
   5.2.1 Constancy of Runner Temperature 208
   5.2.2 Heat Flow in Shaft 212
5.3 THE BEARING SURFACE 216
5.4 THE EDGE CONDITIONS 218
5.5 THE CAVITATION REGION 220
   5.5.1 Streamlets Cavitation 220
   5.5.2 Possible Modes of Cavitation 221

Part II. THE THERMAL WORKSHOP 231

6.0 TURBULENCE 233
6.1 TURBULENCE AND THE FLUID FILM 233
6.2 TURBULENCE AND THD ANALYSIS 236
6.3 THE TRANSITION REGIME 240
6.4 SLIDERS 243
6.5 JOURNAL BEARINGS 248
6.6 THRUST BEARINGS 253

7.0 ELASTIC AND THERMAL DISTORTIONS 261
   7.1 ELLIPTICAL CONTACTS 263
      7.1.1 Quasi-Hertzian Pressure Field 263
      7.1.2 Thin Non-Hertzian Contacts 266
      7.1.3 EHD Films with Full Energy Equation 279
      7.1.4 General Elliptical Contacts 277
   7.2 INFINITELY LONG ROLLERS 277
   7.3 SLIDERS 284
   7.4 THRUST BEARINGS 294
      7.4.1 Line Supported Pad 294
      7.4.2 Effect of Support Geometry 301
   7.5 JOURNAL BEARINGS 304
      7.5.1 Tilting Pad Bearings Including Deformations ... 307
      7.5.2 Thermal Growth of Bearing and Journal 309
      7.5.3 Shaft Seizure 311

8.0 TWO-PHASE REGIMES 317
   8.1 LIQUID-SOLID PHASES 319
   8.2 LIQUID-GAS MIXTURES 322
   8.3 CHANGE OF PHASE 326
      8.3.1 General Characteristics 326
      8.3.2 Isothermal and Adiabatic Cases 328
      8.3.3 Heat Transfer Effects 334
   8.4 MELT LUBRICATION 340
   8.5 CAVITATION 351

9.0 TIME-DEPENDENT FILMS 357
   9.1 SQUEEZE FILMS 357
   9.2 TRANSIENT PROCESSES 366
      9.2.1 Thermal Transients in Sliders 366
      9.2.2 Thermal Transients in Journal Bearings 373
      9.2.3 Thermoelastic Effects 379
      9.2.4 Experimental Correlations 382
   9.3 NON-NEWTONIAN FLUIDS 386
      9.3.1 Traction under Thermal Conditions 387
      9.3.2 Elasto-Plastic Model 391

10.0 INSTRUMENTATION TECHNIQUES 403
   10.1 FILM THICKNESS MEASUREMENTS 404
      10.1.1 Microtransducers 404
      10.1.2 Optical Interferometry 407
      10.1.3 Laser Fluorescence Technique 413
   10.2 TEMPERATURE MEASUREMENTS 415
      10.2.1 Special Thermocouples 415
      10.2.2 Bisignal Transducer 420
      10.2.3 Infrared Radiation Emission Technique 422
   10.3 OTHER MEASUREMENTS 426

      10.3.1 Implanted Radiation Method 426

      10.3.2 Flow Visualization 429

11.0 THE NEEDS OF TECHNOLOGY 433

  11.1 ADVANCED TRIBOSYSTEMS 434

  11.2 GAPS AND NEEDS 438

  11.3 CONCLUSIONS AND RECOMMENDATIONS

     OF THE THERMAL WORKSHOP 440

APPENDIX A: Workshop Specifics 443

APPENDIX B: Thermophysical Properties 455

APPENDIX C: Bibliography 471

AUTHORS INDEX 497

SUBJECT INDEX 501

FOREWORD

Thermal phenomena are pervasive in tribology. They are synonymous with friction and viscous dissipation. The resistance to motion, that is friction, is among other things the result of converting mechanical work into thermal energy which, while being dissipated, gives rise to temperature increases in tribo-elements. This is the case whether we are concerned with boundary friction, where the surfaces of the bodies actually touch, or with hydrodynamic lubrication, where the viscous energy dissipation is distributed throughout the volume of the lubricant. In either case, the temperature increases, leading to changes in the properties of the tribo-materials. These property changes affect the tribo-element and, when large enough, can greatly complicate our attempts to predict performance with engineering confidence.

Historically tribo-engineering has, by and large, developed successfully ignoring thermal effects. Isothermal models, in and of themselves or when tempered with engineering judgement, have been adequate to predict performance. However, as advanced mechanical system development has evolved, requirements have become more severe and thermal effects must be dealt with in a more rigorous manner than was necessary in the past. Not only must the nonlinear effects associated with material property changes be addressed, but the system thermal effects which couple the fluid film with the overall mechanical system and its environment also must be taken into account. Perhaps, not surprisingly, these additional complications have made it difficult to provide clear guidance for the design engineer.

The current difficulties in thermal analysis can be traced to several generic causes. One is the problem of extracting from the plethora of variables and boundary conditions inherent in the thermal problem those few parameters which embody the essentials of the process. Another factor is the difficulty of manipulating the complex set of governing differential equations—the Reynolds and energy equations, and other auxiliary expressions of material property equations and boundary conditions—so as to produce nonisothermal solutions simple enough for the engineer to use. And lastly, there is the issue of formulating the problem such that the solutions will not be limited to a specific configuration but will have some general validity.

In order to deal with the above issues, a workshop organized by myself and Hooshang Heshmat was held at Georgia Tech in May 1988 on the subject of thermal problems in tribology. Approved by the ASME Research Committee on Tribology and sponsored by the U. S. Air Force and the National Science Foundation, the workshop objectives were "to identify the sources of confusion in the field... and to formulate steps to clarify them..." No formal papers were presented, the aim of the working sessions being to discuss and elucidate the nature of the problem and to chart guidelines for future theoretical and experimental work. The workshop program is presented in Appendix A.

This monograph is an outgrowth of the workshop. However, while utilizing the material discussed at the workshop, the book also draws on previously published works to offer a synthesis of the widely scattered material available on the subject. In intent and content, the book has a three-fold objective. A major goal is to present a unified treatment of the subject of thermal phenomena in lubrication, bringing together in a systematic way those parts of the knowledge which have been established as the foundation of thermohydrodynamic analysis. Along with the theory the book also offers solutions to thermohydrodynamic problems in bearings, seals, gears, and similar devices which can be used with a fair degree of reliability. Next, by comparing the various analytical results and examining these in the light of available experimental data, an attempt is made to show to what extent there is quantitative or even qualitative agreement between the various approaches. Finally, by drawing on the conclusions and recommendations of the workshop, the book outlines suggested directions for future analytical and experimental activity to fill the gaps and resolve uncertainties in the body of knowledge on thermal phenomena in tribology.

This book would not have come into being without the efforts of the participants, sponsors, and organizers of the workshop. The individual contributions of the various researchers are cited throughout the text, and a list of the workshop attendees is given in Appendix A. Special acknowledgement is due to B. D. McConnell of the U. S. Air Force and to E. L. Marsh of the National Science Foundation for providing technical and financial support; to Georgia Institute of Technology for hosting the workshop; to J. Yang, a Ph.D. student in tribology at Georgia Tech who prepared the electronic version of all the equations and the camera-ready copy of the text; to Kaye Fuller, my secretary, who input all of the text and tables; and to Mechanical Technology, Inc. and the Woodruff School Drafting Office, which prepared the book's numerous figures. We gratefully acknowledge our indebtedness to those persons and to the support of the George W. Woodruff School of Mechanical Engineering at Georgia Institute of Technology, without which we might not have been able to complete this project. We also express appreciation to the ASME Press, especially in the person of Caryl Dreiblatt whose encouragement and persistent urgings helped complete the publication of the book.

Atlanta, Georgia   Ward O. Winer
August 1990

NOMENCLATURE

Unless otherwise noted, the following symbols are used through the text.

A	Area
B	Damping coefficient; Extent of slider
$\overline{B}$	$(\pi B)/\mu_{0,1}L(R/C)^3$
C	Radial clearance
D	Diameter of bearing or journal
E	Young's modulus; Energy;
	$(2\mu_{0,1}\alpha\omega/cw)(R/C)^2$, adiabatic constant for journal bearings
	$(2\mu_{0,1}\alpha\omega/cw)(R_2/h_2)^2$, adiabatic constant for thrust bearings
Ec	Eckert number, see Table 1.4
F_r	Frictional force
Fo	Fourier number, see Table 1.4
G	Turbulence coefficients; Mass flow rate
H	Power dissipation
$\overline{H}$	$(H/\mu_{0,1}N^2DL^2)(C/R)$, for journal bearings
	$(H/\mu_{0,1}\omega^2 R_2^3)(h_2/R_2)$, for thrust bearings
J	Mechanical equivalent of heat
K	Constant in Equ 2.21
K	Spring coefficient
$\overline{K}$	$K/2\mu_{0,1}NL(R/C)^3$
L	Width of bearing in the z direction for journal bearings
	Width of bearing in the r direction for thrust bearings
M	Moment
$\overline{M}$	$M/\mu_{0,1}NDL^2(R/C)$
M_{CR}	Critical mass for journal bearing stability
$\overline{M}_{CR}$	$M_{CR}N/2\mu_{0,1}L(R/C)^3$
N	Revolutions per unit time
Nu	Nusselt number, see Table 1.4
P	(W/LD), in journal bearings
	(W/A), in thrust bearings
Pe	Peclet number, see Table 1.4
Pe^*	$Pe(h_2/B)$, for sliders
	$Pe(C/R)$, for journal bearings
Pr	Prandtl number, see Table 1.4
Q	Volumetric lubricant flow; Rate of heat flow
Q_s	Side leakage of lubricant
$\hat{Q}_s$	Starvation index, see Sec 5.2.2

$\overline{Q}$	$Q/(\pi/2)(NDLC)$, in journal bearings
	$Q/\omega h_2 R^2$, in thrust bearings
R	Radius of bearing or journal
$\mathcal{R}$	Perfect gas constant
R_1	Inner radius
R_2	Outer radius
Re	$(\rho R\omega h/\mu)$, Reynolds number in journal bearings
	$(\rho r\omega h/\mu)$, Reynolds number in thrust bearings
Re^*	$Re(C/R)$; $Re(h_2/R_2)$
$Re_{0,1}$	$(\rho R\omega C/\mu_{0,1})$, inlet or reference Reynolds number for journal bearings
	$(\rho R_2\omega h_2/\mu_{0,1})$, inlet or reference Reynolds number for thrust bearings
S	Sommerfeld number $(LD\mu_{0,1}N/W)(R/C)^2 = 1/\overline{W}$
T	Temperature
ΔT	Temperature rise, $(T - T_{0,1})$
$\overline{T}$	$\alpha\Delta T$
ΔT_{max}	$(T_{max} - T_{0,1})$
$\overline{\overline{T}}$	$(T/T_{0,1})$
$\Delta\overline{\overline{T}}$	$\left(\frac{T-T_0}{T_0}\right)$
$\hat{T}$	$\frac{\rho c_p(T-T_{0,1})}{\mu_{0,1}\omega(R/C)^2}$, for journal bearings
	$\frac{\rho c_p(T-T_{0,1})}{\mu_{0,1}\omega(R_2/h_2)^2}$, for thrust bearings
$\tilde{T}$	$\rho c_p h_2^2(T-T_{0,1})/6\mu_{0,1}UB$
U	Circumferential velocity (in x direction)
V	Normal velocity (in y direction); Volume
W	Load
W_{CR}	Critical load for journal bearing stability
$\overline{W}_{CR}$	$(W_{CR}/W)(C\omega^2/g)$
$\overline{W}$	$(W/LD\mu_{0,1}N)/(C/R)^2$, in journal bearings
	$(W/R_2^2\mu_{0,1}\omega)(h_2/R_2)^2$, in thrust bearings
a	Major axis
b	Minor axis; Tapered fraction of pad
c	Concentration fraction; Specific heat
c_p	Specific heat at constant pressure
c_v	Specific heat at constant volume
d	Preload or Ellipticity
e	Eccentricity; Energy
f	Friction coefficient, (F_τ/W)
g	Gravitational constant
h	Film thickness; Coefficient of heat transfer
$\overline{h}$	(h/C), (h/h_2)
$\overline{h}_1$	(h_1/h_2)
h_c	Film at center of axes; Coefficient of surface convection
i	Enthalpy

k	Thermal conductivity
l	Length
m	Preload ratio, (d/C)
p	Pressure
q	Rate of heat transfer; (Q/L)
r	Radial coordinate
$\bar{r}$	(r/R_2)
s	Slip, $(U_2 - U_1)$
t	Thickness of pad or runner; Time
u	Longitudinal velocity
$\bar{u}$	(u/U)
v	Transverse velocity
w	Lubricant specific weight; Axial velocity
x, y, z	Rectangular coordinates
$\bar{x}$	(x/B)
$\bar{y}$	(y/h)
$\bar{z}$	$[z/(L/2)]$

α	Angular misalignment; Temperature-viscosity coefficient
β	Pressure viscosity coefficient; Angular extent of bearing arc
δ	Displacement; Distortion; Thermal compressibility; Particle size
δ_{12}	Radial taper, $(h_{10} - h_{12})$
$\bar{\delta}_{12}$	(δ_{12}/h_2)
ϵ_m	Eddy diffusivity
ϵ	Eccentricity ratio, (e/C)
θ	Angular coordinate
θ_S	Start of bearing pad
θ_E	End of bearing pad
θ_1	start of hydrodynamic film
θ_2	End of hydrodynamic film
λ	Mixing coefficient; Quality
μ	Lubricant viscosity
μ_0	Supply viscosity
μ_1	Mixing viscosity
$\bar{\mu}$	$\mu/\mu_{0,1}$
ν	Poisson's ratio; (μ/ρ)
ρ_g	Specific gravity
ρ	Density of lubricant
σ	Diffusivity; Surface tension
τ	Shear stress; ωt
ϕ	Attitude angle; Dissipation
ϕ_L	Load angle
ω	Angular frequency
ω_i	Threshold instability frequency

$\overline{\omega}_i$	(ω_i/ω)
Λ	Relaxation time

Subscripts

a	Ambient
av	Average
c	Center of pressure, cavitation
e	Effective, equivalent
E	End
f	Liquid
fg	Vaporization, saturation
G	Groove
g	Gas, vapor
i	Initial
m	Mean
max	Maximum
min	Minimum
p	Pivot
R	Runner
s	Stationary, bearing side, average
S	Start
w	Wall
x	In the x direction
y	In the y direction
z	In the z direction
θ	In the θ direction
xx	Force in the x direction due to a displacement in the x direction
xy	Force in the x direction due to a displacement in the y direction
yx	Force in the y direction due to a displacement in the x direction
yy	Force in the y direction due to a displacement in the y direction
0	Supply, inlet
1	Mixing inlet

Superscripts

$(-)$	To the left of

(+)	To the right of
·	First derivative
¯	Second derivative
(t)	Time dependent
(0)	At t=0

Abbreviations

EHD	Elastohydrodynamics
HT	Heat transfer
J	Joules
Pa	Pascals
PTFE	Polytetrafluoroethylene
THD	Thermohydrodynamics
TW	Thermal Workshop
N	Newtons
cps	Centipoises
cks	Centistokes
l	litres
ml	milliliter
m	Meter
mil	10^{-3} inch
μm	Micron, 10^{-6} m

PART I

THERMOHYDRODYNAMICS

Chapter 1

PERVASIVENESS OF THERMAL EFFECTS

The hydrodynamic branch of tribology can be subdivided into a number of areas which from an engineering standpoint at least can be delineated with a fair degree of certainty. Phenomena such as compressible films, elastohydrodynamics, viscoelasticity, turbulence, instability, and others are all confined to specific regimes of operation and appropriate criteria exist to delimit the range of parameters within which each is likely to occur. Not so with the thermal dimension. In the domain of fluid film lubrication there is practically no analysis, experiment, or application which does not involve variations in the viscosity of the fluid film. In most cases this variation exerts a profound effect on the process of lubrication. Furthermore, any change in dimensions or operating conditions involves immediately a reorientation of the viscosity field. This pervasiveness is not in the nature of a second order effect or a mere perturbation but very frequently its impact exceeds the effects of the primary variable being considered and can even reverse results obtained on the basis of isothermal analysis.

Variations in the temperature field are due to several causes. First is the heat generated by viscous dissipation, and it is worth pointing out that a given percentage rise in temperature will often cause a higher percentage change in viscosity. The other mechanism is heat transfer to and from the fluid film. If, for example, the shaft carries a cryogenic compressor, heat would be extracted from the film; if it carries a turbine driven by superheated steam, heat might be pumped into the film. Thus in addition to the heat convected by the lubricant, conduction via the runner,* bearing, bearing supports, and various other heat sources and sinks all interact to produce a variable temperature field. The third mechanism which af-

---

* Here and subsequently the term runner denotes the surface travelling relative to the pressure-temperature field of the fluid film, applicable to both thrust and journal bearings and other smilar tribological devices.

fects the thermal map is the mixing inlet temperature. Bearings and seals usually consist of a number of pads arrayed in tandem. The hot fluid discharging from an upstream element mixes with the cold lubricant delivered from an outside source and this mixing is reflected in the inlet boundary temperature. Other tribosystems will have additional heat sources, such as Coulomb friction in seals and pumping rings. There is hardly a lubrication process that does not involve one or more of the above heat generating mechanisms.

Interest in the thermal aspects of tribology can be ascribed to their importance in the following general areas:

[a] Effect on performance – film thickness, power, loss, flow rates, etc.
[b] Magnitude and location of the maximum film temperature
[c] Effects of thermal gradients on the geometry of the fluid film, that is thermal distortion.
[d] Heat flow to and from components of the system.

Interest in the above results is not uniform. It can, for example, be said that item [d] is of concern only in rare cases whereas the other items are all significant. Aside from determining performance, thermal behavior affects bearing reliability and can cause outright failure. This it can do via excessive temperatures leading to the degradation of the bearing material or lubricant; via steep temperature gradients that may cause cracking or, in bearing parlance, ratcheting; and via unequal thermal growth of components which can cause seizure. Fig. 1a shows a bearing that failed because of high temperatures, while Fig. 1b shows a surface that cracked as a consequence of high thermal gradients. A seizure caused by the unequal thermal expansion of bearing and journal during a rapid start up is discussed in Chapter 7 (see Fig. 7.44).

A rigorous solution of the thermo-hydrodynamic (THD) conditions prevailing in the fluid film requires an effort of a very high order. The question that the investigator is often confronted with is not only whether a solution is possible, but what is the correspondence between the effort invested in analysis and the results it may produce. In the thermal field, a vast exertion may yield only a small refinement on solutions obtained by simple, if inexact, methods. One of the considerations, therefore, in solving thermal problems is an economy of approach vis-a-vis the likely results.

1.1 ANALYTICAL RAMIFICATIONS

1.1.1 Impact on Calculated Results

A straightforward isothermal solution normally yields performance on the basis of the lubricant supply viscosity μ_0. The discrepancy between such a set of results and those obtained from an anisothermal analysis will vary with a change in the geometrical and operational parameters. This point

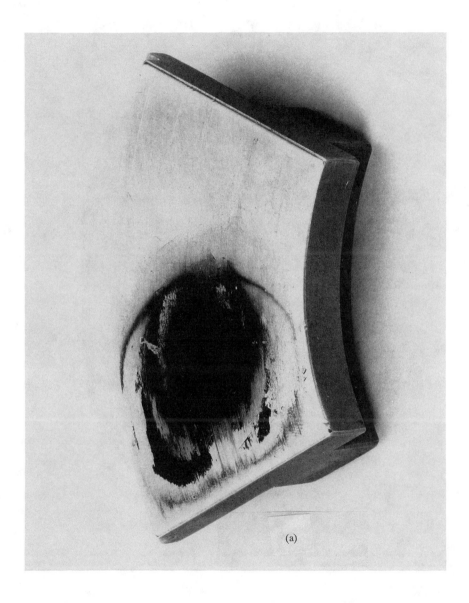

(a)

Fig. 1.1 Photographs of bearing failures due to thermal causes (Kingsbury, Inc), (a) thrust pad failure due to excessive temperatures, (b) ratcheting of thrust bearing due to steep temperature gradients

Fig. 1.1 Photographs of bearing failures due to thermal causes (Kingsbury, Inc), (a) thrust pad failure due to excessive temperatures, (b) ratcheting of thrust bearing due to steep temperature gradients

TABLE 1.1

Effect of viscosity variation in journal bearing

Variable		ϵ		HP		K_{xx} (lb/in.) x 10^{-6}		Thermal Variables	
		$\mu=\mu_0$	$\mu=\mu_e$	$\mu=\mu_0$	$\mu=\mu_e$	$\mu=\mu_0$	$\mu=\mu_e$	ΔT, °F	μ_e, csks
C in.	0.003	.287	.598	14.7	4.34	2.54	4.28	72	9.97
	0.0045*	.469	.681	10.5	4.69	2.38	3.59	50	14.4
	0.006	.593	.739	8.86	4.87	2.38	3.25	38	17.7
L in.	2	680	.822	8.69	3.91	4.39	6.95	53	13.5
	3*	469	.681	10.5	4.69	2.38	3.59	50	14.4
	4	297	.549	13.1	5.19	1.71	2.33	51	14.0
	6	131	.372	19.9	5.82	1.48	1.39	62	11.7
N rpm	2,000	.595	.740	3.66	2.02	3.24	4.35	38	17.8
	3,600*	.469	.681	10.5	4.69	2.38	3.59	50	14.4
	5,000	.394	.647	19.4	7.48	2.02	3.22	58	12.5
	10,000	.245	.577	75.3	19.6	1.63	2.66	80	8.97
ϕ_L	-30°	.533	.782	11.6	3.39	4.33	6.49	83	8.51
	0°*	.469	.681	10.5	4.69	2.38	3.59	50	14.4
	30°	.566	.653	10.2	5.08	1.43	1.72	35	18.8
	45°	.761	.889	12.6	7.00	1.96	2.00	36	18.5

Standard conditions: D x L x C = 6 in. x 3 in. x 0.0045 in.

ϕ_L = 0, W = 4,000 lbs; N = 3600 rpm; T_0 = 100°F @ μ_0 = 41 csks;

ρ_g = 0.87; β = 150°

*Reference conditions

can be illustrated by subjecting a standard two-axial groove bearing to a variation in its basic parameters—clearance, length, speed, or load angle. Table 1.1 shows two sets of solutions: one on the basis of inlet viscosity μ_0; the other based on an average or effective viscosity. A scrutiny of this table

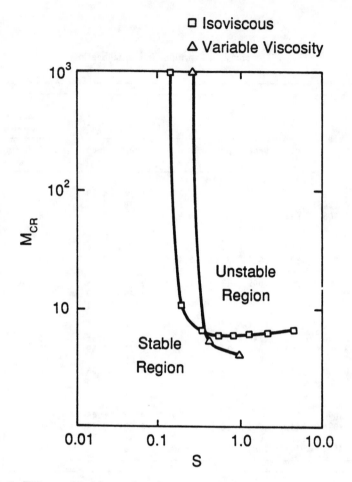

Fig. 1.2 Effects of isothermal and anisothermal solutions on journal bearing stability (Craighead, 1980)

shows that the increase in ϵ due to μ_e, as compared to μ_0, can be more than double; that the effect on power loss can be a reduction to less than one third; and that in the case of K_{xx} the effect can be either an increase or a decrease in the colinear stiffness. This inversion of variable viscosity effect on the dynamic characteristics of journal bearings is corroborated by an example taken from Craighead (1980) and illustrated in Fig. 1.2.

The effective viscosity μ_e is based on an average of flow exiting the bearing at various local temperatures, or

$$\Delta T_e = (T_e - T_0) = \frac{\int_0^B q(T - T_0)d\xi}{Q}$$

where $\overline{OB}$ represents the periphery along which lubricant leaves the clearance space with q and T functions of some coordinate ξ. The μ_e based on this T_e offers a simple and extremely useful tool for obtaining performance characteristics. In fact, this method is often superior to more sophisticated, and certainly more laborious, THD approaches, for reasons that will be explained in Chapter 4. However, this method of effective temperature fails to yield one of the most important tribological quantities, namely the maximum film temperature. The importance of knowing the value of the film maximum temperature, T_{max}, is often on a par with that of h_{min}. While too small an h_{min} can cause damage by physical contact, excessive temperatures can cause failure by softening or melting the surfaces which can occur even when there is an ample film thickness. And ΔT_e is no indicator of the magnitude of T_{max}. As an illustration consider Fig. 1.3 representing two different temperature profiles. While their effective temperatures T_e would be the same, the two temperature distributions have radically different values of T_{max}. Effective viscosity remains a useful strategy for calculating performance but it cannot provide the crucial quantity T_{max}.

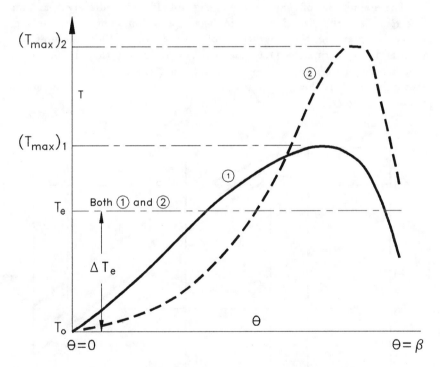

Fig. 1.3 Relationship between effective and maximum temperatures

In addition to the direct effect on performance, variable temperatures also exert an indirect impact on the behavior of the fluid film. Most notable among these is turbulence. The Reynolds and power loss equations accounting for both laminar and turbulent operation can be written as

$$\frac{\partial}{\partial x}\left[G_x\frac{h^3}{\mu}\left(\frac{\partial p}{\partial x}\right)\right] + \frac{\partial}{\partial z}\left[G_z\frac{h^3}{\mu}\left(\frac{\partial p}{\partial z}\right)\right] = 6U\left(\frac{\partial h}{\partial x}\right) + 12\left(\frac{\partial h}{\partial t}\right) \qquad (1.1)$$

$$Q = \int_{-L/2}^{L/2}\left[\frac{Uh}{2} - \frac{G_x h^3}{12\mu}\left(\frac{\partial p}{\partial x}\right)\right]dz \qquad (1.2)$$

$$H = U\int_{\theta_1}^{\theta_2}\int_{-L/2}^{L/2}\left(\frac{\mu U}{hG_\tau} + \frac{h}{2}\frac{\partial p}{\partial x}\right)dz\,dx \qquad (1.3)$$

where G_x, G_z, G_τ are the turbulence coefficients. These depend on the Reynolds number

$$Re \equiv \frac{\rho r \omega h}{\mu}$$

The magnitudes of G are shown in Fig. 1.4. It is immediately clear that since $Re \sim \mu^{-1}$ the viscosity variation exerts a profound effect on the level of turbulence and on the magnitudes of the G factors. While the presence of G_x and G_z in equation (1.1) does not reveal directly their influence on the pressure, the G_τ factor, given the fact that the first right hand term in equation (1.3) is the dominant one, yields a power loss almost inversely proportional to G_τ.

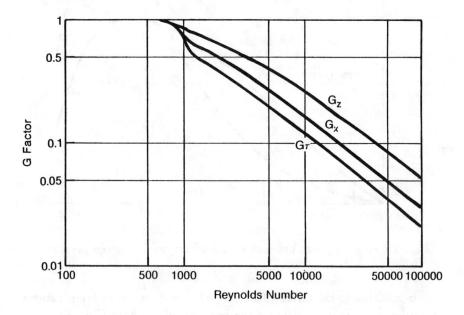

Fig. 1.4 Values of the linearized turbulence coefficients

In fact the influence of variable viscosity on turbulence is quite complex, as can be seen from the following:

- Load Capacity - In general, turbulence can be said to increase load capacity, that is, for a given load, turbulent conditions will produce a higher film thickness. However, the boost in load capacity is not as much as is usually portrayed in isothermal solutions because turbulence is accompanied by higher losses and thus lower operating viscosities.

- Power Loss - Losses go up appreciably with a rise in turbulence. Here, too, there exists an interlocking effect of turbulence and viscosity. While higher turbulence lowers the viscosity due to heating, which yields lower power loss, the lower viscosity simultaneously raises the Reynolds number and intensity of turbulence. Still, the net effect is an increase in power loss above laminar.

- Maximum Temperature - Turbulence always produces higher temperatures, including higher maximum temperatures.

- Heat Transfer - In those cases where film coefficients are employed to account for heat transfer between dissimilar bodies, these coefficients are substantially higher for turbulent films, that is, heat transfer rates are enhanced. In turn this will affect the resulting viscosity field in the fluid films.

- Stability - It can be surmised that the effect of turbulence on stability is bound to be deleterious. This is so because the film thicknesses are higher, the viscosities lower, and the locus of shaft center has a larger horizontal component (higher attitude angles).

Another effect worth mentioning is that variations in the temperature field generate thermal gradients. These may cause thermoelastic distortions in the geometry of the film so as to affect its hydrodynamic operation. Often this can be of a fundamental nature. It is an old paradox in the field of lubrication that parallel surfaces and centrally pivoted pads, which by the very principles of hydrodynamic theory ought to be outright failures, operate without particular difficulty. Theories such as the "density wedge," relying on the compressibility of the fluid, or the "thermal wedge," relying on the variation of viscosity, have been propounded to explain the generation of hydrodynamic pressures in parallel films. But these have by now been sufficiently discounted and, even if not, these theories lead to the generation of pressures nowhere near those encountered in reality. What may, however, explain the operation of initially parallel films is crowning of the surfaces due to transverse thermal gradients. Fig. 1.5 shows results for a typical babbitt-steel construction in which the babbitt deformed due to a combination of a bimetallic effect and calculated transverse temperature variation. The deflections are seen to range from $\frac{1}{4}$ to $\frac{1}{2}$ mil which are of the same order as the film thickness and are thus capable of imparting a proper geometric wedge on the originally flat surface.

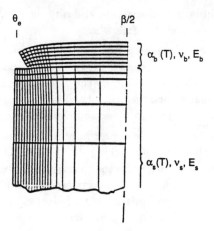

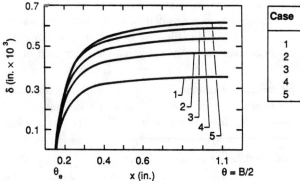

Case	T_0	T_1	T_{max}
1	167	185	194
2	194	210	231
3	210	232	253
4	222	237	278
5	225	242	270

Subscripts:

b ≡ babbitt

s ≡ steel

Fig. 1.5 Thermal crowning of an initially flat surface (Heshmat, 1987)

1.1.2 Computational Procedures

All of the previously discussed items can be viewed to be the result of physical factors. There are occasions, however, when the difficulties are ones of methodology. These are often due to the fact that, as pointed out in the introductory remarks, THD analysis has not as yet established a sufficiently clear correlation between its methods and the validity of the sought after results. A case in point are the questions arising in the calculation of the dynamic coefficients K_{ij} and B_{ij}. The value of K, for example, is obtained from the relationship

$$K = \lim_{\Delta e \to 0} \left(\frac{\Delta W}{\Delta e} \right)$$

Now one might think that when a small perturbation is used it would be permissible to ignore the variation of viscosity with Δe. However, since

$W = W(e, \mu)$, we have

$$\frac{dW}{de} = \left(\frac{\partial W}{\partial e}\right) + \left(\frac{\partial W}{\partial \mu}\right)\left(\frac{\partial \mu}{\partial e}\right)$$

and since Δe is small, the last term cannot be automatically neglected. This is illustrated in Table 1.2. In those cases where under the heading (dT/de) the column reads "Yes," it means that, for any perturbation Δe, a corresponding ΔT and $\Delta \mu$ were calculated for the process; where the column reads "No," it means that during the perturbation the temperature was kept at its original value. The table includes two methods of solution, one using the full energy equation, the other employing the Couette approximation (explained in Chapter 2). The striking errors, often as much as 100%, that can result from neglecting the perturbed change in ΔT, attest to the care that has to be taken not only in the use of approximate equations, but also in the method of solution of the full energy equation.

1.1.3 Boundary Conditions

The complexities in formulating the thermal equations are paralleled and often exceeded by difficulties in ascribing boundary conditions for the solution of these equations. Generally, the boundary temperatures of the fluid film are not simply boundary conditions, but rather solutions of the thermal problem of the entire system. Nor are these conditions fixed. For example, experiments on journal bearings indicate that at low speeds a significant portion of the viscous heat is conducted away through the shaft since the time available for relaxing temperature differences is long and the heat generated is small compared to the thermal capacity of the shaft. At high velocities, however, heat production in the film is large in comparison with the heat capacity of the shaft, and the time available to absorb this energy is short. Under these conditions, one would expect approximately an adiabatic runner surface.

Another common complication is that of lubricant mixing at the entrance, as shown in Fig. 1.6. The general impact of this phenomenon will be treated more fully later on; here only its formal presence as a boundary condition will be mentioned. The runner leaving an upstream bearing or seal element carries with it an adhering layer of hot lubricant which, along with new cold lubricant, is dragged into the downstream pad. Thus, the inlet temperature of the fluid is neither known, nor necessarily constant across and along the leading edge. The initial statement that the boundary conditions often constitute the essence of the problem can be visualized from the consequences of the mixing inlet temperature proposition. If we take a journal bearing running at $\epsilon = 0$, or nearly so, then there is no side leakage and a given amount of lubricant recirculates ad infinitum in the clearance space. This would mean that with time the temperature of the lubricant would reach infinity, a thing that in practice never happens. It

TABLE 1.2

Effects of method of solution in starved journal bearings

$$L/D = 1, \ \overline{W} = 10, \ \overline{Q}_0 = 0.1, \ T_0 = 120°F, \ SAE \ 20$$

	Full energy equation			Couette approximation			
	(dT/dε)			(dT/dε)			
	Yes	No		Yes		No	
	Value*	Value	%Δ+	Value	%Δ	Value	%Δ
ϵ	0.924			0.923	-0.1		
ϕ	9.9°			9.6°	-3		
$\overline{Q}_z$	0.0106			0.0100	-5.7		
$\overline{T}_s$	0.2623			0.2871	9.5		
$\overline{T}_{max}$	0.8182			0.9208	12.5		
θ_1	166°			167°	0.6		
θ_2	198°			199°	0.5		
$\overline{M}$	4.885			5.11	10.4		
$\overline{K}_{xx}$	7.3	13.4	83	8.4	15	13.5	85
$\overline{K}_{xy}$	77.4	-1.5	-102	73.8	-5	-1.1	-101
$\overline{K}_{yx}$	-220	-176	-20	-250	14	-186	-15
$\overline{K}_{yy}$	1280	530	-59	1480	16	-537	-58
$\overline{B}_{xx}$	2.4	1.7	-31	2.3	-6	-1.7	-31
$\overline{B}_{xy}$	8.5	-9.5	-21	6.5	-24	-9.9	-216
$\overline{B}_{yx}$	4.4	-9.5	115	-4.1	-7	-10	123
$\overline{B}$	369	181	-51	419	14	189	-49

*Reference solution   Pinkus, 1987
+Δ - deviation

is the process of lubricant mixing and of heat transfer that eliminates the above paradox in the operation of concentric journal bearings.

By ignoring conduction within the lubricant, a knowledge of the surface temperatures may be sufficient to solve the energy equation. However, when fluid conduction terms

$$\left(\frac{\partial^2 T}{\partial x^2}\right), \quad \left(\frac{\partial^2 T}{\partial z^2}\right), \quad \left(\frac{\partial^2 T}{\partial y^2}\right)$$

are included, then these second order terms require additional information. Yet the fluid temperatures at the seal or bearing edges, $z = \pm(L/2)$, are not known. These have to be supplied by the history of the lubricant subsequent to its discharge from the interface, as portrayed in Fig. 1.7. The

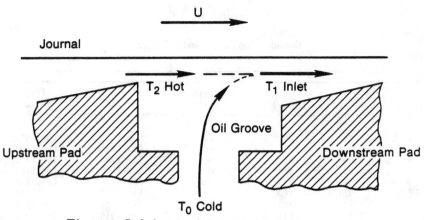

Fig. 1.6 Lubricant mixing at inlet to bearing pad

problem to be solved is that of a free or semi-attached jet with the upstream condition that of the exiting lubricant and the downstream condition given, theoretically, by

$$(x, z) \to \infty, \quad T \to T_a$$

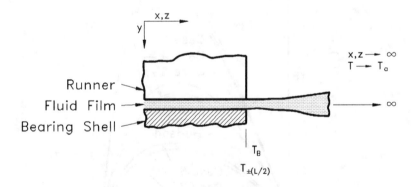

Fig. 1.7 Lubricant discharging at edge of bearing pad

Additional modifiers of the boundary conditions are heat sources and sinks and insulated or cooled surfaces. The effect that various boundary conditions may have on bearing performance can be gleaned from Fig. 1.8, where the labeled pressure profile numbers correspond to those given in Table 1.3. The profound difference in results as a consequence of assuming different boundary conditions is evident.

1.1.4 Gas Lubrication

There are a number of fundamental differences between gas and liquid

TABLE 1.3

Boundary conditions used in Fig. 1.8

Case	μ(T)	μ(p)	Compression work	T(x) bearing	T(x) runner
1	No	No	No	Constant	Constant
2	Yes	No	No	Constant	Constant
3	Yes	No	No	Constant	Constant
4	Yes	Yes	No	Constant	Constant
5	Yes	Yes	No	Constant	Constant
6	Yes	Yes	Yes	Variable	Constant
7	Yes	Yes	Yes	Variable	Variable
8	Yes	Yes	Yes	Variable	Variable
9	No	No	No	Constant	Constant

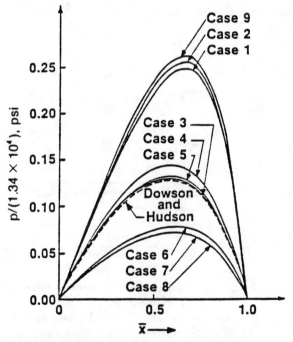

Fig. 1.8 Effect of thermal boundary conditions on pressure in slider bearing (Hahn and Kettleborough, 1967)

fluid films as far as their THD behavior is concerned. These are as follows:
- Viscous dissipation in gases is minimal primarily because, as shown in Fig. 1.9, their viscosity is several orders of magnitude lower than those of liquid lubricants. At 0°F the viscosity of air is nearly 10^{-5} that of a light oil; at 300°F it is 10^{-2}.

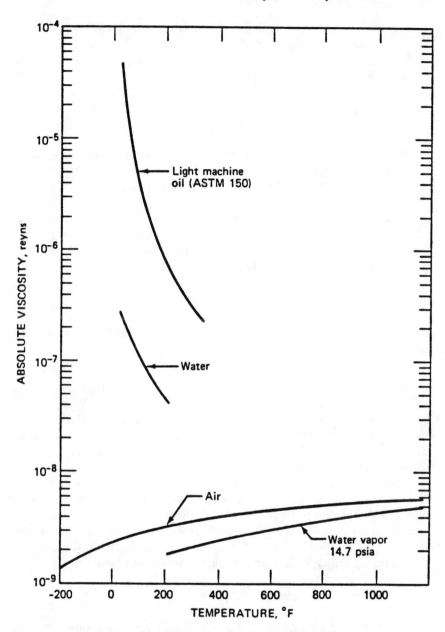

Fig. 1.9 Viscosity of liquids and gases

- Because of its low density the heat capacity of a gaseous film is likewise small. Consequently, whatever heat is generated is readily absorbed by the bounding surfaces leaving the fluid film close to its entrance temperature.
- Most important is the fact that the variation of viscosity with tem-

perature in gases is almost negligible as compared with that of liquid lubricants. A change from 0°F to 300°F produces a 600-fold drop in the viscosity of a light oil; in air the corresponding change is an increase of about one half over its original value.

- Even the simplest equation of state, namely the perfect gas equation

$$\rho = \left(p \big/ \mathcal{R}T\right)$$

introduces non-linearities in both the Reynolds and the energy equations genetrating computational entanglements out of proportion to the possible results. Therefore, there has been very little work invested in THD gas analysis.

The above qualitative argument can be supported by quantitative considerations. Elrod and Burgdorfer, 1959, have shown that for a gas film flowing between walls, each having a constant temperature of T_0, the maximum temperature rise is given by

$$\frac{\Delta T_{max}}{T_0} = O(M^2)$$

where M is the Mach number. Now the speed of sound in a perfect gas is given by

$$c = \sqrt{(k\mathcal{R}T/m)}$$

where m is the molecular weight of the gas. Under standard atmospheric conditions the above yields for air a speed of sound of 1100 ft/sec; and for hydrogen 4200 ft/sec. A 1 in diameter gas bearing running at the extremely high speed, say, of 100,000 rpm would have an average speed of $U/2 = 109$ ft/sec, or $M = 0.1$. Thus ΔT_{max} for air would be of the order of 1% or some 5°F. For hydrogen the increase would be of the order of 1°F.

Consequently, since gas films can be considered to operate under close to isothermal conditions, the bulk of our attention throughout the present text will be devoted to liquid lubricated films. However, the latter can on occasion, such as in the case of cryogenic fluids or under very high pressures, exhibit compressibility effects. Whenever appropriate, then, the effects of such compressibility variations are taken into account later on.

1.1.5 Thermal Parameters

A number of parameters and dimensionless groupings enter the analysis of thermal problems. Of these, of course, viscosity is the paramount variable. One of the difficulties encountered in dealing with energy relations stems from the fact that there are no exact analytical expressions for viscosity as a function of temperature and pressure. And yet in the usual range of operation viscosity is a very strong function of temperature. Numerous empirical expressions of $\mu(T, p)$ have been developed, and some of them are given below.

If the lubricant viscosity at a pressure p_0 and a temperature T_0 is given by μ_0 the viscosity at an arbitrary pressure p and temperature T can be represented by

$$\ln \frac{\mu}{\mu_0} = \beta(p - p_0) + A \left(\frac{1}{T + a} - \frac{1}{T_0 + a} \right) \qquad (1.4a)$$

This equation has a theoretical basis in the kinetic theory of liquids besides being verifiable experimentally over a reasonable range. The pressure and temperature scales can be so chosen that p_0 and T_0 are both zero. Equation (1.4a) then reduces to

$$\ln \frac{\mu}{\mu_0} = \beta p + A' \left(\frac{1}{T + a'} - \frac{1}{a'} \right) \qquad (1.4b)$$

The temperature rise likely to be encountered in practical problems is small enough so that for most lubricants the second term in the above equation is nearly linear in T. Therefore an approximation to the above equation is the widely used expression

$$\mu = \mu_0 e^{\beta p} e^{-\alpha T} \qquad (1.4c)$$

If the pressure and temperature range is large, the following expression holds:

$$\mu = \mu_0 \exp \left(\frac{\alpha}{T} + \beta p + \frac{\gamma p}{T} \right) \qquad (1.4d)$$

When viscosity is a function of temperature only, Equ. (1.4e) is often used

$$\mu = \frac{\gamma}{1 + \alpha T + \beta T^2} \qquad (1.4e)$$

Above, α, β and γ are all constants whose values differ in the last five expressions.

The density of most lubricating òils varies linearly with temperature, decreasing about 2% for a 50°F temperature rise. For a reasonable range the following formulas are very close approximations to this relationship:

$$\rho = \rho_0 e^{-\lambda T} \qquad (1.5a)$$

$$\rho = \rho_0 \left[1 + \gamma(T - T_1) \right] \qquad (1.5b)$$

The internal energy is assumed to be independent of pressure and linearly related to temperature

$$e = e_0 + c\,T \qquad (1.6)$$

Some of the other major thermal properties that may be required in THD analysis are as follows:

$$H_{fg} = \frac{(110.8 - 0.09T)}{\rho_g} \quad \text{BTU/lb} \qquad (1.7)$$

where H_{fg} is the heat of evaporation. The specific heat and the thermal conductivity of petroleum oils can be approximated by the equations

$$c_p = \frac{(0.388 + 0.00045T)}{\sqrt{\rho_g}} \quad \frac{\text{BTU}}{\text{lb} \, {}^\circ\text{F}} \tag{1.8}$$

$$k = \frac{0.812}{\rho_g}[1 - 0.0003(T - 32)] \quad \frac{\text{BTU}}{\text{hr} \, \text{ft}^2 \, {}^\circ\text{F/in}} \tag{1.9}$$

In the course of analysis the above variables often appear in groupings, or dimensionless numbers and these along with their names are given in Table 1.4. In addition to their role as dimensionless numbers these factors by their relative magitudes often determine the form of the differential equation to be used and it is thus of importance to investigate prior to analysis the regime of applicability of one or the other of these groupings. It is also of some interest to compare the magnitude levels of some of these variables in the case of liquids and gases. In a very general way their relative standing is as given in Table 1.5.

Two examples of the importance played by these dimensionless numbers will be shown. The first, given in Fig. 1.10, portrays theoretical values of T_{max} vs measured values. One set is plotted in terms of the dimensionless groupings $(\Delta T_{max}/\Delta T_e)$; the other in terms of the parameter

$$\frac{2\omega}{c_p \rho (C/R)^2}(\mu_e/\Delta T_e) = \frac{E}{\alpha \Delta T_e}$$

The contraction of the scatter achieved with the second approach testifies to the effectiveness of selecting dimensionless groupings which best reflect the ongoing thermal process.

The second example concerns regimes in which certain parameters play only a subsidiary role, so that they may, perhaps, be ignored. The case, taken from Cameron, 1967, is that of a parallel gap in which the moving surface is at a temperature ΔT higher than that of the stationary surface. Both transversely and along x the temperatures are taken to be linear with distance, so that

$$T(x, y) = (x/B)(y/h)\Delta T$$

At any station x the rate of heat conducted along an element dx is

$$dQ_k = k(dT/dy) \, dx = k(x/B)(\Delta T/h) \, dx$$

Integrated, the heat conducted along B is

$$Q_k = k \int_0^B (\Delta T/Bh)x \, dx = \frac{kB\Delta T}{2h}$$

TABLE 1.4

Dimensionless thermal groupings

Group	Symbol	Name
$k/\rho c_p$	σ	Diffusivity
$\dfrac{2\omega\alpha\mu_0\ (R/C)^2}{c_p\rho}$	E	Adiabatic Number (in x,z coordinates)
$\dfrac{U^2}{c_p\Delta T}$	Ec	Eckert number
$\Delta p/\rho U^2$	Eu	Euler number
Gt/r_0	F_0	Fourier number
$(L/D)\ (k/UD\rho c_p)$	Gz [= L/D)/Re·Pr]	Graetz number
$g\beta(\Delta T)\ L^3\rho^2/\mu^2$	Gr	Grashof number
λ/L	Kn	Knudsen number
$h_c L/k$	Nu	Nusselt number
$Uh\rho c_p/k$	Pe (= Re·Pr)	Peclet number
$c_p\mu/k$	Pr	Prandtl number
$g\beta(\Delta T)L^3\rho^2 c_p/\mu k$	Ra (= Gr·Pr)	Rayleigh number
$\rho Uh/\mu$	Re	Reynolds number
$h_c/c_p G$	St (= Nu/Re·Pr)	Stanton number

The convected heat is the product of mass flow, the specific heat and the average temperature rise, which is $\Delta T/2$. Thus the heat convected by the lubricant is

$$Q_c = c_p\rho \left(\frac{Uh}{2}\right) \left(\frac{\Delta T}{2}\right) = \frac{c_p\rho Uh\Delta T}{4}$$

The ratio of conducted to convected heat is thus given by

$$\frac{(kB\Delta T/2h)}{\frac{1}{4}Uhc_p\rho\Delta T} = \left(\frac{k}{c_p\rho}\right)\left(\frac{2B}{Uh^2}\right) \tag{1.10}$$

TABLE 1.5

Orders of magnitude of thermal parameters
in lubrication films

Fluid	ρ	p	h	μ	k	c_p	Re	$(k/c_p\rho)$
Gas*	10^{-3}	10^7	10^{-4}	10^{-4}	10^{-4}	10^7	10	10^{-1}
Liquid*	1	10^8	10^{-3}	10^{-1}	10^{-3}	10^7	100	10^{-3}
Liquid/gas ratio	1,000	10	10	1,000	10	1	10	100

*In absolute cgs units   Cope, 1949

The group in the first bracket on the right-hand side is the diffusivity of the fluid. Its value for various substances is given in Table 1.6. Now for a bearing and a gear the following operational parameters may be typical.

Parameter	Bearing	Gear
B, in.	5	0.04
h, in.	10^{-3} to 10^{-2}	10^{-5}
U, ft/sec	200	30

The conduction to convection ratio for the bearing is thus 0.07 and nearly all the heat is taken away from a bearing by convection; for the gear the ratio is about 300 showing that virtually all the heat is removed by conduction, which is typical of EHD problems. More generally we can write the energy equation in the form of

$$u\frac{\partial T}{\partial x} + w\frac{\partial T}{\partial z} + v\frac{\partial T}{\partial y} = \frac{k}{\rho c_p}\left[\frac{\partial^2 T}{\partial x^2} + \frac{\partial^2 T}{\partial z^2} + \frac{\partial^2 T}{\partial y^2}\right] + \frac{\mu}{\rho c_p}\Phi \qquad (1.11)$$

The above expression in words reads

$$\text{Convection}\left(\sim \frac{U\Delta T}{L}\right) = \text{Diffusion}\left(\sim \frac{k\Delta T}{\rho c_p h^2}\right)$$

$$+ \text{ViscousDissipation}\left[\sim \left(\frac{\mu}{\rho c_p}\right)\left(\frac{U}{h}\right)^2\right]$$

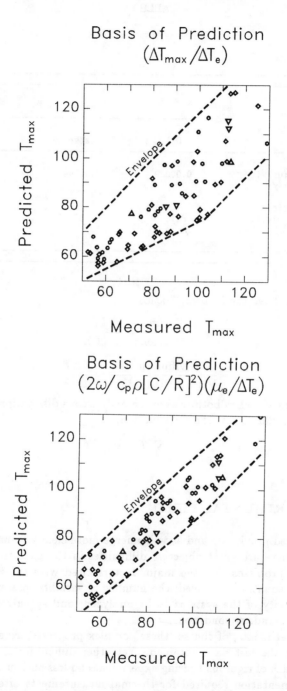

Fig. 1.10 Scatter of experimental data when plotted against two different parameters. All T's are in °C. (Martin, 1980)

TABLE 1.6

Diffusivity of various fluids

	Fluid		
	Oil	Water	Air
Diffusivity cm^2/s	0.00089	0.00144	0.187
Ratio of diffusivities	Ref.	1.62	210

and we then have the following ratios

$$\frac{\text{Diffusion}}{\text{Convection}} = \frac{kL}{\rho c_p U h^2}$$

$$\frac{\text{Diffusion}}{\text{ViscousDissipation}} = \frac{k\Delta T}{\mu U^2}$$

For the case when both convection and viscous dissipation are zero, we have the ordinary heat transfer equation,

$$\nabla^2 T = 0$$

1.2 EXPERIMENTAL FIELD

As noted previously and as will emerge more clearly from the material to follow there are great objective difficulties in the analytical treatment of thermal problems. Of the major factors cited were the nature of the differential equations involved; the number of variables present; the physical complexity of the systems to be analyzed; and speculation as to the prevailing boundary conditions.

In experiments, of course, these complex processes are automatically reflected in the test data obtained. Yet other difficulties arise and these make the task of rigorous bearing experiments no less onerous. In addition, the instrumentation required for thermal measurements often becomes a research project in itself, consuming much of the effort designated originally for the THD problem.

1.2.1 Viscosity

Of all the factors that make lubrication experiments difficult that of variable viscosity constitutes the most troublesome element. It is almost impossible to vary any parameter during testing without simultaneously altering the viscosity field of the lubricant. With any change in diameter, clearance, speed, etc., there is a parasitic intrusion of viscosity changes which often vitiates the results of the most careful experiment.

To illustrate the degree to which viscosity changes pervert experimental data we may resort to the example discussed at the opening of the chapter, Table 1.1. The clearance of the reference bearing was 0.0045 in. for which case the effective operating viscosity was 14.4 csks. Let us say we want to determine experimentally the effect of changing the clearance on bearing performance. This means that to obtain a correct set of data for bearing performance as a function of clearance, tests would have to be conducted at $\mu_e = 14.4$ csks. However, changing the clearance would immediately and drastically alter the viscosity field and thus the value of μ_e. Fig. 1.11 shows two sets of results: one, the changes in ϵ, H, K_{xx} and B_{xx} that would follow a change in clearance from the reference value of 0.0045 in. when μ_e =const. for all cases; and two, the changes an experimenter would obtain as a result of simultaneous changes in both clearance and effective viscosity (due to a change in C). The changes in μ_e due to a variation in clearance are as follows:

C mils	μ_e csks
3.0	9.97
4.5	14.4 (Reference)
6.0	17.7

What is particularly worth noting in Fig. 1.11 is that while B is little affected, both ϵ and K can show either a positive or a negative departure from the correct curve (based on 14.4 csks). The most striking upset, however, is shown by the power loss, for it exhibits a reverse effect from that due to a change in clearance. Whereas the $\mu_e =$ const. curve shows an increase in power loss when C is decreased, the curve that includes a change in μ_e shows a decrease in power loss; and where the $\mu_e =$ const. curve shows a drop in H with an increase in C the $\mu_e = 14.4$ csks shows an increase in H. A routine set of test data would, of course, correspond to the variable μ_e curve and would not reflect correctly the change of bearing performance with a change in C. The same distortion would occur in testing for variation of any other bearing parameter.

1.2.2 Transverse measurements

In the case of THD experiments a particularly difficult aspect is that

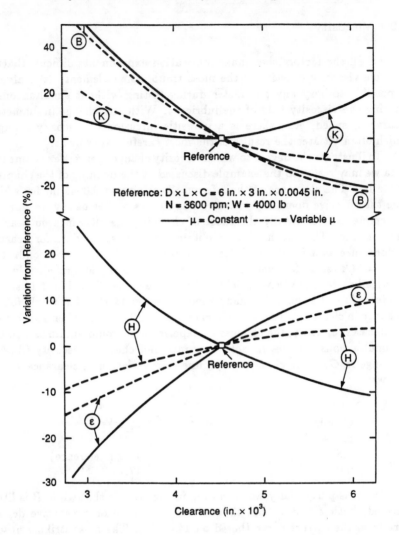

Fig. 1.11 Effect of variable viscosity during a parametric change of clearance

of obtaining transverse measurements across the fluid film. As will be seen later, unlike the pressure field, the temperature distribution and the associated thermal mapping are very strong functions of the transverse variable. Since the height of a fluid film is of the order of mils, and often microns, it is very difficult to measure temperatures and temperature gradients across it.

Even to measure the temperatures of the surfaces adjacent to the fluid film, particularly that of the runner, is a difficult task. This can be seen in a relatively simple case of measuring the planar temperature in a stationary journal bearing. In a discussion by Vohr to a paper by Heshmat and Pinkus

(1986) the discusser showed analytically that what was measured was a surface temperature, whereas the authors claimed to have measured the bulk temperature of the lubricant at a given point (x, z). Fig. 1.12 shows the topography of thermocouple location and mounting vis-a-vis the fluid film and adjacent surfaces. The picture is drawn to scale, with the dimensions as indicated. Note should be taken of the fact that thermal insulation separates the thermocouple from the bearing metal; that the thermocouple head is closer to the runner surface than it is to the babbitt; and that the only thing it is in contact with is a pool of fluid. This fluid is carried into the indentation housing the thermocouple, so that a circulation is set up in the quasi-spherical well about the thermocouple. What then is the temperature registered by the thermocouple? Is it the surface temperature at the bottom of the hydrodynamic film or some bulk temperature of the surrounding fluid? Such uncertainties arose when a single measurement was at issue, but in a THD experiment what would be required is a temperature profile across the film, perhaps five to ten such measurements. To accomplish this with any degree of accuracy and confidence by conventional means seems unlikely. Some advanced, non-intrusive instrumentation would have to be utilized to meet this task.

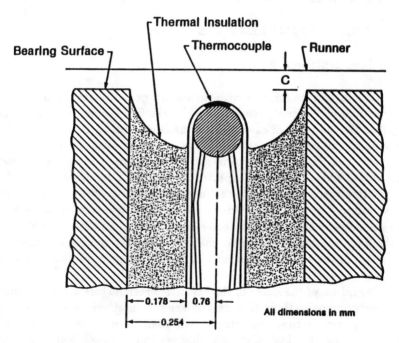

Fig. 1.12 The topography of a thermocouple

1.2.3 Methodology

Most experiments are conducted in circumstances specific to a particular test rig. Since the peripheral equipment and the mode of assembly influence the heat transfer pattern, it is difficult to generalize such results to cases where the environment is different. While load capacity measurements are little affected by such diversities, thermal effects in fluid films are very sensitive to these factors. The above is partly the reason that, compared with the body of test results on other facets of tribology such as load capacity, flow, etc., good experiments on temperature mapping and heat flow are scarce. While objective factors are certainly responsible for this quandary, it seems that to some extent the fault is also due to the fact that most experiments on thermal effects involve too many variables. A more modest but more rigorous investigation of single elements of the problem in which the scope of the investigation and the experimental setup are kept as simple as possible, may provide more insight.

Thus in parallel to the analytical field where the researcher is challenged to isolate the essential THD parameters and to generate basic models of thermal systems—the experimenter must develop a programmatic strategy on how to simplify his experimental setup and utilize sophisticated instrumentation so as to be able to observe and record the essential features of the thermal process.

1.3 SPECIAL COMPLEXITIES

1.3.1 Non-Isotropic Films

Tribologists are familiar with the occurrence of cavitation in the trailing portions of a hydrodynamic film where the film geometry diverges. Less familiar are the cases of cavitation at the beginning of the film, due either to a divergent geometry; to insufficient lubricant supply; to a restrictive geometry that impedes the formation of a full film; or to so-called flooded lubrication where the fluid is expected to enter the clearance from the sides. In addition there is uncertainty as to the form the cavitating film assumes. As shown in Fig. 1.13, if the fluid striates to form transverse filaments then the geometry of the region will correspond to that of model A in Fig. 1.13; if the fluid adheres sheet-like to one or both of the mating surfaces, then the space in the cavitating region will resemble respectively models B and C.

Regardless of the shape of the cavitating region there are discrepancies between conventional predictions and experimental evidence as far as the thermal picture is concerned. It has long been observed that beyond h_{min} the temperatures of hydrodynamic films tend to decrease. Rationally there seems to be no reason for this. Even though the rate of heat generation may fall once h_{min} is passed, there continues to be energy dissipation which should further raise the film temperature. Fig. 1.14 shows the results of an experiment in which a thrust bearing was run backwards, i.e. the motion took place against a diverging film. The temperatures over this cavitating

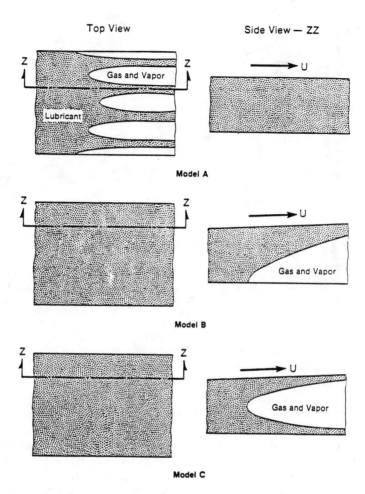

Fig. 1.13 Models of the cavitation zone

surface are seen to decrease regardless of the speed, load, and temperature levels involved. This is corroborated by many other examples. Thus the usual equations for calculating the dissipation and temperature variation need a new formulation in order to make them account for the specific conditions prevailing in this mixed liquid-gas regime.

An even more complex situation arises in the case of a change of state, or two-phase flow as might occur in steam lubrication or in cryogenic devices. In addition to the formulation of the energy dissipation and corresponding variation in temperature and viscosity there are also uncertainties as to whether the heat of vaporization required is taken from the solid surfaces or from the fluid itself, two modes which would yield strikingly different results for the fluid film temperatures.

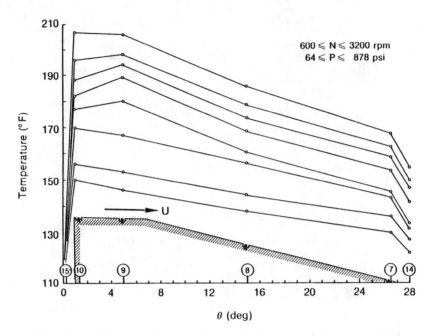

Fig. 1.14 Temperatures in the cavitation zone (Heshmat, 1987)

1.3.2 Reverse Flows

Another complexity arises in the case of reverse flow. Under conditions of high film convergence or high upstream boundary pressures fluid instead of being dragged into the clearance may actually be exiting from the pad entrance. A sketch of such reverse flow is shown in Fig. 1.15 along with some of the temperature interactions occurring in the inlet zone. As seen, one part of fluid entering the pad is the hot layer Q_2 at the high exit temperature T_2; another part, Q_0, is the supply lubricant at the cold temperature T_0; and finally there is a recirculating portion Q_R which enters at a temperature T_0 and exits at some indeterminate temperature T_R. This temperature rise $\Delta T = (T_R - T_0)$ is due to several sources; shearing of the fluid along its circulation path; heating by the stationary surface; and conduction of heat between the Q_0 and Q_R streams. While it is shown later that as compared with convection, and other heat transfer modes, conduction within the lubricant is small, for the reverse flow regime this may not be true. In cases where reverse flow is a major component of total flow, such as in pumping rings or seals, this would make the generally postulated energy equations invalid.

1.3.3 Rheodynamic Models

So far the discussion has been confined to fluids whose viscosity is

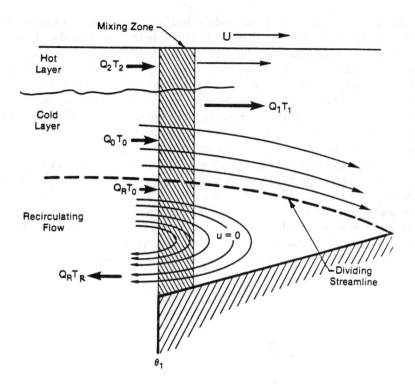

Fig. 1.15 Reverse flow at inlet to pad

primarily a function of temperature or $\mu = \mu(T)$. There are, of course, tribosystems where this would be far from adequate. Thus in gears and many other elastohydrodynamic systems the prevailing high pressures would dominate the temperature effects so that we would have to write $\mu = \mu(T, p)$. This eventually leads us to the general case of non-Newtonian fluids where the viscosity in addition to pressure and temperature is also a function of the rate of shear, or in general

$$\mu = \mu\left(T, p, \frac{du}{dy}, \ldots\right) \tag{1.11}$$

Similar to two-phase flow, however, there are no generally agreed rheodynamic models to correctly represent the functional form of Equ. (1.11). In such cases it would be futile to labor on thermal solutions in which a simple temperature-dependent viscosity was employed to formulate the problem.

1.3.4 System Interaction

In contrast to isothermal solutions where one can more or less ignore the surrounding hardware, a THD solution is intimately linked to the thermal history of the system of which the seal or bearing is only one component. This introduces both methodological and mathematical difficulties. The first refers to the issue that in his attempts to provide seal or bearing solutions the tribologist not be faced with a syndrome that is bound to overwhelm the specific tribological assignment. Thus, in order to calculate the characteristics of the fluid film one must not be embroiled in the task of calculating the temperatures, stresses, and deflections of bearing shell, housing, pedestals, shaft, rotor, including perhaps also such items as radiation and forced cooling. The oil ring bearing offers a glimpse of this problem even in the most mundane of cases. A schematic configuration of an oil ring bearing assembly is shown in Fig. 1.16. To arrive at the temperature map of the fluid film in the journal bearing one would have to consider the heat transfer loop in the sump; the rates of oil lift-off, shearing losses, and temperatures of the oil ring; the heat interchange with the thrust bearing, shaft, thrust runner, housing, and pedestals. Such an undertaking would vitiate the very purpose of arriving at a manageable and useful THD solution of a simple oil ring journal bearing.

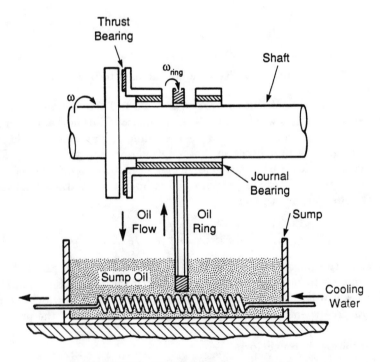

Fig. 1.16 Oil ring bearing assembly

The other problem with system interaction is analytical, in that an attempt to account mathematically for all the elements in the system would introduce cumulative errors and approximations to a degree that would deny the very purpose of having included the system in the first place. Heat transfer equations are semi-empirical to start with and the components of heat exchangers are mechanically complex in the extreme and a THD solution for the fluid film incorporating all these factors would probably be less accurate than one confined to the bearing itself but based on plausible simplifications. The problem here is linked to the previously discussed subject of boundary conditions. For the tribologist the history of the system as such is of little concern except to the extent that it affects the fluid film. Thus the efforts must lie in reducing common mechanical systems to generic models which once solved would provide envelopes of boundary conditions to be used in the differential equations directly applicable to the fluid film.

Chapter 2

APPROACHES TO THE
ADIABATIC PROBLEM

The intent of the following three chapters is to present the various conceptual approaches to the solution of THD problems. The present chapter will be confined to the adiabatic field. This is done, first of all, as a logical step in the build-up of analytical complexity in dealing with THD analysis, but also for other cogent reasons. One of these is that the adiabatic problem confines attention to the fluid film proper without encumbering and obscuring the subject by extraneous factors. By making the bounding surfaces adiabatic, any assumptions or approximations will reflect exclusively on the thermal behavior of the fluid film. The full three-dimensional aspects of thermal analysis will be considered in the subsequent chapter.

2.1 THE ENERGY EQUATION

Many investigators have derived the energy equation for the specific conditions of hydrodynamic lubrication. The one obtained by Cope (1949) is written in the following form

$$
\rho c_p \left[u\left(\frac{\partial T}{\partial x}\right) + w\left(\frac{\partial T}{\partial z}\right) \right] + p\, dV + k\left[\frac{\partial^2 T}{\partial x^2} + \frac{\partial^2 T}{\partial z^2} \right]
$$
$$
= \mu\left[\left(\frac{\partial u}{\partial y}\right)^2 + \left(\frac{\partial w}{\partial y}\right)^2 \right]
$$

(2.1)

with some of the nomenclature shown on the basic tribological element of Fig. 2.1. Spelled out, the above equation balances the following energy fluxes

$$ConvectiveHeat + DilatationWork + FluidConduction$$
$$= ViscousDissipation$$

35

where by convection here we mean forced as opposed to free heat convection.

It should be emphasized that Equ. 2.1. adheres to the basic hydro-dynamic lubrication postulate of no transverse pressure or temperature variations. Since all changes with y are ignored, the resulting temperature field will be a function of the planar coordinates (x, z) or (r, θ) and will thus be referred to as the two-dimensional solution. The adiabatic approach encompasses thus two related yet not exclusive of each other premises, namely,

- No heat transfer to the mating surfaces
- No variations across the fluid film.

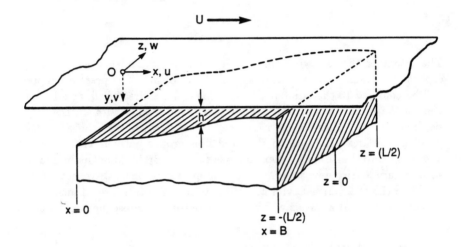

Fig. 2.1 Basic tribological element

In most cases fluid conduction can be ignored when compared with convection. Also, given the essential incompressibility of liquid lubricants, dilatation work is negligible. We then have

$$k \left[\frac{\partial^2 T}{\partial x_i^2} \right] \ll \rho c_p u_i \left(\frac{\partial T}{\partial x_i} \right)$$
$$dV \simeq 0$$

and the adiabatic energy equation reduces to

$$\rho c_p \left(u \frac{\partial T}{\partial x} + w \frac{\partial T}{\partial z} \right) = \mu \left\{ \left(\frac{\partial u}{\partial y} \right)^2 + \left(\frac{\partial w}{\partial y} \right)^2 \right\} \tag{2.2}$$

Substituting for u and w from

$$u = \frac{1}{2\mu} \frac{\partial p}{\partial x} y(y - h) + \frac{h - y}{h} U \tag{2.3a}$$

$$w = \frac{1}{2\mu} \frac{\partial p}{\partial z} y(y - h) \tag{2.3b}$$

we obtain the two-dimensional adiabatic energy equation in rectangular coordinates in the form of

$$6U\rho c_p h\left[\left(1 - \frac{h^2}{6\mu U}\frac{\partial p}{\partial x}\right)\left(\frac{\partial T}{\partial x}\right) - \frac{h^2}{6\mu U}\left(\frac{\partial p}{\partial z}\right)\left(\frac{\partial T}{\partial z}\right)\right]$$
$$= \frac{12\mu U^2}{h}\left\{1 + \frac{h^4}{12\mu^2 U^2}\left[\left(\frac{\partial p}{\partial x}\right)^2 + \left(\frac{\partial p}{\partial z}\right)^2\right]\right\}$$

(2.4)

and in polar coordinates

$$\frac{1}{2}\omega\rho c_p h\left\{\left[1 - \frac{h^2}{6\mu\omega r^2}\left(\frac{\partial p}{\partial\theta}\right)\right]\frac{\partial T}{\partial\theta} - \frac{h^2}{6\mu\omega}\left(\frac{\partial p}{\partial r}\right)\left(\frac{\partial T}{\partial r}\right)\right\}$$
$$= \frac{\mu\omega^2 r^2}{h} + \frac{h^3}{12\mu}\left[\frac{1}{r^2}\left(\frac{\partial p}{\partial\theta}\right)^2 + \left(\frac{\partial p}{\partial r}\right)^2\right]$$

(2.5)

This is a differential equation for T in which the coefficients ρ and c_p are considered constant. The equation contains the pressure derivatives $(\partial p/\partial x)$ and $(\partial p/\partial z)$ both unknown functions of x and z. These have to be obtained from the Reynolds equation which in rectangular coordinates reads

$$\frac{\partial}{\partial x}\left(\frac{h^3}{\mu}\frac{\partial p}{\partial x}\right) + \frac{\partial}{\partial z}\left(\frac{h^3}{\mu}\frac{\partial p}{\partial z}\right) = 6U\left(\frac{\partial h}{\partial x}\right) + 12\left(\frac{\partial h}{\partial t}\right)$$

(2.6)

and in polar coordinates

$$\frac{\partial}{\partial r}\left(\frac{rh^3}{\mu}\frac{\partial p}{\partial r}\right) + \frac{1}{r}\frac{\partial}{\partial\theta}\left(\frac{h^3}{\mu}\frac{\partial p}{\partial\theta}\right) = 6U\left(\frac{\partial h}{\partial\theta}\right) + 12r\left(\frac{\partial h}{\partial t}\right)$$

(2.7)

To solve the problem for both the pressure and temperature fields, the energy and the Reynolds equations must be solved simultaneously either by iterative or some other suitable techniques.

2.2 METHOD OF EFFECTIVE VISCOSITY

Normally, in methods that ignore the thermal problem completely, the inlet viscosity μ_0 is used as the value for μ. The method of effective viscosity attempts to borrow the simplicity of the constant μ approach, but instead of μ_0 it uses a value for the viscosity that bears some relation to the thermal process in the fluid film. Based on an effective film temperature $T_e = T_0 + \Delta T$, this μ_e is then used to calculate performance.

From a simple energy balance we have

$$H = c_p Q_s \Delta T$$

or

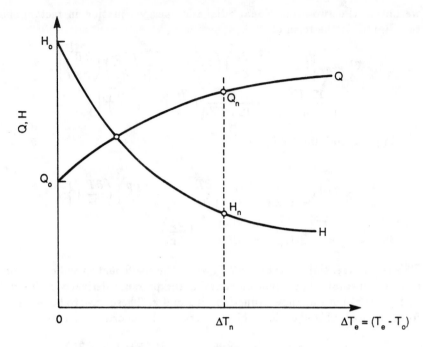

Fig. 2.2 Method of effective viscosity

$$\Delta T = \left(\frac{H}{c_p Q_s} \right)$$

where Q_s represents the lubricant discharge, or side leakage, and ΔT represent the bulk temperature increase of the discharged lubricant. Since H decreases, while Q_s increases, with a rise in ΔT, a plot of H and Q_s versus assumed values of ΔT would yield the curves shown in Fig. 2.2. Thus the effective temperature rise ΔT would be that location where

$$\left(\frac{H_n}{c_p Q_{sn}} \right) = \Delta T_n \tag{2.8}$$

This approach represents an extreme in thermal analysis, since by this method the viscosity varies neither with y, nor with x or z. Still, the method yields generally fair results in terms of overall performance and by its simplicity and adaptability to a wide range of problems constitutes in fact a very useful tool in dealing with THD problems.

The usefulness of the method is enhanced by the fact that any nondimensional solution of the isothermal case can serve as a solution for the effective viscosity method. It suffices simply to find the correct value of the effective viscosity in order to be able to utilize the various isothermal results by simply substituting in the normalized quantities the proper μ_e for the viscosity term appearing in these parametric groupings. The results of

Table 1.1 in the previous chapter were obtained by this method and reflect the case of full bearings. Similar results for partial journal bearings are given in Fig. 2.3. As seen, even in the present simple example the interplay between the high ϵ for the shorter pads and the longer heating paths for the higher $\beta's$ makes the results unpredictable so that over a certain range the short pads and in another the long pads exhibit higher temperature rises.

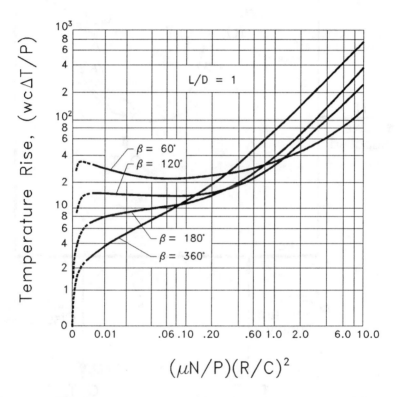

Fig. 2.3 Effective temperature as function of S and β for $L/D = 1$ (Raimondi and Boyd, 1958)

It is also worth stressing that the effective viscosity method automatically fulfills one of the major demands regarding the inlet boundary condition. As was shown in Fig. 1.6, the lubricant entering a pad is not at μ_0 but at some lower viscosity, due to the mixing of the hot and cold lubricant streams. A simple adding of the two streams yields the equation

$$Q_0 T_0 + Q_2 T_2 = (Q_0 + Q_2) T_1 \tag{2.9}$$

The use of an effective or bulk temperature T_e can be shown to be equivalent to employing the above relationship. In the effective viscosity method outlined above we essentially consider the bearing enclosed in a box across which pass three energy fluxes; the power supplied to the bearing, H; the cold supply oil Q_0 at T_0; and the side leakage Q_S at some effective T_e, as shown in Fig. 2.4.

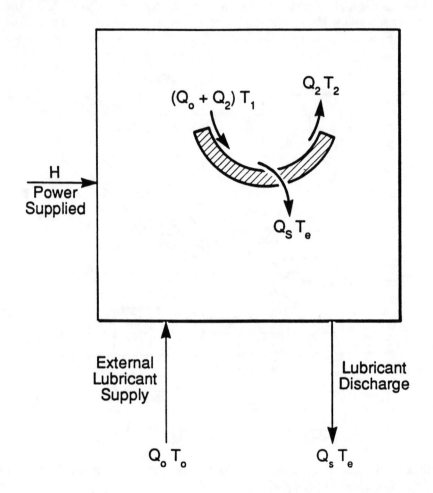

Fig. 2.4 Energy balance for effective viscosity

For this system we have

$$H + c_p Q_0 T_0 = c_p Q_s T_e \qquad (2.10)$$

Since $Q_s = Q_0$ we have

$$H = c_p Q_s (T_e - T_0)$$

Now in terms of the flows within the bearing, the energy balance reads

$$H = c_p [Q_s (T_e - T_1) + Q_2 (T_2 - T_1)]$$

Equating the two expressions for H, we obtain

$$Q_s(T_e - T_0) = Q_s(T_e - T_1) + Q_2(T_2 - T_1)$$

T_e is eliminated and we obtain

$$Q_s T_0 + Q_2 T_2 = (Q_s + Q_2) T_1$$

which is the same as Equ. (2.9) with $Q_s = Q_0$.

The above approach can be generalized by requiring that it be dictated not by Equ. (2.8) but by the more valid constraint that μ_e be determined by the results of a genuine THD analysis. By running two solutions, one isothermal, and one with a variable viscosity, two sets of, say, load capacity are obtained. In order to reconcile the incorrect isothermal W_0 solution with the correct THD result, W, one can reduce the inlet viscosity μ_0 by the ratio of the two values of load, namely

$$\frac{\mu_e}{\mu_0} = \frac{W}{W_0}$$

Similarly, one can find an effective viscosity for calculating the correct friction, as shown in Fig. 2.5. It is immediately clear that since the ratio H/H_0 is different from W/W_0, the two values of μ_e, the one for determining load and the other for determining friction, will not be the same.

Representative differences for the two approaches in calculating W and H are given in Table 2.1. Values of the effective viscosity in terms of its dependence on eccentricity and supply temperature T_0 are given in Fig. 2.6, where the variation of μ_e/μ_0 is seen to range from 0.9 to 0.3. Similar discrepancies in μ_e/μ_0 will occur if a representative μ were to be established for calculating spring and damping coefficients, or some other performance item. In short, while the notion of effective viscosity may partly mitigate the consequences of isothermal analysis for one particular item of interest, as a general concept μ_e becomes untenable. Finally, while an effective viscosity may offer shortcuts to an approximate determination of performance, this does not hold for T_{max}, which can be obtained only from an anisothermal solution.

2.3 THE COUETTE APPROXIMATION

As pointed out previously, due to the presence of pressure terms, equations (2.4) and (2.5) can be solved only in conjunction with the Reynolds equation. This demands iterative or other computational procedures which make the process laborious. If the terms containing the pressure gradients $\partial p/\partial x$ and $\partial p/\partial z$ could be considered small vis-a-vis the other terms, the energy equation could be uncoupled from the Reynolds equation and solved independently. Thus for purposes of temperature evaluation the flows and

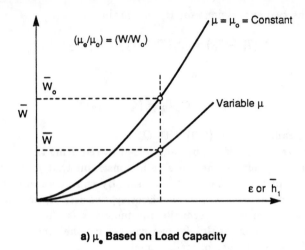

a) μ_e **Based on Load Capacity**

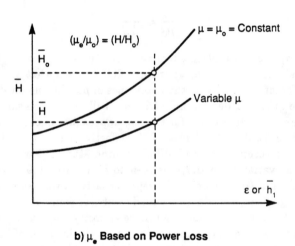

b) μ_e **Based on Power Loss**

Fig. 2.5   Effective viscosity for different performance parameters

viscous losses would be based on Couette type of flow; hence the name of the method.

The importance of the pressure terms vis-a-vis the other terms in the energy equation has been examined by McCallion et al. (1970). Figs. 2.7 and 2.8 show the relative discrepancies of the Couette approximation from an adiabatic solution that does retain the pressure terms. As seen the differences are small in the lower range of eccentricities (where the pressure gradients are small) and even at moderate values of ϵ they are not too far apart. From their numerical results McCallion et al. extracted the following range of validity of the Couette method for E values up to about 0.4.

TABLE 2.1

Values of effective viscosities

β	ϵ	(μ_e/μ_0) for W	(μ_e/μ_0) for H	$\dfrac{\mu_{eW}}{\mu_{eH}}$
150°	0.78	0.65	0.88	0.74
120°	0.78	0.67	0.88	0.76
100°	0.78	0.68	0.91	0.74
80°	0.78	0.71	0.91	0.78
60°	0.784	0.735	0.90	0.81
40°	0.852	0.69	0.78	0.88
20°	0.942	0.35	0.41	0.85
10°	0.981	0.11	0.18	0.61

Pinkus and Wilcock, 1985

$(L/D) = 1/3$; $Re_0 = 2400$; W = constant for all μ's.
Isothermal $\mu_0 = 20$ cps; THD solution based on E = 0.12.

Validity of Couette approach

L/D	ϵ
0.25	0 – 0.9
0.5	0 – 0.8
1.0	0 – 0.6

The simplification achieved via the Couette method in handling the momentum and energy equations is, of course, considerable.

When the pressure gradients are eliminated from the energy equation

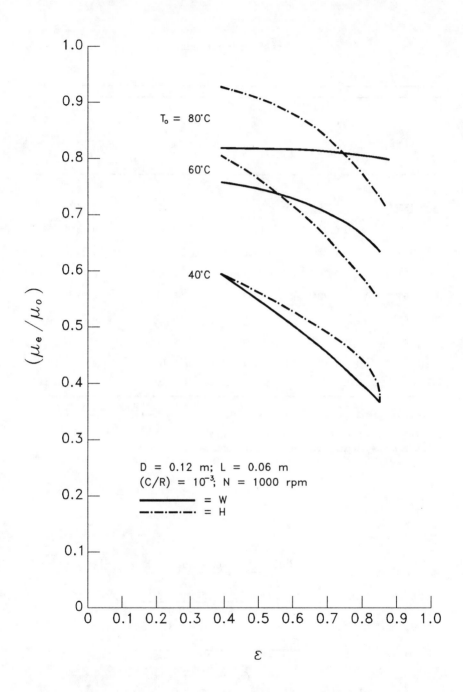

Fig. 2.6 Effective viscosity as function of eccentricity and T_0 for $L/D=\frac{1}{2}$, (Motosh, 1964)

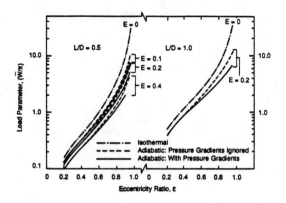

Fig. 2.7 Effect of ignoring pressure gradients in energy equation on load (McCallion et al., 1970)

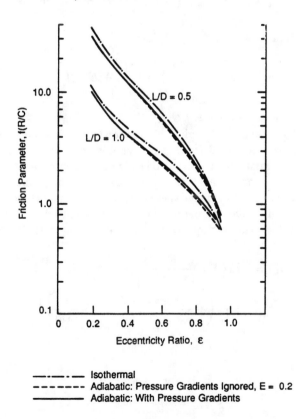

Fig. 2.8 Effect of ignoring pressure gradients in energy equation on friction (McCallion et al., 1970)

(2.4), we obtain

$$\frac{\partial T}{\partial x} = \frac{2\mu U}{\rho c_p h^2}$$

Using the exponential viscosity-temperature relationship

$$\mu = \mu_0 e^{-\alpha(T-T_0)} \tag{2.11}$$

we have

$$-\alpha \frac{dT}{dx} = \frac{1}{\mu}\left(\frac{d\mu}{dx}\right)$$

and in normalized form

$$\frac{d}{d\theta}\left(\frac{1}{\mu}\right) = \left(\frac{E}{\bar{h}^2}\right)$$

The above integrates into

$$\alpha(T - T_0) = \ln\left(\frac{1}{\bar{\mu}}\right) = \ln\left\{1 + E \int \frac{d\theta}{\bar{h}^2}\right\} \tag{2.12a}$$

for x, z coordinates with $\bar{h} = (h/C)$, and

$$\alpha(T - T_0) = \ln\left(\frac{1}{\bar{\mu}}\right) = \ln\left\{1 + E\bar{r}^2 \int \frac{d\theta}{\bar{h}^2}\right\} \tag{2.12b}$$

for (r, θ) coordinates with $\bar{h} = h/h_2$.

Equs. (2.12) thus yield explicit expressions for the variation of temperature, which are uncoupled from the Reynolds equation. While in thrust bearings and in misaligned journal bearings a dependence of viscosity on shear flow only will produce a two-dimensional variation in the temperature field, in aligned journal bearings there would be a temperature variation in the θ direction only. The temperature distribution in such a journal bearing obtained by integration of Equ. (2.12a) is then given by the function

$$(T - T_0) = \frac{1}{\alpha} \ln\left\{1 + \frac{E}{(1-\varepsilon^2)}\left[\frac{\delta}{(1-\varepsilon^2)^{1/2}} \times \right.\right.$$
$$\left.\left.\arccos\left(\frac{\varepsilon + \cos\theta}{1 + \varepsilon\cos\theta}\right) - \frac{\varepsilon\sin\theta}{(1 + \varepsilon\cos\theta)}\right]\right\}_{\theta_0}^{\theta} \tag{2.13}$$

where

$$\delta = 1 \quad \text{for} \quad \sin\theta > 0,$$
$$\delta = -1 \quad \text{for} \quad \sin\theta < 0.$$

The present method provides a rough guide to the temperature map of bearings in which the Couette component of flow is the predominant mode.

However, for bearings with a large film convergence when the bulk of the lubricant exits in the form of side leakage $(h_2/h_0 \to 0 \; or \; \epsilon \to 1)$ inordinately high temperature rises result from this method of solution.

A number of adiabatic solutions for journal and thrust bearings based on the above approach are given for the common bearing designs shown in Figs. 2.9 and 2.10. These results are compiled in Figs. 2.11 and 2.12 and in Table 2.2 for circular and elliptical bearings. Table 2.3 gives results for tapered land thrust bearings with both circumferential and radial tapers, as given by the expression

$$\bar{h} = \bar{h}_{11} - (\bar{h}_{11} - 1)\frac{\theta}{b\beta} - \bar{\delta}_r \left[\frac{\bar{r} + (L/R - 1)}{L/R_2}\right]\left(1 - \frac{\theta}{b\beta}\right); \quad 0 \le \theta \le b\beta$$

$$\bar{h} = 1; \quad b\beta \le \theta \le \beta \tag{2.14}$$

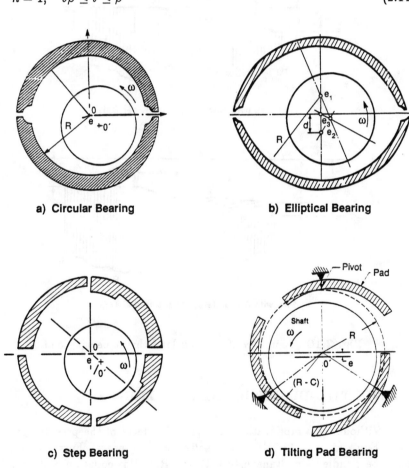

a) Circular Bearing   b) Elliptical Bearing

c) Step Bearing   d) Tilting Pad Bearing

Fig. 2.9 Common journal bearing configurations

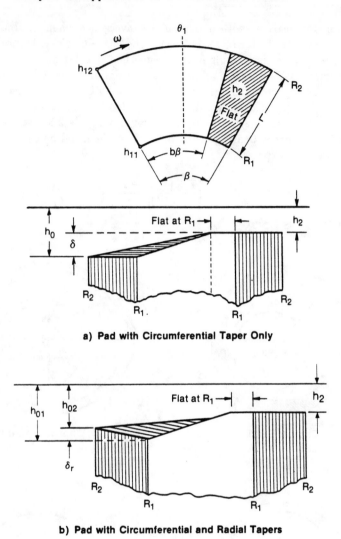

a) Pad with Circumferential Taper Only

b) Pad with Circumferential and Radial Tapers

Fig. 2.10 Geometry of tapered land thrust bearing pad

2.4 THE TWO-DIMENSIONAL SOLUTION

Without the simplification of neglibible effects of the pressure gradients, Equ. (2.4) or (2.5) must be solved as it appears. Furthermore, it must be handled simultaneously with the Reynolds equation and usually by some iterative scheme whereby initally the pressures are ignored and then entered into the scheme until the momentum and energy equations yield a consistent set of pressure and temperature distributions.

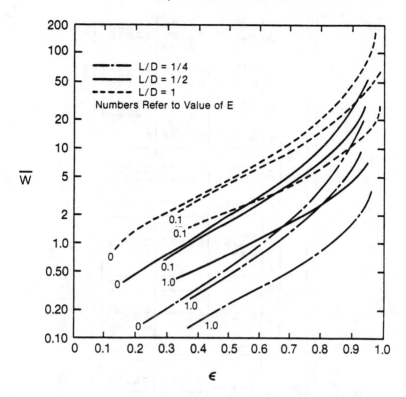

Fig. 2.11 Couette approximation load capacity in two-axial- groove circular journal bearing (Pinkus and Bupara, 1979)

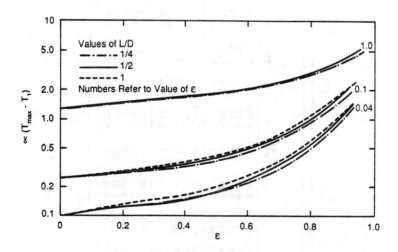

Fig. 2.12 Maximum temperatures in couette approximation solution for two-axial-groove journal bearing (Pinkus and Bupara, 1979)

TABLE 2.2

Couette approximation solution for elliptical bearings

E	ε	φ deg	h_min	θ^min deg	W	Q̄ In.	Q̄ Sides	(R/C) f Brg.	GRVS*	ΔT_max	K̄ XX	K̄ XY	K̄ XX	K̄ YY	B̄ XX	B̄ XY	B̄ YX	B̄ YY	w_i	M_cr
0	0.1	93.5	0.48	350	0.537	1.45	0.442	47.3	0.322	0	12.4	4.91	-7.16	-0.069	19.7	-7.67	-7.67	5.62	0.267	0.202
	0.2	90.8	0.46	200	1.24	1.45	0.495	21.1	0.145	0	14.1	7.77	-7.28	0.218	24.9	-7.67	-7.67	6.86	0.206	0.411
	0.3	86.6	0.40	207	2.45	1.45	0.581	11.0	0.071	0	16.9	9.92	-6.63	2.63	27.2	-3.02	-3.02	7.98	0.135	1.15
	0.4	77.1	0.29	215	5.26	1.43	0.674	5.55	0.033	0	30.4	19.0	-2.11	6.94	38.6	4.65	4.65	7.59	ST	ST
	0.45	61.1	0.18	207	12.3	1.40	0.682	2.84	0.014	0	95.3	58.6	2.71	2.00	112	25.6	25.6	16.5	0.03	14.4
0.1	0.1	94.8	0.48	350	0.439	1.46	0.444	45.6	0.395	0.39	10.2	4.16	-6.21	-0.076	16.1	-6.72	-6.72	5.04	0.299	0.146
	0.2	92.1	0.45	330	0.994	1.46	0.490	20.3	0.175	0.43	11.4	6.22	-6.37	0.044	19.7	-6.78	-6.78	4.98	0.235	0.284
	0.3	88.0	0.405	207	1.88	1.45	0.584	10.9	0.093	0.52	12.9	7.32	-6.02	1.74	20.4	-3.69	-3.69	6.71	0.162	0.732
	0.4	83.0	0.32	215	3.45	1.44	0.679	6.14	0.051	0.71	18.1	14.0	-5.45	3.61	29.8	-0.117	-0.152	10.1	0.121	1.52
	0.45	75.1	0.245	215	5.53	1.42	0.728	3.96	0.032	0.94	28.8	21.2	-2.56	7.70	39.4	6.93	6.89	9.22	0.087	2.65
0.5	0.1	98.6	0.48	350	0.255	1.47	0.452	46.7	0.68	1.21	6.18	2.74	-4.22	-0.27	9.65	-4.65	-4.65	3.68	0.359	0.07
	0.2	95.6	0.44	335	0.571	1.47	0.505	20.6	0.30	1.28	6.61	4.06	-4.43	-0.08	12.1	-4.42	-4.43	4.29	0.298	0.134
	0.3	92.5	0.405	330	1.03	1.47	0.593	11.5	0.17	1.42	7.14	4.00	-4.27	0.626	11.1	-3.26	-3.26	4.47	0.219	0.282
	0.4	89.0	0.35	220	1.73	1.46	0.696	6.74	0.10	1.68	8.96	6.62	-4.36	1.21	14.5	-2.47	-2.47	6.12	0.187	0.511
	0.45	86.4	0.305	220	2.28	1.45	0.755	5.21	0.08	1.83	9.74	7.39	-3.01	2.69	14.1	0.264	0.264	4.80	0.193	0.581

Pinkus, 1985.

β = 150°; (L/D) = m = 0.5; φ_L = 0.

*Frictional loss in grooves.

TABLE 2.3

Couette approximation solution for thrust bearings

All results are per single pad.

Re_0	E	$\dfrac{W}{\mu_0 R_1^2 \omega}\left(\dfrac{h_2}{R_1}\right)^2$	$Q/R_1^2\,h_2\omega$ At $\theta=0$	At R_1	At R_2	$(T_{max}-T_0)$	$\dfrac{H}{W\omega h_2}$
LAMINAR	0.0	1.92E-01	1.58	2.96E-01	4.37E-01	0.0	1.76E+01
	0.20	1.82E-01	1.58	2.94E-01	4.46E-01	5.27E-01	1.83E+01
	0.35	1.75E-01	1.59	2.93E-01	4.51E-01	8.62E-01	1.85E+01
	0.50	1.68E-01	1.59	2.92E-01	4.56E-01	1.16	1.87E+01
	1.00	1.51E-01	1.60	2.90E-01	4.69E-01	2.00	1.91E+01
1500	0.0	3.37E-01	1.61	3.24E-01	4.55E-01	0.0	2.15E+01
	0.20	3.07E-01	1.62	3.21E-01	4.65E-01	1.02	2.16E+01
	0.35	2.89E-01	1.62	3.19E-01	4.71E-01	1.63	2.18E+01
	0.50	2.75E-01	1.63	3.18E-01	4.76E-01	2.16	2.19E+01
	1.00	2.40E-01	1.64	3.14E-01	4.90E-01	3.60	2.23E+01
3500	0.0	5.67E-01	1.63	3.39E-01	4.62E-01	0.0	2.32E+01
	0.20	4.99E-01	1.64	3.34E-01	4.75E-01	1.74	2.32E+01
	0.35	4.63E-01	1.64	3.31E-01	4.82E-01	2.71	2.33E+01
	0.50	4.35E-01	1.65	3.29E-01	4.88E-01	3.53	2.34E+01
	1.00	3.69E-01	1.66	3.25E-01	5.04E-01	5.64	2.37E+01

Pinkus and Wilcock, 1985

$(L/R_2) = 0.5$; $\beta = 40°$; $\overline{h}_{11} = 3$; $\delta_{12} = 0.5$; $b = 0.8$

2.4.1 Journal Bearings

Before a solution of the energy equation is attempted, the latter is usually normalized so as to yield non-dimensional results in terms of a few basic parameters. In rectangular coordinates one such normalized form is given by

$$\left[1 - \frac{\overline{h}^2}{\overline{\mu}}\left(\frac{\partial\overline{p}}{\partial\theta}\right)\right]\frac{\partial\overline{T}}{\partial\theta} - \frac{\overline{h}^2}{\overline{\mu}}\left(\frac{\partial\overline{p}}{\partial\varsigma}\right)\cdot\frac{\partial\overline{T}}{\partial\varsigma}$$
$$= \frac{E\overline{\mu}}{\overline{h}^2}\left[1 + 3\frac{\overline{h}^2}{\overline{\mu}}\left(\frac{\partial\overline{p}}{\partial\theta}\right)^2 + 3\frac{\overline{h}^2}{\overline{\mu}}\left(\frac{\partial\overline{p}}{\partial\varsigma}\right)^2\right] \qquad (2.15)$$

where $\overline{T} = \alpha(T - T_0)$ and $\varsigma = (z/R)$.

The above is convenient when the exponential viscosity function given by Equ. (2.11) is used. In this case the solution to Equ. (2.15) is a function of the usual isothermal parameters plus the adiabatic constant E, or for a given geometry

$$\chi = \chi\left[\epsilon, \phi_0, (L/D), E\right]$$

where $E = 2(\mu_0\alpha/c_p\rho)(\omega R^2/C^2)$ contains all of the lubricant properties needed to solve the problem. If some other $\mu - T$ relationship is used which may contain more than one arbitrary constant or one which is not amenable to analytic representation, then the energy equation, by dividing out Equ. (2.15) by E, becomes

$$\left[1 - \frac{\overline{h}^2}{\overline{\mu}}\left(\frac{\partial \overline{p}}{\partial \theta}\right)\right]\frac{\partial \widehat{T}}{\partial \theta} - \frac{\overline{h}^2}{\overline{\mu}}\left(\frac{\partial \overline{p}}{\partial \varsigma}\right) \cdot \frac{\partial \widehat{T}}{\partial \varsigma}$$
$$= \frac{\overline{\mu}}{\overline{h}^2}\left[1 + 3\frac{\overline{h}^2}{\overline{\mu}}\left(\frac{\partial \overline{p}}{\partial \theta}\right)^2 + 3\frac{\overline{h}^2}{\overline{\mu}}\left(\frac{\partial \overline{p}}{\partial \varsigma}\right)^2\right] \quad (2.16)$$

where

$$\widehat{T} = \frac{c_p\rho(T - T_0)}{\mu_0\omega(R/C)^2}$$

The relation between the two normalized temperatures is simply

$$\overline{T} = \frac{1}{2}E\widehat{T}$$

Thus while Equ. (2.15) requires the use of the exponential $\mu - T$ relationship, Equ. (2.16) is free to use any arbitrary viscosity function.

To continue our scrutiny of results obtained by the various approaches to the THD problem, Table 2.4 gives data for the same set of cases used in Table 1.1, but this time based on Equ. (2.15). The temperature field is shown in Table 2.5 and in Fig. 2.13, the latter also showing the effective bearing temperature based on side leakage. Two things are worth noting:

- The average temperature T_s based on Equ. (2.8) falls below the average temperature obtained from the adiabatic THD solution.
- There is practically no variation in the axial (z) direction in temperature, and if anything there is a slight rise toward the edges of the bearing $(z = \pm L/2)$.

That the latter is not an isolated case is corroborated by the results of Fig. 2.14, which shows a similar $T(z) \simeq constant$ for a different L/D ratio bearing and for several different values of the adiabatic constant E.

Comparing further the present results with those catalogued in Table 1.1 we see that the adiabatic solution falls closer to the isothermal $\mu_0 = const.$ results than to those obtained on the basis of an effective temperature. These results are due to the profound effect of the mixing inlet temperature T_1. As was shown in Section 2.2 the use of Equ. (2.8) is tantamount to the employment of an inlet temperature T_1 much higher

TABLE 2.4

Two dimensional adiabatic solutions

Variable		ϵ	HP	K_{XX} (lb/in.) X 10^{-6}	ΔT °F	ΔT_{max} °F
C	0.003	0.405	6.94	2.53	24	59
in.	0.0045	0.55	6.21	2.27	13	45
	0.006	0.65	5.75	2.25	9	37
L	2	0.77	4.47	4.21	18	78
in.	3	0.55	5.23	2.27	13	45
	6	0.17	11.5	1.52	12	24
N	2,000	0.65	2.38	3.03	9	37
rpm	3,600	0.55	6.22	2.27	13	45

$D = 6$ in.; $W = 4{,}000$ lb; $\phi_L = 0$; $T_0 = 100$°F @ 41 cs/cs

$\beta = 150$°. Standard conditions: $C = 0.0045$ in.; $L = 3$ in.; $N = 3{,}600$ rpm.

then T_0. Thus while both the T_0 and the two-dimensional solutions are based on the same inlet temperature T_0, the effective viscosity approach uses $T_1 > T_0$.

A bar chart comparing bearing performance as calculated from the various approaches considered thus far is given in Fig. 2.15. As seen not only does anisothermal analysis make an important difference but the matter of mixing inlet temperature too has a telling effect. As may have been expected, the largest effect is on the resulting maximum temperatures and the viscous dissipation.

Table 2.6 and Fig. 2.16 give non-dimensional solutions for journal bearings based on the exponential viscosity-temperature relationship. Additional non-dimensional solutions for journal bearings are contained in Figs. 2.7 and 2.8 where they are compared with the results of the Couette approximations as well as that of an isothermal solution.

2.4.2 Thrust Bearings

As with the case of rectangular coordinates the energy equation in polar coordinates must first be normalized before a solution can be attempted.

TABLE 2.5

Two-dimensional
temperature field in 2-axial groove
journal bearing - bottom lobe
(for bearing data see Table 1.1)

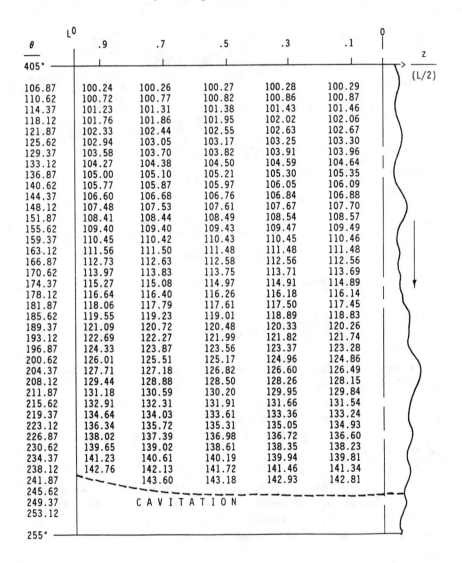

θ	L^0 .9	.7	.5	.3	.1	0
405°						(L/2)
106.87	100.24	100.26	100.27	100.28	100.29	
110.62	100.72	100.77	100.82	100.86	100.87	
114.37	101.23	101.31	101.38	101.43	101.46	
118.12	101.76	101.86	101.95	102.02	102.06	
121.87	102.33	102.44	102.55	102.63	102.67	
125.62	102.94	103.05	103.17	103.25	103.30	
129.37	103.58	103.70	103.82	103.91	103.96	
133.12	104.27	104.38	104.50	104.59	104.64	
136.87	105.00	105.10	105.21	105.30	105.35	
140.62	105.77	105.87	105.97	106.05	106.09	
144.37	106.60	106.68	106.76	106.84	106.88	
148.12	107.48	107.53	107.61	107.67	107.70	
151.87	108.41	108.44	108.49	108.54	108.57	
155.62	109.40	109.40	109.43	109.47	109.49	
159.37	110.45	110.42	110.43	110.45	110.46	
163.12	111.56	111.50	111.48	111.48	111.48	
166.87	112.73	112.63	112.58	112.56	112.56	
170.62	113.97	113.83	113.75	113.71	113.69	
174.37	115.27	115.08	114.97	114.91	114.89	
178.12	116.64	116.40	116.26	116.18	116.14	
181.87	118.06	117.79	117.61	117.50	117.45	
185.62	119.55	119.23	119.01	118.89	118.83	
189.37	121.09	120.72	120.48	120.33	120.26	
193.12	122.69	122.27	121.99	121.82	121.74	
196.87	124.33	123.87	123.56	123.37	123.28	
200.62	126.01	125.51	125.17	124.96	124.86	
204.37	127.71	127.18	126.82	126.60	126.49	
208.12	129.44	128.88	128.50	128.26	128.15	
211.87	131.18	130.59	130.20	129.95	129.84	
215.62	132.91	132.31	131.91	131.66	131.54	
219.37	134.64	134.03	133.61	133.36	133.24	
223.12	136.34	135.72	135.31	135.05	134.93	
226.87	138.02	137.39	136.98	136.72	136.60	
230.62	139.65	139.02	138.61	138.35	138.23	
234.37	141.23	140.61	140.19	139.94	139.81	
238.12	142.76	142.13	141.72	141.46	141.34	
241.87		143.60	143.18	142.93	142.81	
245.62						
249.37		C A V I T A T I O N				
253.12						
255°						

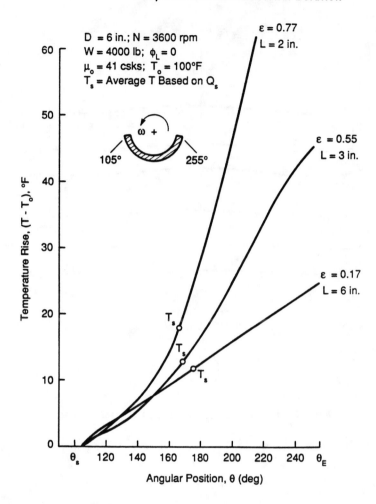

Fig. 2.13 Temperature distribution at bearing centerline for two-dimensional solution

Using the non-dimensional quantities of

$$\bar{p} = \frac{p(h_2/R_2)^2}{\mu_0 \omega}$$

$$\hat{T} = \frac{\rho c_p (T - T_0)}{\mu_0 \omega (R_2/h_2)^2}$$

(2.17)

we obtain from Equ. (2.5)

$$\left[6 - \frac{\bar{h}^2}{\bar{\mu}\bar{r}^2}\left(\frac{\partial\bar{p}}{\partial\theta}\right)\right]\left(\frac{\partial\hat{T}}{\partial\theta}\right) - \frac{\bar{h}^2}{\bar{\mu}}\left(\frac{\partial\bar{p}}{\partial\bar{r}}\right)\left(\frac{\partial\hat{T}}{\partial\bar{r}}\right)$$

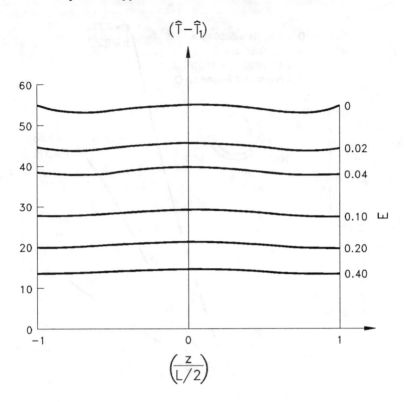

Fig. 2.14 Axial temperatures in journal bearings

$$= \frac{12\overline{\mu r}^2}{\overline{h}^2} + \frac{\overline{h}^2}{\overline{\mu}}\left[\left(\frac{1}{\overline{r}}\frac{\partial\overline{p}}{\partial\theta}\right)^2 + \left(\frac{\partial\overline{p}}{\partial\overline{r}}\right)^2\right] \tag{2.18}$$

In terms of the adiabatic constant E the above equation transforms into

$$\left[6 - \frac{\overline{h}^2}{\overline{\mu r}^2}\left(\frac{\partial\overline{p}}{\partial\theta}\right)\right]\left(\frac{\partial\overline{T}}{\partial\theta}\right) - \frac{\overline{h}^2}{\overline{\mu}}\left(\frac{\partial\overline{p}}{\partial\overline{r}}\right)\left(\frac{\partial\overline{T}}{\partial\overline{r}}\right)$$

$$= E\left\{\frac{6\overline{\mu r}^2}{\overline{h}^2} + \frac{\overline{h}^2}{2\overline{\mu}}\left[\left(\frac{1}{\overline{r}}\frac{\partial\overline{p}}{\partial\theta}\right)^2 + \left(\frac{\partial\overline{p}}{\partial\overline{r}}\right)^2\right]\right\} \tag{2.19}$$

When this equation is solved by finite difference methods the sectorial element is usually subdivided in the manner given in Fig. 2.17 in which case the expression for a nodal temperature $T_{i,j+1}$ is given by the finite difference equation

$$\widehat{T}_{i,j+1} = \frac{2\Delta\overline{\theta}\left(\frac{\overline{r}^2}{3\overline{h}}\right)_{i,j} + \left.\frac{\overline{h}^3}{\overline{\mu}}\right|_{i,j}\left[\left(\frac{\overline{p}_{i,j+1}-\overline{p}_{i,j-1}}{2\overline{r}\Delta\overline{\theta}}\right)^2 + \left(\frac{\overline{p}_{i+1,j}-\overline{p}_{i-1,j}}{2\Delta\overline{r}}\right)^2\right]}{\left[\left.\frac{\overline{h}}{\overline{h}}\right|_{i,j} - \left.\frac{\overline{h}^3}{\overline{\mu r}^2}\right|_{i,j}\left(\frac{\overline{p}_{i,j+1}-\overline{p}_{i,j+1}}{2\overline{\theta}}\right)\right]}$$

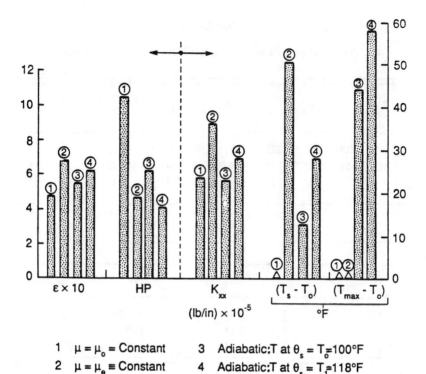

Fig. 2.15 Journal bearing solutions according to four different methods of thermal approach

$$
+ \frac{\left.\dfrac{\bar{h}^3}{\bar{\mu}}\right|_{i,j} \left(\dfrac{\bar{p}_{i+1,j}+\bar{p}_{i-1,j}}{2\Delta\bar{r}}\right)\left(\dfrac{\widehat{T}_{i+1,j}-\widehat{T}_{i-1,j}}{2\Delta\bar{r}}\right)}{\left[\left.\bar{h}\right|_{i,j} - \left.\dfrac{\bar{h}^3}{\bar{\mu}r^2}\right|_{i,j}\left(\dfrac{\bar{p}_{i,j+1}-\bar{p}_{i,j-1}}{2\Delta\bar\theta}\right)\right]}
$$
$$
+ \frac{\left[\left.\bar{h}\right|_{i,j} - \left.\dfrac{\bar{h}^3}{\bar{\mu}r^2}\right|_{i,j}\left(\dfrac{\bar{p}_{i,j+1}-\bar{p}_{i,j-1}}{2\Delta\theta}\right)\right]\dfrac{\widehat{T}_{i,j-1}}{2\Delta\bar\theta}}{\left[\left.\bar{h}\right|_{i,j} - \left.\dfrac{\bar{h}^3}{\bar{\mu}r^2}\right|_{i,j}\left(\dfrac{\bar{p}_{i,j+1}-\bar{p}_{i,j-1}}{2\Delta\bar\theta}\right)\right]}
\tag{2.20}
$$

The geometry $\bar{h}$ above can be that of Equ. (2.14) or any other chosen film shape.

An example of pressure and temperature distributions resulting from a two- dimensional adiabatic solution for the kind of geometry represented by Equ. (2.14) is shown in Fig. 2.18. As seen, unlike journal bearings where the maximum temperature profile occurs at the bearing centerline, here the maximum temperatures occur at the outer radius with the peak

TABLE 2.6

Two-dimensional adiabatic solutions for two axial groove bearing

$\bar{W}$	ϵ	ϕ_0	υ_2	$\bar{M}$	$\bar{Q}_S$	$\left(\dfrac{T_S - T_0}{T_0}\right)^*$	$\left(\dfrac{T_{max} - T_0}{T_0}\right)^*$
0.201	.124	84°	255°	9.27	.206	.121	.24
1.611	.550	51°	240°	9.74	.694	.13	.45
4.03	.750	40°	230°	10.5	.797	.18	.77

$\bar{K}$				$\bar{B}$			
yy	yx	xy	xx	yy	yx	xy	xx
0.767	- 3.36	1.26	0.454	3.31	-0.106	-.137	1.26
8.22	-10.6	2.00	4.74	9.57	-2.25	-2.46	2.95
33.06	-27.63	0.61	11.48	22.90	-4.95	-5.83	4.72

$$L/D = 0.5; \quad \phi_L = 0$$

*In (°F/°F) at $T_0 = 100°F$.

temperature lying at the (R_2, β) coordinate. To remedy this, the clearances in thrust bearings are often opened up at the outer radius so as to reduce the dissipation and permit the ingress of more lubricant at the cold supply temperature T_0. A comprehensive set of solutions for such thrust bearings with a radially divergent clearance and no flat $(b = 1)$ were obtained by Strömberg, 1971. The film geometry of these bearings is given by

$$\bar{h} = 1 + K\bar{r}\sin(\beta - \theta) \tag{2.21}$$

with $K > 0$, a constant. The range of parameters spanned by these solutions is as follows:

$$(L/R_2) = 0.25, \quad 0.5, \quad ,0.75$$
$$\beta = 30°, \quad 45°, \quad 60°, \quad 90°$$
$$K = 0.5 - 7.0$$
$$E = 0 - 4$$

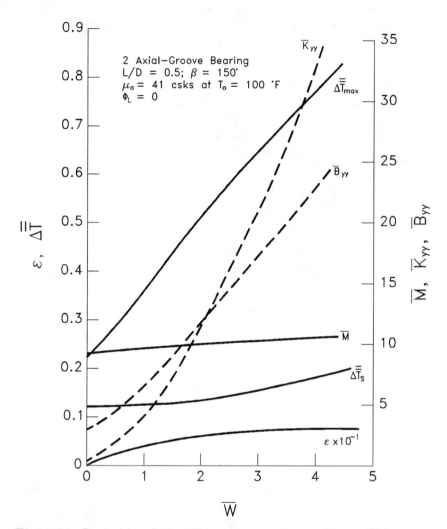

Fig. 2.16 Journal bearing performance according to the two-dimensional solution

The solutions are given in Figs. 2.19, 2.20, and 2.21, all for an exponential $\mu - T$ relationship. These solutions contain the effect of a mixing inlet temperature as given by Equ. (2.9). The plots showing the variation of T_{max} include the case of $E = 0$. This is a somewhat fictitious curve, representing the temperature rise for a hypothetical fluid whose viscosity remains constant regardless of the temperature. However, the curve is of practical use when values of E lying between 1 and 0 are being sought. By defining the upper limit of T_{max} for $E = 0$ it is then possible to interpolate for all $0 < E < 1$ cases. Finally Table 2.7 provides average temperature

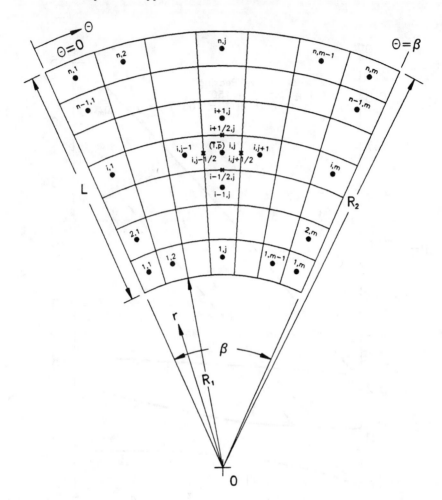

Fig. 2.17 Discretization of sectorial pad for finite difference thermal solution

rises at both the inlet and outlet of the fluid film based on a value of T_a obtained from

$$T_a = \frac{1}{Q_r} \int_{R_1}^{R_2} q(r) \left[T(r) - T_0 \right] dr \qquad (2.22)$$

taken at either $\theta = 0$ or at $\theta = \beta$.

2.4.3 Pumping Rings

Whatever the difficulties in solving the set of energy and Reynolds equations, these have been at least conceptually straightforward. The next case will be examined in some detail because it exhibits a number of complexities which mirror many of the difficulties inherent in THD analysis.

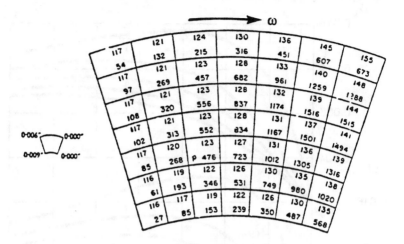

a) With Radial Taper

Lower Number = Pressure in psi
Upper Number = Temperature in °F

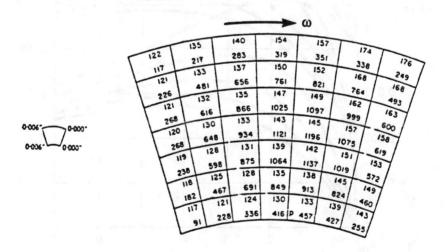

b) Without Radial Taper

$\beta = 40°$; $L/R_2 = 0.56$; $\mu_0 = 23.2$ csks; $h_2 = 0.002$ in.;
$P = 566$ psi; $T_1 = 115°F$; $N = 3600$ rpm

Fig. 2.18 Pressure and temperatures in thrust bearing pad based on the two-dimensional solution. (Sternlicht, 1957)

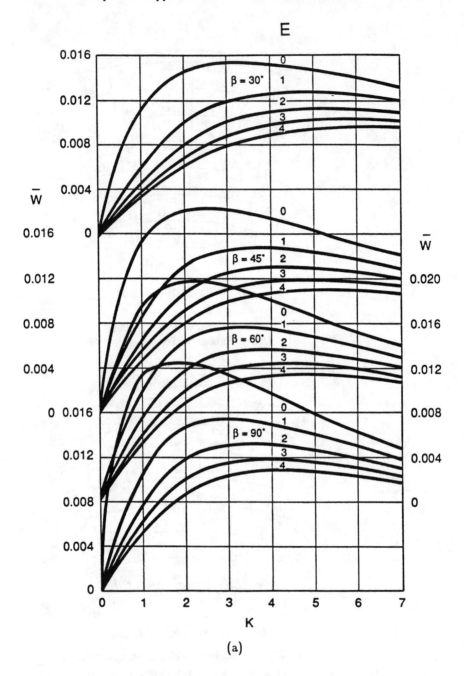

Fig. 2.19 Effect of the parameter E on load capacity of tapered land thrust bearing pad (Strömberg, 1971), (a) $L/R_2 = 0.25$, (b) $L/R_2 = 0.50$, (c) $L/R_2 = 0.75$

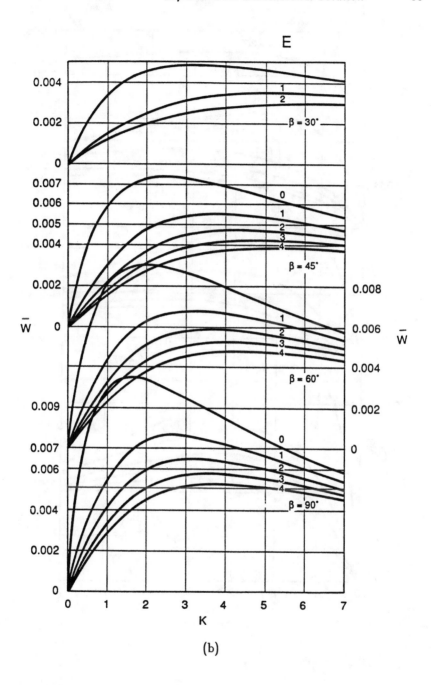

Fig. 2.19 Effect of the parameter E on load capacity of tapered land thrust bearing pad (Strömberg, 1971), (a) $L/R_2 = 0.25$, (b) $L/R_2 = 0.50$, (c) $L/R_2 = 0.75$

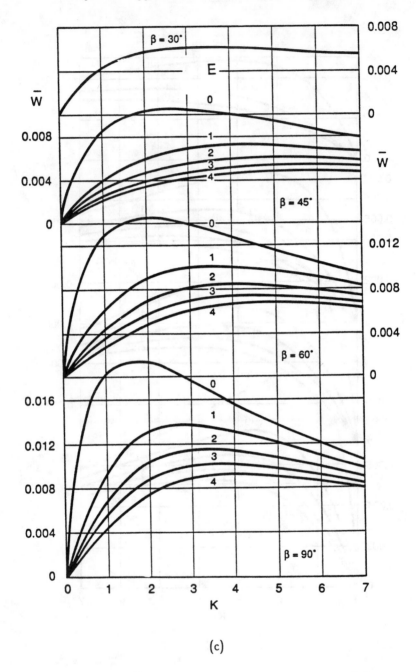

(c)

Fig. 2.19 Effect of the parameter E on load capacity of tapered land thrust bearing pad (Strömberg, 1971), (a) $L/R_2 = 0.25$, (b) $L/R_2 = 0.50$, (c) $L/R_2 = 0.75$

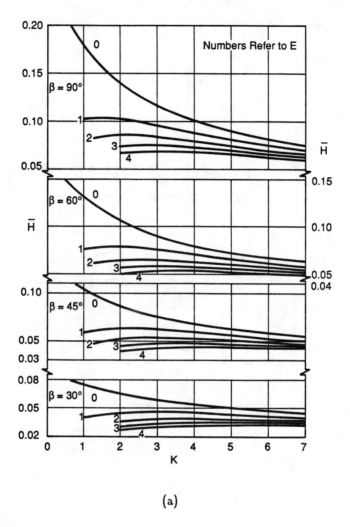

(a)

Fig. 2.20 Effects of the parameter E on power loss in tapered land thrust bearing pad (Strömberg, 1971), (a) $L/R_2 = 0.25$, (b) $L/R_2 = 0.50$,(c) $L/R_2 = 0.75$

This is the case of a reciprocating seal, commonly called a pumping ring, shown schematically in Fig. 2.22. The pumping ring is a particularly suitable candidate for scrutiny because, due to its axial symmetry, it can be treated as an infinitely long slider, both simplifying and clarifying the arguments that will arise in the course of the analysis.

The coordiante system used for the pumping ring is shown in Fig. 2.23. With the z derivatives eliminated from Equ. (2.4) the energy equation for

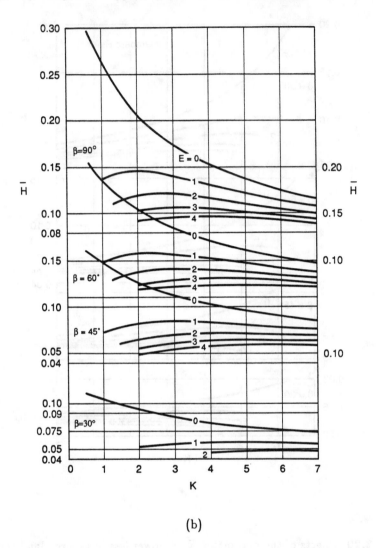

(b)

Fig. 2.20 Effects of the parameter E on power loss in tapered land thrust bearing pad (Strömberg, 1971), (a) $L/R_2 = 0.25$, (b) $L/R_2 = 0.50$,(c) $L/R_2 = 0.75$

the pumping ring becomes

$$c_p\rho\left[\frac{Uh}{2} - \frac{h^3}{12\mu}\left(\frac{dp}{dx}\right)\right]\frac{dT}{dx} = \frac{\mu U^2}{h} + \frac{h^3}{12\mu}\left(\frac{dp}{dx}\right)^2 \qquad (2.23)$$

Utilizing the one-dimensional Reynolds equation

$$\frac{d}{dx}\left(\frac{h^3}{\mu}\frac{dp}{dx}\right) = 6U\left(\frac{dh}{dx}\right)$$

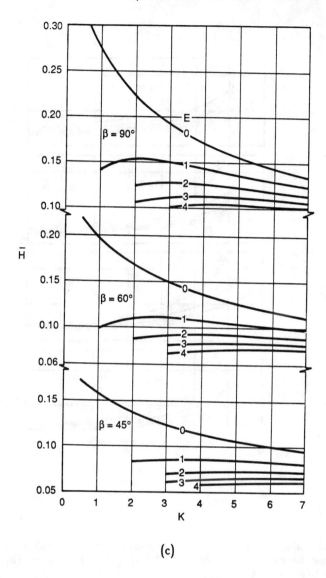

(c)

Fig. 2.20 Effects of the parameter E on power loss in tapered land thrust bearing pad (Strömberg, 1971), (a) $L/R_2 = 0.25$, (b) $L/R_2 = 0.50$,(c) $L/R_2 = 0.75$

we obtain, with K a constant of integration, in normalized form the energy equation

$$\frac{d\widetilde{T}}{d\overline{x}} = \frac{\overline{\mu}}{K}\left[\left(\frac{\overline{h}-K}{\overline{h}^3}\right) + \frac{1}{3\overline{h}}\right]$$

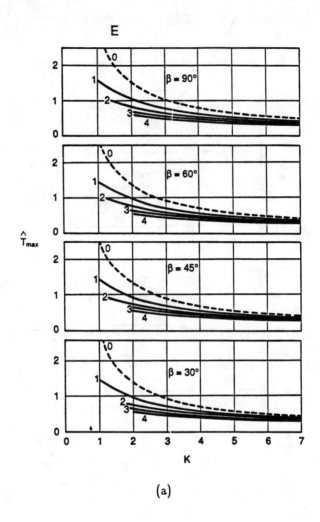

(a)

Fig. 2.21 Effect of the parameter E on maximum temperature in tapered land thrust bearing pad (Strömberg, 1971), (a) $L/R_2 = 0.25$, (b) $L/R_2 = 0.50$, (c) $L/R_2 = 0.75$

where

$$\widetilde{T} = \frac{\rho c_p C^2 (T - T_0)}{6\mu_0 U B}$$

with C denoting the original, undeformed radial clearance between the ring and shaft.

The flow pattern at the interface of a pumping ring is somewhat problematic, consisting as it does of three flow layers possessing the following characteristics:

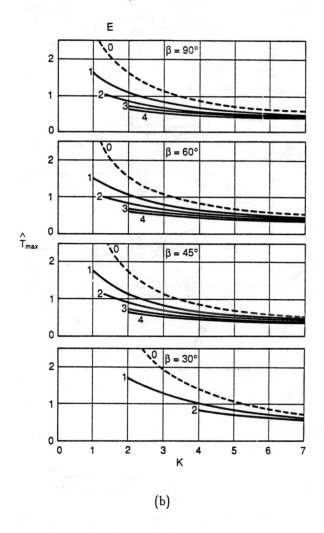

(b)

Fig. 2.21 Effect of the parameter E on maximum temperature in tapered land thrust bearing pad (Strömberg, 1971), (a) $L/R_2 = 0.25$, (b) $L/R_2 = 0.50$, (c) $L/R_2 = 0.75$

- Forward flow - This flow is along the moving surface. It enters at a temperature T_0 and is heated on its way to the reservoir to some T_{max} at $x = L$.
- Recirculating flow - This flow also enters at a temperature T_0, but various portions of that flow penetrate only part way into the film before they are reversed and returned to their source. It undergoes relatively little viscous shearing, resulting in low energy dissipation to the fluid.
- Back flow - This flow originates in the reservoir entering at a tem-

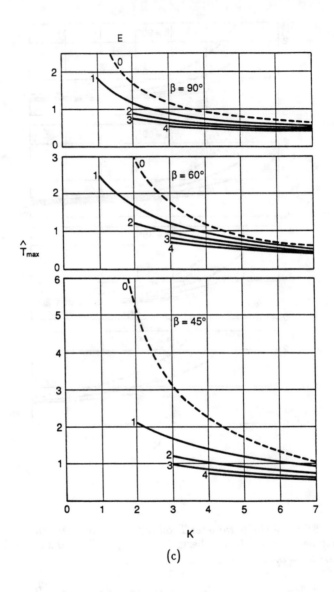

Fig. 2.21 Effect of the parameter E on maximum temperature in tapered land thrust bearing pad (Strömberg, 1971), (a) $L/R_2 = 0.25$, (b) $L/R_2 = 0.50$, (c) $L/R_2 = 0.75$

perature T_f and is heated while traveling upstream. Its maximum temperature is reached at the entrance to the pumping ring, at $x = 0$. Thus we are faced here with an extreme case of complications resulting

TABLE 2.7

Mean inlet and outlet temperatures in
adiabatic thrust bearings

$(L/R_2) = 0.5$

K	E 0 (1)*	(2)*	1 (1)	(2)	2 (1)	(2)	3 (1)	(2)	4 (1)	(2)
30° 0.5	10.09	10.91								
1	4.587	5.310								
2	1.953	2.540	1.051	1.357						
3	1.091	1.577	0.7259	1.035						
4	0.7354	1.163	0.5362	0.8368	0.4388	0.6772				
5.5	0.4622	0.8267	0.3343	0.5710	0.3159	0.5533				
7	0.3195	0.6392	0.2714	0.5347	0.2469	0.4771	0.2158	0.4169		
45° 0.5	5.996	6.925								
1	2.575	3.347	1.264	1.632						
2	1.022	1.610	0.6796	1.059	0.5316	0.8259	0.4464	0.6920	0.3910	0.5039
3	0.5858	1.071	0.4318	0.7869	0.3584	0.5487	0.3097	0.5586	0.2749	0.4947
4	0.3698	0.7822	0.3009	0.6283	0.2595	0.5369	0.2297	0.4729	0.2082	0.4264
5.5	0.2252	0.5697	0.1969	0.4888	0.1738	0.4294	0.1583	0.3879	0.1466	0.3561
7	0.1527	0.4518	0.1375	0.4004	0.1254	0.3613	0.1165	0.3321	0.1081	0.3068
60° 0.5	4.864	5.942								
1	1.999	2.851	1.053	1.499						
2	0.7462	1.3711	0.5230	0.9513	0.4180	0.7574	0.3566	0.5446	0.3135	0.5672
3	0.4045	0.9052	0.3181	0.7049	0.2688	0.5904	0.2362	0.5169	0.2117	0.4630
4	0.2577	0.5515	0.2160	0.5643	0.1892	0.4895	0.1691	0.4361	0.1546	0.3959
5.5	0.1541	0.5049	0.1364	0.4393	0.1229	0.3934	0.1130	0.3580	0.1048	0.3297
7	0.1035	0.4072	0.0942	0.3648	0.0867	0.3321	0.0808	0.3062	0.0758	0.2859
90° 0.5	4.536	5.912								
1	1.754	2.798	0.9404	1.493						
2	0.6144	1.333	0.4364	0.9380	0.3511	0.7519	0.3807	0.6420	0.2645	0.5637
3	0.3201	0.8807	0.2548	0.6943	0.2159	0.5831	0.1907	0.5125	0.1719	0.4619
4	0.1991	0.6675	0.1676	0.5553	0.1474	0.4831	0.1318	0.4294	0.1215	0.3952
5.5	0.1169	0.5017	0.1027	0.4344	0.0929	0.3901	0.0857	0.3570	0.0792	0.3255
7	0.0780	0.4116	0.0702	0.3655	0.0647	0.3341	0.0597	0.3044	0.0551	0.2850

*(1) Mean Inlet Temperture $\hat{T}_{a1}$ at $\theta = 0$ Stromberg, 1971

(2) Mean Outlet Temperature $\hat{T}_{a2}$ at $\theta = \beta$

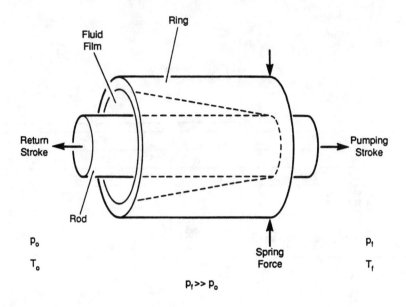

Fig. 2.22 Schematic of pumping ring seal

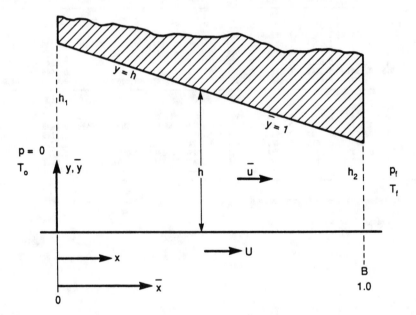

Fig. 2.23 Pumping ring coordinate system

from both recirculating and back flows. However, since the bulk of the intermediate layer recirculates near the entrance where the temperature differential is relatively small, its effect may be left out of the heat balance.

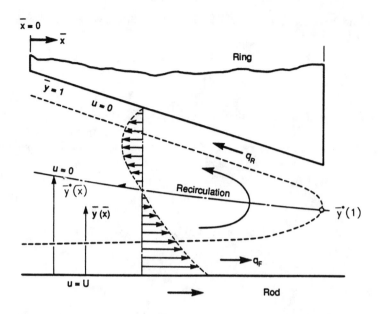

Fig. 2.24 Flow regimes in pumping ring

This treatment represents a conservative approach because inclusion of the recirculating flow would yield lower temperatures and thus safer operating conditions.

The first task is to find expressions for the three flow regimes in terms of ring geometry and its operating conditions. The details of these calculations are not very pertinent to the thermal aspects of the problem; they can be found in Walowit and Pinkus (1986). It will be sufficient to give the resulting expressions.

Referring to Fig. 2.24 any streamline at the interface is given by

$$\psi(\overline{y}, \overline{x}) = \overline{h}\overline{y}(\overline{y}-1)^2 + K\overline{y}^2\left(\frac{3}{2} - \overline{y}\right) \tag{2.24}$$

where for $\mu = \mu(x)$ the value of K is

$$K = \frac{-\tilde{p}_f + \int_0^1 \frac{\overline{\mu}d\overline{x}}{\overline{h}^2}}{\int_0^1 \frac{\overline{\mu}d\overline{x}}{\overline{h}^2}}$$

and

$$\tilde{p}_f = \frac{p_f C^2}{6\mu U L}$$

The individual streamlines are obtained by assigning different constants to $\psi(\overline{y}, \overline{x})$ in Equ. (2.24). Back flow will commence when the dividing streamline between the forward and recirculating streams reaches $\overline{x} = 1$. This occurs at

$$\overline{y}^* = \frac{1}{3\left[1 - \frac{K}{\overline{h}_2^2}\right]} = 1$$

or at $K/\overline{h}_2 = 2/3$. At values of $K/\overline{h}_2^2 < 2/3$ the tangent point of the recirculating envelope at $\overline{x} = 1$ will lie below $\overline{y} = 1$, opening up a passage for back flow. Below the recirculating envelope the fluid will, as shown in Fig. 2.26, flow forward at a rate of

$$\overline{q}_F = \psi[\overline{y}^*(1), 1] \qquad (2.25a)$$

whereas above the envelope there will be back flow at a rate of

$$q_B = q_F - q_{NET} = \psi[\overline{y}^*, 1] - \left(\frac{K}{2}\right) \qquad (2.25b)$$

Having determined the flow boundaries, thermal mapping can be attempted. By the nature of the imposed constraints namely,

- no temperature variation across the film
- no heat transfer to the boundaries
- no fluid heat conduction

the only variation will be in the x direction.

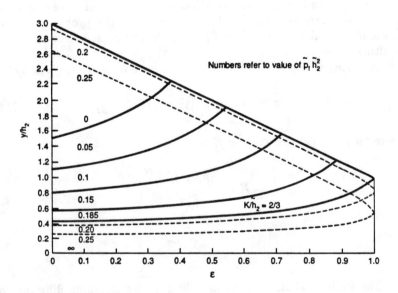

Fig. 2.25 Fluid film streamlines for $a = 3$ (Walowit and Pinkus, 1986)

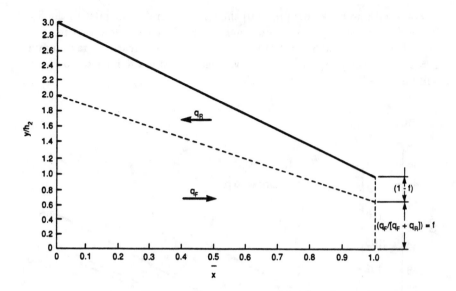

Fig. 2.26 Subdivision of clearance space for convective heat flow

As stated previously, the viscous heating is assumed to be convected away only by the fluid that transits the interspace, be it to the right or to the left. For this purpose, the interspace is modelled so as to be filled with these two streams, i.e., a forward flow q_F and a back flow q_B as shown in Fig. 2.25. Designating by

$$f = \frac{q_F}{q_F + q_B}$$

$f \cdot \Delta T(x)$ will be the fraction of heat convected to the right and $(1-f)\Delta T(x)$ the fraction convected to the left. The above ΔT's will be averaged across the film (i.e. in the y but not in the x direction) with $T_F(x)$ and $T_B(x)$ representing the temperature profiles generated in the two flow layers, q_F and q_B. The overall average temperature profile will be obtained from

$$T(x) = fT_F(x) + (1 - f)T_B(x) \qquad (2.26)$$

The expression for the viscous heating, given by the right hand side of Equ. (2.23), is exact in the sense that the losses are calculated over the exact velocity profile, including the region of recirculating flow, namely

$$\Psi(x) = \mu(x) \int_0^h \left(\frac{du}{dy}\right)^2 dy \qquad (2.27)$$

where $u(x, y)$ is the velocity profile shown in Fig. 2.24. Consistent with the present flow model, $\mu(x)$ in the calculation of this viscous shear will be that corresponding to the average temperature $T(x)$. It should be noted that, whereas with no back flow $(q_B = 0)$ the temperature at the inlet to the film is a constant, say T_0, this is no longer true when back flow sets in. The back flow, starting at an initial temperature, T_f, will reach a maximum at $x = 0$ and, since temperatures are averaged across y, the inlet temperature will be some $T_1 > T_0$.

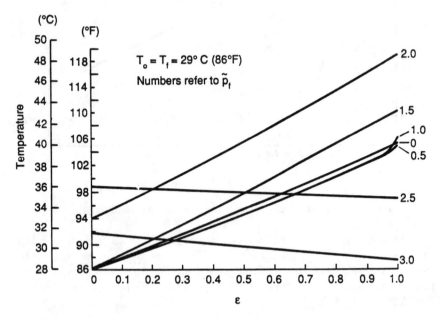

Fig. 2.27 Pumping ring temperature profiles for several different values of $\tilde{p}_f$ (Walowit and Pinkus, 1986)

A sample solution for the temperature profile in the fluid film for various values of upstream pressure is given in Fig. 2.27. The solutions are for the case of equal upstream and downstream boundary temperatures. Curves which start at $T = 86°F$ are those without back flow and thus have what may be called a conventional profile. However, at $\tilde{p}_f > 1.5$, there is back flow and, due to the averaging of temperatures across y, the inlet temperature is higher than 86°F. It is interesting to note that due to the cooling effect of the back flow, the maximum temperatures in the film actually decrease as the back flow increases above a certain level. Thus, for $\tilde{p}_f = 0$, $T_{max} = 107°F$, but for $\tilde{p}_f = 3.0$, $T_{max} = 90°F$ and occurs not at the trailing but at the leading edge of the film. The highest temperatures occur at some intermediate combination of forward and back flows; in this particular example it happens at $p_f = 2.0$ with T_{max} reaching 120°F.

The problem is still more difficult when the reverse stroke is to be considered. This situation shown in Fig. 2.28 is complicated by the presence of cavitation and often by direct contact between ring and rod at $x = B$.

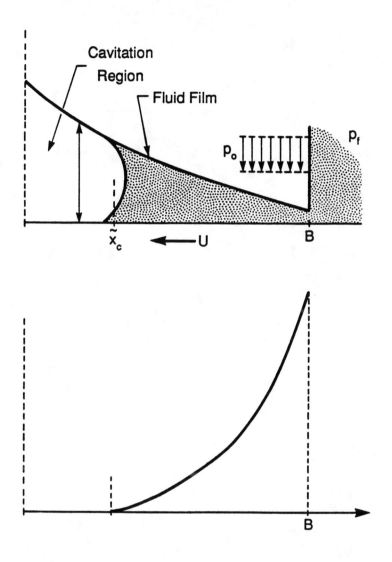

Fig. 2.28 Fluid film during reverse stroke in pumping ring

Then, in addition to viscous dissipation, contact frictional resistance must be taken into account.

Chapter 3

TRANSVERSE TEMPERATURE VARIATIONS

The adiabatic treatment presented in the previous chapter applies to a number of practical tribological applications and can, therefore, be used with a fair degree of confidence to obtain realistic solutions. This pertains for example to large bearings where the rate of viscous dissipation is high relative to the heat capacity of the bounding surfaces, or to films where lubricant flow is high, and naturally, to systems with insulated surfaces. But most tribological applications do not fall within these categories and most seals or EHD systems do not resemble adiabatic conditions at all.

Moreover, from a rigorous thermal standpoint, an adiabatic solution with an invariant transverse temperature presents a contradiction in terms. It was pointed out that the Reynolds equation ignores variations in the y direction. That transverse pressure variations are negligible in tribological phenomena can be shown to be both theoretically and practically valid, but the hypothesis that the temperature too remains constant across y leads to serious difficulties. To start with, the frictional losses incurred across a lubricant film are given by

$$dH = \tau u dx$$

where

$$\tau = \mu \left(\frac{\partial u}{\partial y} \right)$$

It is clear that since both the gradient $(\partial u/\partial y)$ and the velocity u differ at various locations of y, the energy dissipation too will differ, and so will the local temperatures.

More serious complications follow. In the adiabatic approach, there is no heat transfer to or from the bounding surfaces. For the latter to hold, the temperature gradients at the interface of the two surfaces must

79

be zero. One of two things must then be true: either the runner and bearing have at each (x, z) location identical temperatures, in which case $T(y)$ can conceivably be constant; or, if $T_R \neq T_S$, there must be a variation in T with y. Now, it is contrary to experience to say that the $T(x, z)$ variation of the stationary surface is identical to that of the journal or runner, thus excluding the first hypothesis. The conclusion is that even adiabaticity of the lubricant film implies a transverse variation in temperature.

3.1 THE BASIC DIFFERENTIAL EQUATIONS

3.1.1 The Energy Equation

In order to assess the character of the basic differential equation of thermohydrodynamic lubrication, and particularly the extent to which certain terms are retained and others ignored, the three-dimensional energy equation will here be derived from first principles. We shall consider a cubic element of dimension Δx, Δz, and Δy and account for the various heat and work fluxes flowing in and out of this elementary volume of fluid. Under steady state conditions these are simply

$$E_i - E_0 = H_i - H_0$$

where

E_i = energy transported into the control volume

E_0 = energy transported out of the control volume

H_i = work done on the fluid volume by the surroundings

H_0 = work done by the fluid volume on the surroundings

The above equation ignores the following modes of energy transfer
- radiation
- effect of external body forces
- gravity effects

The basic modes of energy transport into the fluid are by fluid conduction and by convection of intrinsic energy, i.e., of fluid possessing both kinetic and internal energies. The transported energies and mechanical work involved are indicated separately on the control volumes of Fig. 3.1a and b; so as not to encumber the sketches, not all components are specified. The differential changes in energies are all taken about the midpoint 0 of the control volumes.

The transported energies of Fig. 3.1a summed over the surfaces of the control volume are

$$E_0 - E_i = \left\{ \left[\frac{\partial(\rho u e)}{\partial x} + \frac{\partial(\rho v e)}{\partial y} + \frac{\partial(\rho w e)}{\partial z} \right] - \left[\frac{\partial}{\partial x}\left(k\frac{\partial T}{\partial x}\right) \right. \right.$$
$$\left. \left. + \frac{\partial}{\partial y}\left(k\frac{\partial T}{\partial y}\right) + \frac{\partial}{\partial z}\left(k\frac{\partial T}{\partial z}\right) \right] \right\} \Delta x \Delta y \Delta z \qquad (3.1a)$$

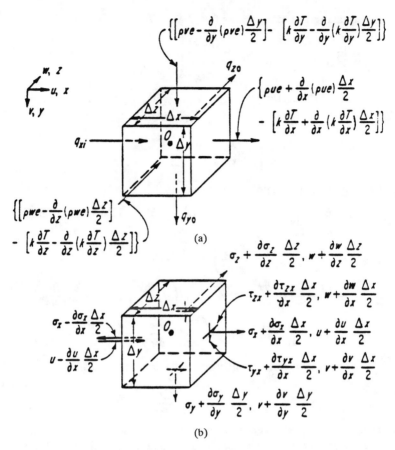

Fig. 3.1 Control volumes for transported and mechanical energies

where the intrinsic energy is given by

$$e = \frac{u^2 + v^2 + w^2}{2} + c_v T$$

The elements of mechanical work, too, must be summed over the volume surfaces. Here work done by the fluid will be taken to be the work performed by the fluid on the upstream surfaces of the control volume, i.e., where the velocity components are in a direction opposite to the stress components. With this approach, the net work is given by

$$H_i - H_0 = \left[\frac{\partial}{\partial x} \left(u\sigma_x + v\tau_{yx} + w\tau_{zx} \right) + \frac{\partial}{\partial y} \left(u\tau_{xy} + v\sigma_y + w\tau_{zy} \right) \right.$$
$$\left. + \frac{\partial}{\partial z} \left(u\tau_{xz} + v\tau_{yz} + w\sigma_z \right) \right] \Delta x \Delta y \Delta z \tag{3.1b}$$

By equating the expressions for E and H we obtain

$$\left[\frac{\partial(\rho u e)}{\partial x} + \frac{\partial(\rho v e)}{\partial y} + \frac{\partial(\rho w e)}{\partial z}\right]$$

$$-\left[\frac{\partial}{\partial x}\left(k\frac{\partial T}{\partial x}\right) + \frac{\partial}{\partial y}\left(k\frac{\partial T}{\partial y}\right) + \frac{\partial}{\partial z}\left(k\frac{\partial T}{\partial z}\right)\right]$$

$$= \frac{\partial}{\partial x}(u\sigma_x + v\tau_{yx} + w\tau_{zx}) + \frac{\partial}{\partial y}(u\tau_{xy} + v\sigma_y + w\tau_{zy})$$

$$+ \frac{\partial}{\partial z}(u\tau_{xz} + v\tau_{yz} + w\sigma_z)$$

and by rearranging some of the terms,

$$\rho\left[u\frac{\partial e}{\partial x} + v\frac{\partial e}{\partial y} + w\frac{\partial e}{\partial z}\right] - \left[\frac{\partial}{\partial x}\left(k\frac{\partial T}{\partial x}\right) + \frac{\partial}{\partial y}\left(k\frac{\partial T}{\partial y}\right) + \frac{\partial}{\partial z}\left(k\frac{\partial T}{\partial z}\right)\right]$$

$$= u\left(\frac{\partial\sigma_x}{\partial x} + \frac{\partial\tau_{xy}}{\partial y} + \frac{\partial\tau_{xz}}{\partial z}\right) + v\left(\frac{\partial\tau_{yz}}{\partial x} + \frac{\partial\sigma_y}{\partial y} + \frac{\partial\tau_{yz}}{\partial z}\right)$$

$$+ w\left(\frac{\partial\tau_{zx}}{\partial x} + \frac{\partial\tau_{zy}}{\partial y} + \frac{\partial\sigma_z}{\partial z}\right) + \left(\sigma_x\frac{\partial u}{\partial x} + \sigma_y\frac{\partial v}{\partial y} + \sigma_z\frac{\partial w}{\partial z}\right)$$

$$+ \tau_{xy}\left(\frac{\partial u}{\partial y} + \frac{\partial v}{\partial x}\right) + \tau_{yz}\left(\frac{\partial v}{\partial z} + \frac{\partial w}{\partial y}\right) + \tau_{zx}\left(\frac{\partial w}{\partial x} + \frac{\partial u}{\partial z}\right) \quad (3.1c)$$

where use was made of the continuity equation

$$\frac{\partial(\rho u)}{\partial x} + \frac{\partial(\rho v)}{\partial y} + \frac{\partial(\rho w)}{\partial z} = 0$$

The equilibrium equations of fluid flow for steady state conditions are given by the expressions (see, for example, Pinkus and Sternlicht, 1961, p. 4)

$$\rho\left(u\frac{\partial u}{\partial x} + v\frac{\partial u}{\partial y} + w\frac{\partial u}{\partial z}\right) = \frac{\partial\sigma_x}{\partial x} + \frac{\partial\tau_{xy}}{\partial y} + \frac{\partial\tau_{xz}}{\partial z}$$

$$\rho\left(u\frac{\partial v}{\partial x} + v\frac{\partial v}{\partial y} + w\frac{\partial v}{\partial z}\right) = \frac{\partial\tau_{yx}}{\partial x} + \frac{\partial\sigma_y}{\partial y} + \frac{\partial\tau_{yz}}{\partial z}$$

$$\rho\left(u\frac{\partial w}{\partial x} + v\frac{\partial w}{\partial y} + w\frac{\partial w}{\partial z}\right) = \frac{\partial\tau_{zx}}{\partial x} + \frac{\partial\tau_{zy}}{\partial y} + \frac{\partial\sigma_z}{\partial z}$$

Using for σ_i and τ_{ij} their equivalents in terms of pressure and velocity gradients, and substituting the last three expressions into Equ. (3.1c), we obtain

$$\rho\left(u\frac{\partial e}{\partial x} + v\frac{\partial e}{\partial y} + w\frac{\partial e}{\partial z}\right) - \left[\frac{\partial}{\partial x}\left(k\frac{\partial T}{\partial x}\right) + \frac{\partial}{\partial y}\left(k\frac{\partial T}{\partial y}\right) + \frac{\partial}{\partial z}\left(k\frac{\partial T}{\partial z}\right)\right]$$

$$= \rho\left\{u\frac{\partial[(u^2 + v^2 + w^2)/2]}{\partial x} + v\frac{\partial[(u^2 + v^2 + w^2)/2]}{\partial y}\right.$$

$$\left. + w\frac{\partial[(u^2 + v^2 + w^2)/2]}{\partial z}\right\} - p\left(\frac{\partial u}{\partial x} + \frac{\partial v}{\partial y} + \frac{\partial w}{\partial z}\right) + \Phi \quad (3.1d)$$

From $e = (u^2 + v^2 + w^2)/2 + c_v T$ we have

$$\rho\left[u\frac{\partial(c_v T)}{\partial x} + v\frac{\partial(c_v T)}{\partial y} + w\frac{\partial(c_v T)}{\partial z}\right] + p\left[\frac{\partial u}{\partial x} + \frac{\partial v}{\partial y} + \frac{\partial w}{\partial z}\right]$$
$$= \left[\frac{\partial}{\partial x}\left(k\frac{\partial T}{\partial x}\right) + \frac{\partial}{\partial y}\left(k\frac{\partial T}{\partial y}\right) + \frac{\partial}{\partial z}\left(k\frac{\partial T}{\partial z}\right)\right] + \Phi \quad (3.1)$$

where

$$\Phi = \mu\left[2\left(\frac{\partial u}{\partial x}\right)^2 + 2\left(\frac{\partial v}{\partial y}\right)^2 + 2\left(\frac{\partial w}{\partial z}\right)^2 - \frac{2}{3}\left(\frac{\partial u}{\partial x} + \frac{\partial v}{\partial y} + \frac{\partial w}{\partial z}\right)^2\right.$$
$$\left. + \left(\frac{\partial u}{\partial y} + \frac{\partial v}{\partial x}\right)^2 + \left(\frac{\partial v}{\partial z} + \frac{\partial w}{\partial y}\right)^2 + \left(\frac{\partial w}{\partial x} + \frac{\partial u}{\partial z}\right)^2\right]$$

The terms in the first bracket of Equ. (3.1) are the convection of internal energy of the fluid. The terms of the second bracket are the rate of work done by a differential volume of fluid in expansion against the surrounding pressure. The terms in the third bracket are the rate of heat conduction in the fluid, and Φ is the rate at which kinetic energy is dissipated into heat. In Equ. (3.1) only a $\partial(c_v T)/\partial t$ need be added within the first bracket to make the equation applicable also to transient states.

As derived, the energy equation is still valid for compressible as well as incompressible fluids. By restricting ourselves to the latter we can assume the following:

• Dilatation work as represented by the term $p\nabla V$ is zero.
• The thermal coefficients c_v and k can be treated as constants; moreover, $c_v = c_p$.
• All velocity derivatives are negligible when compared to the two transverse gradients $\frac{\partial u}{\partial y}$ and $\frac{\partial w}{\partial y}$.
• Compared with the conductive term across the film the two other conductive terms are small. This also is consistent with the previous assumption about the velocity gradients.

With these simplifications entered in Equ. (3.1) we obtain

$$\rho c_p\left[u\frac{\partial T}{\partial x} + v\frac{\partial T}{\partial y} + w\frac{\partial T}{\partial z}\right] = k\frac{\partial^2 T}{\partial y^2} + \mu\left[\left(\frac{\partial u}{\partial y}\right)^2 + \left(\frac{\partial w}{\partial y}\right)^2\right] \quad (3.2)$$

As seen, two new terms appear here when compared with the adiabatic equations, namely $v(\partial T/\partial y)$ and $(\partial^2 T/\partial y^2)$, both due to transverse variations. It could be supposed that in the convective term where v is much smaller than either u or w, $v\partial T/\partial y$ could be ignored. But due to the smallness of y we have $(\Delta v/\Delta y)\Delta T \sim 1 \cdot \Delta T$ and it is, therefore, not obvious that this term will necessarily be small.

Equ. (3.2) is a partial differential equation of the second order in T yielding a three-dimensional temperature field $T(x, y, z)$ containing three convective velocity terms, u, v, and w. As in the adiabatic case these velocities have to be obtained from an appropriate momentum equation, in our case the Reynolds equation. If in Equ. (3.2) we normalize the velocities, say, by $U = R\omega$, distances by R, y by C, and the temperatures by some reference temperature T_r, we can write

$$\frac{\rho c_p \omega C^2}{k} \left[\tilde{u}\left(\frac{\partial \tilde{T}}{\partial \tilde{x}}\right) + \left(\frac{R}{C}\right)\tilde{v}\frac{\partial \tilde{T}}{\partial \tilde{y}} + \tilde{w}\left(\frac{\partial \tilde{T}}{\partial \tilde{z}}\right) \right]$$

$$= \left(\frac{\partial^2 \tilde{T}}{\partial \tilde{y}^2}\right) + \frac{\mu \omega^2 R^2}{k T_r} \left[\left(\frac{\partial \tilde{u}}{\partial \tilde{y}}\right)^2 + \left(\frac{\partial \tilde{w}}{\partial \tilde{y}}\right)^2 \right]$$

The group of variables $(\rho c_p \omega C^2/k)$ multiplying the convection terms is known as the Peclet number. It dominates the thermal behavior in a manner siminar to that of the Sommerfeld number in the Reynolds equation. But wheras $S = (\mu N/P)(R/C)^2$ yields similar solutions for geometrically similar bodies, that is, once the (L/D) and (R/C) are fixed, that is not the case here. The Peclet number reads

$$Pe = \frac{\rho c_p \omega}{k} \cdot C^2$$

and thus the actual bearing clearance C will dictate the solution. There is thus no geometric similarty in THD analysis. The situation here is similar to that prevailing in turbulent flow. Here the Reynolds number,

$$Re = \frac{\rho U C}{\mu} = \frac{\rho \omega}{\mu}\left(\frac{R}{C}\right) \cdot C^2$$

which determines the level of turbulence, is also seen to depend on C. In fact the Peclet number can be rewritten as

$$Pe = Pr \cdot Re\left(\frac{C}{R}\right) \sim C^2$$

where $Pr = \mu c_p/k$ depends only on the lubricant properties while the rest depends on C^2. The laminar thermal solution has thus a similar dynamics as the turbulent isothermal solution.

In subsequent analyses the transverse variable y will, in the usual manner, be normalized by h to give $\bar{y} = (y/h)$. This makes $\bar{y}$ a function of the other coordinates, just as h is. One must now account for this fact whenever derivatives are formulated, as follows:

$$\frac{\partial T(x_i, \bar{y})}{\partial x_i} = \frac{\partial T}{\partial x_i} + \left(\frac{\partial T}{\partial \bar{y}}\right)\left(\frac{\partial \bar{y}}{\partial x_i}\right)$$

Now,

$$\frac{\partial \bar{y}}{\partial x_i} = \frac{\partial (y/h)}{\partial x_i} = \frac{\bar{y}}{h}\left(\frac{\partial h}{\partial x_i}\right)$$

Therefore

$$\frac{\partial T(x_i, \overline{y})}{\partial x_i} = \frac{\partial T}{\partial x_i} - \left(\frac{\overline{y}}{h}\right)\left(\frac{dh}{dx_i}\right)\left(\frac{\partial T}{\partial \overline{y}}\right) \tag{3.3}$$

If h is not a function of the variable x_i – as would be the case with z for aligned journal bearings – then $(\partial h/\partial x_i) = 0$, and

$$\frac{\partial T(x_i, \overline{y})}{\partial x_i} = \left(\frac{\partial T}{\partial x_i}\right)$$

3.1.2 Compressibility Effects

It was stated in Chapter 1 that liquid films exhibit on occasion compressibility effects. The influence of this compressibility can be accounted for by rewriting the thermodynamic terms in Equ. (3.1d). These terms are made up of

$$\rho\left[u\frac{\partial e}{\partial x} + v\frac{\partial e}{\partial y} + w\frac{\partial e}{\partial z}\right] + p\left[\frac{\partial u}{\partial x} + \frac{\partial v}{\partial y} + \frac{\partial w}{\partial z}\right]$$

or in terms of the total derivatives

$$\rho\frac{De}{Dt} + p\frac{DV}{Dt}$$

We now consider the above expressions for the case when V is not invariable. The enthalpy of a fluid is given by $i = (e + pV)$ and we can then write

$$\rho\frac{Di}{Dt} = \rho\frac{D(e + pV)}{Dt} = \rho\frac{De}{Dt} + p\frac{DV}{Dt} + v\frac{Dp}{Dt}$$

or

$$\rho\frac{De}{Dt} + p\frac{DV}{Dt} = \rho\frac{Di}{Dt} - \frac{Dp}{Dt} \tag{3.4}$$

For any arbitrary fluid

$$\frac{Di}{Dt} = c_p\frac{DT}{Dt} + \left[V - \left(\frac{\partial V}{\partial T}\right)\right]\frac{Dp}{Dt} \tag{3.5}$$

Substituting for Di/Dt in Equ. (3.4), we obtain

$$\rho\left[\frac{De}{Dt} + p\frac{DV}{Dt}\right] = \rho\left[c_p\frac{DT}{Dt} - T\left(\frac{\partial V}{\partial T}\right)\right]\frac{Dp}{Dt}$$

Now (DV/DT) is the thermal compressibility of the fluid, which we denote here by δ. The thermodynamic terms, therefore, in Equ. (3.1d) become

$$\rho\left[\frac{De}{Dt} + p\frac{DV}{Dt}\right] = \rho\left[c_p\frac{DT}{Dt} - \delta T\frac{Dp}{DT}\right] \tag{3.6}$$

With the above substitution, the energy equation for a compressible fluid reads (after ignoring $\partial p/\partial y$)

$$\rho\left[u\frac{\partial T}{\partial x} + v\frac{\partial T}{\partial y} + w\frac{\partial T}{\partial z}\right] = k\left(\frac{\partial^2 T}{\partial y^2}\right) + \rho\delta T\left[u\frac{\partial p}{\partial x} + w\frac{\partial p}{\partial z}\right]$$

$$+ \mu\left[\left(\frac{\partial u}{\partial y}\right)^2 + \left(\frac{\partial w}{\partial z}\right)^2\right] \tag{3.7}$$

If the compressibility $\delta = 0$, then Equ. (3.7) reverts to Equ. (3.2).

3.1.3 The Reynolds Equation

The Reynolds equation confining itself only to viscous and pressure forces and making use of the thinness of the fluid film whereby $y \ll x, z$, is essentially derived from the two simple relationships of

$$\frac{\partial}{\partial y}\left(\mu\frac{\partial u}{\partial y}\right) = \frac{\partial p}{\partial x} \qquad (3.8a)$$

$$\frac{\partial}{\partial y}\left(\mu\frac{\partial w}{\partial y}\right) = \frac{\partial p}{\partial z} \qquad (3.8b)$$

Integrating twice with respect to y with $\mu = \mu(x, y, z)$, we obtain

$$u = \left(\frac{\partial p}{\partial x}\right)\int_0^y \frac{\varsigma d\varsigma}{\mu} + C_1\int_0^y \frac{d\varsigma}{\mu} + C_2$$

$$w = \left(\frac{\partial p}{\partial z}\right)\int_0^y \frac{\varsigma d\varsigma}{\mu} + C_3\int_0^y \frac{d\varsigma}{\mu} + C_4$$

where ς is the integration variable used in place of y. Applying the boundary conditions

$$u = U \quad \text{at} \quad y = 0$$

and

$$u = w = 0 \quad \text{at} \quad y = h$$

we obtain

$$C_2 = U$$

$$C_4 = 0$$

$$C_1 = -\frac{U + (\partial p/\partial x)I_1}{I_0}$$

$$C_3 = -\left(\frac{\partial p}{\partial z}\right)I_3$$

where

$$I_0 = \int_0^h \frac{d\varsigma}{\mu}$$

$$I_1 = \int_0^h \frac{\varsigma d\varsigma}{\mu}$$

$$I_3 = (I_1/I_0)$$

The two expressions for the velocities now are

$$u = U\left[1 - \frac{\int_0^y (d\varsigma/\mu)}{I_0}\right] + \left(\frac{\partial p}{\partial x}\right)\left[\int_0^y \frac{\varsigma d\varsigma}{\mu} - I_3\int_0^y \frac{d\varsigma}{\mu}\right] \qquad (3.9a)$$

$$w = \left(\frac{\partial p}{\partial z}\right)\left[\int_0^y \frac{\varsigma d\varsigma}{\mu} - I_3\int_0^y \frac{d\varsigma}{\mu}\right] \qquad (3.9b)$$

From the continuity equation we have

$$v = - \int \left[\frac{\partial u}{\partial x} + \frac{\partial w}{\partial z} \right] dy \qquad (3.9c)$$

and when the two derivatives $(\partial u / \partial x)$ and $(\partial w / \partial z)$ are integrated across the film thickness between $y = 0$ and $y = h$ we obtain a Reynolds equation which accounts for a μ variable in all three directions x, y, and z. Its form is

$$\frac{\partial}{\partial x} \left(I_R \frac{\partial p}{\partial x} \right) + \frac{\partial}{\partial z} \left(I_R \frac{\partial p}{\partial z} \right) = U \left(\frac{\partial I_R}{\partial x} \right) \qquad (3.10)$$

where

$$I_R = \int_0^h \frac{(\varsigma - I_3) d\varsigma}{\mu}$$

Thus the THD Reynolds equation retains its original form with the difference that it contains unevaluated definite integrals I_0, I_1, I_3, and I_R, all of them, after the integrations are performed, functions of (x, z) only. For future use these and other related integrals are defined in Table 3.1.

Two approaches would seemingly bracket the options for treating the problem, namely,

a. Retention of the Conventional Reynolds Equation. Here in the calculation of the transverse temperatures, the energy equation would employ pressures based on a $\mu(x, z)$ field that ignores the transverse viscosity variation altogether. There would still be a need for calculating the transverse velocity v appearing in the energy equation. This transverse velocity can be obtained from the continuity equation (3.9c).

b. Reformulation of the Energy Equation. In this approach the new Reynolds equation would have to be used, as given by Equ. (3.10). Conceptually this does not differ much from method (a) since even for a constant μ both u and w are functions of y and must be integrated across the fluid film. Here, however, along with u and w, μ also has to be integrated across y, and since $\mu(x, y, z)$ is an unknown function, and its determination hinges on a simultaneous solution with the energy equation, the unevaluated integrals, I_R, are bound to burden the entire mathematical procedure.

3.2 INFINITE PARALLEL SURFACES

In considering surfaces which extend infinitely in both the x and the z directions, the only variations left to consider are those in the transverse direction. This, strictly speaking, is not true because energy is continuously being stored in the fluid, altering its state. But since no boundary conditions at entrance and exit are available, the present approach is almost inevitable. Physically, one may view the transverse-variations-only approach as a case in which all the heat dissipated in the film is conducted away via the bounding surfaces.

TABLE 3.1

Transverse viscosity variations integrals

Symbol	Integral	Symbol	Integral
$I_0(y)$	$\displaystyle\int_0^y (d\varsigma/\mu)$	I_3	(I_1/I_0)
		I_4	(I_2/I_0)
$I_1(y)$	$\displaystyle\int_0^y (\varsigma d\varsigma/\mu)$		
$I_2(y)$	$\displaystyle\int_0^y (\varsigma^2 d\varsigma/\mu)$	I_R	$\displaystyle\int_0^h [(y - I_3)\, dy/\mu]$
		I_{00}	$\displaystyle\int_0^I (\overline{dy}/\mu)$
$I_3(y)$	$I_1(y)/I_0(y)$		
$I_4(y)$	$I_2(y)/I_0(y)$	I_{11}	$\displaystyle\int_0^I (\overline{y}dy/\mu)$
I_0	$\displaystyle\int_0^h (dy/\mu)$	I_{22}	$\displaystyle\int_0^I (\overline{y^2}dy/\mu)$
I_1	$\displaystyle\int_0^h (ydy/\mu)$	I_{33}	(I_{11}/I_{00})
		I_{44}	(I_{22}/I_{00})
I_2	$\displaystyle\int_0^h (y^2dy/\mu)$	I_{RR}	$\displaystyle\int_0^I [(\overline{y} - I_{33})\, \overline{dy}/\mu]$

$\mu = \mu(x, y, z)$ or $\mu = \mu(x, \varsigma, z)$.

Probably the earliest attempt to examine transverse thermal effects in parallel surfaces goes back to Kingsbury who in 1933 produced the results given in Fig. 3.2 based on the one-dimensional energy balance

$$k\left(\frac{dT}{dy}\right) = -\mu\left(\frac{du}{dy}\right)(u - u_0)$$

The above equation is somewhat inconsistent with the one that results from a formal reduction of Equ. (3.2), namely

$$k\frac{\partial^2 T}{\partial y^2} = -\mu\left(\frac{\partial u}{\partial y}\right)^2 \tag{3.11}$$

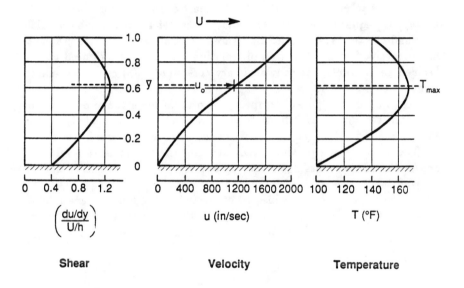

| Shear | Velocity | Temperature |

Fig. 3.2 Transverse temperature solution by Kingsbury of 1933

Solving Equ. (3.11) for the boundary conditions of T_0 at the stationary surface, and T_1 at the surface moving with velocity U in which the expression for du/dy is taken from Equ. (3.8a) for a flow with a constant pressure gradient, we obtain for a solution of Equ. (3.11)

$$
\begin{aligned}
\frac{T - T_0}{T_1 - T_0} =& \bar{y} + Pr \cdot Ec \cdot \left\{ \left[\frac{1}{2}\bar{y}(1 - \bar{y})\right] \right. \\
& + 2G\left[\frac{1}{2}\bar{y}(1 - \bar{y}) - \frac{1}{3}\bar{y}(1 - \bar{y}^2)\right] \\
& \left. + G^2\bar{y}\left[\frac{1}{3}(1 - \bar{y}^3) - \frac{2}{3}(1 - \bar{y}^2) + \frac{1}{2}(1 - \bar{y})\right] \right\} \tag{3.12}
\end{aligned}
$$

Above

$$Pr = \frac{\mu c_p}{k} \equiv \text{Prandtl number}$$

$$Ec = \frac{U^2}{c_p(T_1 - T_0)} \equiv \text{Eckert number}$$

$$G = \frac{h^2}{2\mu U}\left(\frac{dp}{dx}\right)$$

Particular note should be taken of the fact that Equ. (3.11) has here been integrated by considering μ as invariant with temperature. The case where the variation of μ with temperature is taken into account will be taken up later on.

The first term $\bar{y}$ in Equ. (3.12) is the part of the solution with both U and (dp/dx) equal to zero; the first square bracket is the contribution due to U; and the terms multiplied by G are the result of a pressure gradient. One attractive feature of this simple solution is that T_1 and T_0 need not be specified and a solution for any arbitrary $\Delta T = (T_1 - T_0)$ can be evaluated by means of Equ. (3.12).

Fig. 3.3 offers a plot of temperature profiles for a range of values of the parameter $Pr \cdot Ec$ and different pressure gradients. Some of the trends indicated by Eq. (3.12) and also noticeable in the Figure are:

- The presence of a pressure gradient raises the temperature levels.
- As $Pr \cdot Ec$ increases T_{max} goes up and moves closer to the stationary wall.
- Above a certain U, heat instead of flowing to the fluid, reverses direction. By setting $(\partial T/\partial y) = 0$ at $\bar{y} = 1$ the point at which heat starts flowing from the fluid to the moving wall is given by

$$\frac{\mu U^2}{k} > \frac{6(T_1 - T_0)}{G^2 - 2G + 3}$$

The above formally incorrect solution was cited both to provide an upper limit to the problem as well as to juxtapose this case to a solution which does consider the variation of viscosity across the film.

Using the familiar exponential $\mu - T$ relationship we have for the energy equation (still neglecting v)

$$k\frac{d^2T}{dy^2} = -\mu_0 e^{-\alpha(T-T_0)}\left(\frac{du}{dy}\right)^2 \tag{3.13}$$

We now must go back to the momentum equation and obtain an expression for μ with $\mu = \mu(y)$. From the basic relation

$$\left(\frac{dp}{dx}\right) = \frac{d}{dy}\left(\mu\frac{du}{dy}\right)$$

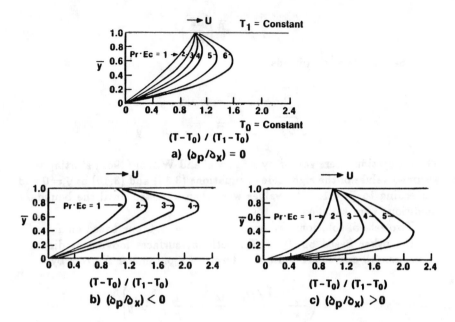

Fig. 3.3 Isoviscous temperature distribution in infinite parallel surfaces (Jain *et al.*, 1976)

we obtain

$$\frac{dp}{dx} = \left(\frac{d\mu}{dy}\right)\left(\frac{du}{dy}\right) + \mu\left(\frac{d^2u}{dy^2}\right)$$

or

$$\frac{d^2u}{dy^2} = \frac{(dp/dx)}{\mu_0}e^{-\alpha(T-T_0)} + \alpha\left(\frac{dT}{dy}\right)\left(\frac{du}{dy}\right) \tag{3.14}$$

Equations (3.13) and (3.14) are two differential equations for $T(y)$ and $u(y)$ which must be solved simultaneously as both are implicit in T and u. The problem is solved with both surfaces in motion, one at U_1 and T_0, the other at U_2 and T_1. Using the substitutions

$$\overline{u} = \left(u - \frac{U_1 + U_2}{2}\right)\left(\frac{\mu_0\alpha}{k}\right)^{\frac{1}{2}}$$

$$\overline{P} = \left(\frac{dp}{dx}\right)h^2\left(\frac{\alpha}{\mu_0 k}\right)^{\frac{1}{2}}$$

$$\overline{U} = (U_1 - U_2)^2\left(\frac{\mu_0\alpha}{k}\right)$$

$$\overline{T}_\Delta = \alpha(T_1 - T_0)$$

we obtain

$$\frac{d^2\overline{u}}{d\overline{y}^2} = \overline{P}e^{\overline{T}} + \left(\frac{d\overline{T}}{d\overline{y}}\right)\left(\frac{d\overline{u}}{d\overline{y}}\right) \tag{3.15}$$

$$\frac{d^2\overline{T}}{d\overline{y}^2} = -e^{\overline{T}}\left(\frac{d\overline{u}}{d\overline{y}}\right)^2 \tag{3.16}$$

with the boundary conditions

$$\overline{u} = \sqrt{\overline{U}/2} \quad \text{and} \quad \overline{T} = 0 \quad \text{at} \quad \overline{y} = 1$$

$$\overline{u} = -\sqrt{\overline{U}/2} \quad \text{and} \quad \overline{T} = \overline{T}_\Delta \quad \text{at} \quad \overline{y} = 0$$

These equations were solved by Aggarwal and Wilson (1980) starting with assumed values of the right side of equations (3.15) and (3.16) at $\overline{y} = 0$ and proceeding to $\overline{y} = 1$. The values were then adjusted until the boundary conditions were satisfied.

Two sets of solutions are presented in Fig. 3.4, one based on $T_1 = T_0$ and the other based on $T_1 \neq T_0$, both for surfaces moving at different velocities. All curves are plotted in terms of a parameter F given by

$$F = \left(\frac{\mu_0 \alpha}{k}\right)\left[\left(\frac{U_1 + U_2}{2}\right) - \frac{1}{h}\int_0^h u\,dy\right]^2$$

which, in essence, represents the deviation from a linear velocity profile due to the presence of a pressure gradient and thermal effects. At low values of F the flow approaches that of an isothermal distribution but as F increases the flow in the center is retarded due to the effect of the relatively cool viscous layers near the surfaces (kept at fixed temperatures). As seen in the Figure in all cases high velocities and peak temperatures occur near the faster, hotter surface.

3.3 HYDROSTATIC FILMS

The essence of hydrostatic lubrication is that the fluid film is externally pressurized and does not rely for its load capacity on relative motion between the surfaces. The basic hydrostatic contrivance then is a set of non-rotating parallel discs to which pressurized fluid is supplied via a central hole. Such a set of disks would have no variation in either pressure or temperature in the circumferential direction and conduction in the radial direction would be small compared to that across the film. Thus the simple axisymmetric energy equation reads

$$\rho c_p w\left(\frac{\partial T}{\partial r}\right) = k\frac{\partial^2 T}{\partial y^2} + \mu\left(\frac{\partial w}{\partial y}\right)^2 \tag{3.17}$$

In the above, the transverse velocity v was neglected; otherwise the equation is equivalent to Equ. (3.2) with $\partial/\partial x = 0$ and the radial variable r replacing z.

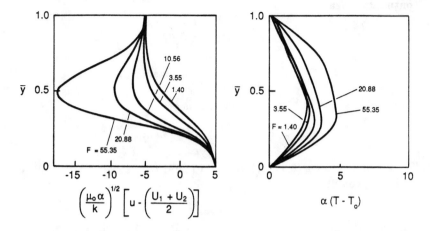

a) Solution with Equal Wall Temperatures, U = 100

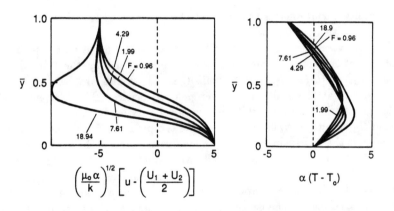

b) Solution with Unequal Wall Temperatures, $\overline{U}$ = 100

Fig. 3.4 Variable viscosity solution for infinite parallel surfaces (Aggarwal and Wilson, 1980)

The anisothermal problem for such a hydrostatic film was solved by Cuenca and Raynor, 1964, for the simple boundary conditions of the fluid at the inlet and at both surfaces remaining at a constant temperature T_0. Physically this is a Poiseuille type of flow with a continually varying pressure gradient in which the fluid, while being heated by the internal viscous shear, has much of this heat conducted away via the walls. The auxiliary equations required to solve the problem are those of momentum and continuity given by

$$\frac{dp}{dr} = \frac{\partial}{\partial y}\left[\mu\left(\frac{\partial w}{\partial y}\right)\right] \tag{3.18}$$

$$2\pi r \int_0^h w\,dy = Q = \text{constant} \tag{3.19}$$

with the very simple boundary conditions of

$$w = 0 \quad \text{at} \quad y = 0 \quad \text{and} \quad y = h$$

The set of equations (3.17), (3.18), and (3.19) were solved for the following operating conditions:

$R_1 = 0.125$ in.

$R_2 = 2.0$ in.

$h = 0.005$ in.

$Q = 10$ in.^{3}/sec

$\mu_0 = 16 \times 10^{-6}$ lb-sec/in.^{2}

$\rho = 8.3 \times 10^{-5}$ lb-sec^2/in.^{4}

$c_p = 0.5$ BTU/lb-°F

$k = 1.76 \times 10^{-6}$ BTU/sec-in.-°F

The viscosity variation was taken to be the exponential function given by Equ. (2.11) with $\alpha = 0.033/°$F.

Fig. 3.5 gives the velocity distribution for both the $\mu = const.$ and anisothermal solutions. The velocities are normalized by the inlet fluid velocity, $V_1 = (Q/2\pi R_1 h)$. The anisothermal profiles are seen to have steeper slopes at the walls, but lower peak velocities. The temperature profiles given in Fig. 3.6 exhibit an inverted behavior due to the imposed constant T_0 along the walls. Normally, the temperatures would be at their maximum at the walls, since, as can be seen from Fig. 3.5, the viscous shear is the highest there. But since the walls are maintained at T_0 the maximum in each profile occurs some distance away from the wall. Also, due to the cooling effects of T_0, the temperatures near the wall actually decrease as the flow progresses. In the center, however, around $h/2$, the temperatures rise in the direction of flow, though not sufficiently to overtake the peak temperatures on the wings. The latter stay throughout the flow region a distance of 10% to 20% from the wall.

The pressures developed under variable viscosity are, as shown in Fig. 3.7, lower than those for constant viscosity. When integrated for the total force, the anisothermal load capacity is 32% lower than based on $\mu_0 =$ constant. When compared to a solution based on an average temperature, the anisothermal force is still some 10% below the solution with $\mu_{av} =$ constant

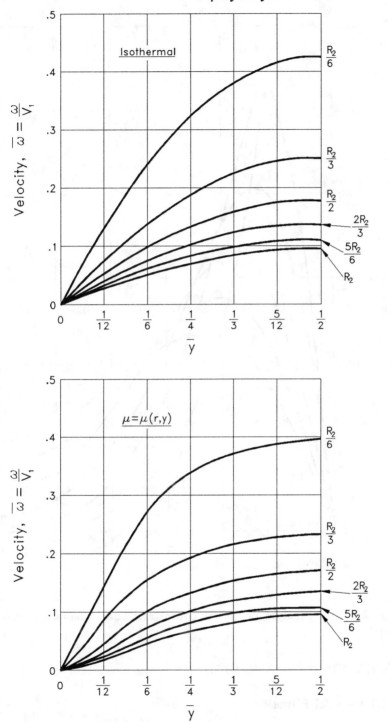

Fig. 3.5 Velocity profiles in hydrostatic film (Cuenca and Raynor, 1964)

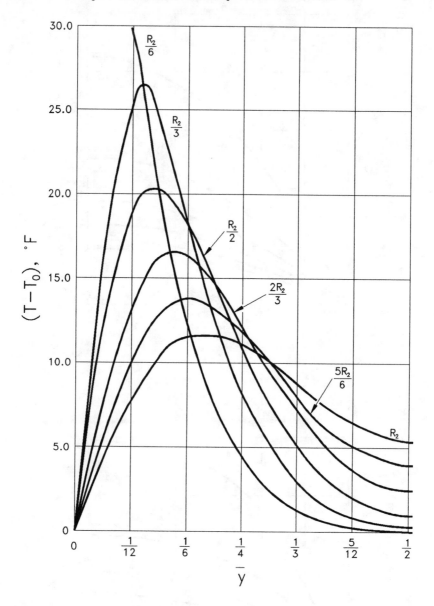

Fig. 3.6 Temperature profiles in hydrostatic film (Cuenca and Raynor, 1964)

3.4 HYDRODYNAMIC SLIDERS

3.4.1 Parallel Films

Parallel surface sliders represent the simplest possible bearing geome-

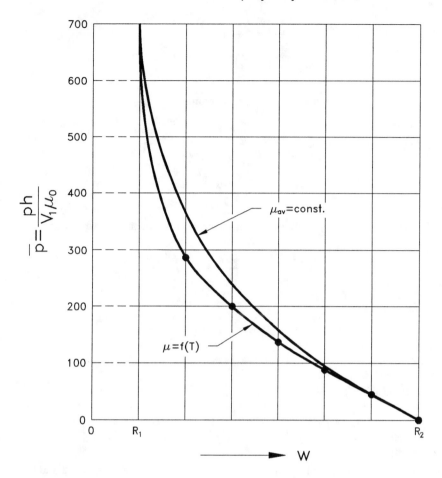

Fig. 3.7 Pressure profile in hydrostatic film with variable viscosity (Cuenca and Raynor, 1964)

try, yet they exhibit most of the essential features of other more elaborate configurations. Furthermore, they represent a number of actual tribological devices such as face seals, hydrostatic bearings, parallel plate thrust bearings, stepped configurations, traction drives, and others. Combined with the relative simplicity of the required analytical effort they represent an important family of geometries in THD analysis.

For sliders infinitely long in the axial(z) direction, we have from Equ. (3.7)

$$\rho c_p u \left(\frac{\partial T}{\partial x} \right) = \mu \left(\frac{\partial u}{\partial y} \right)^2 + k \left(\frac{\partial^2 T}{\partial y^2} \right) \tag{3.20}$$

Using the normalized quantities

$$\hat{p} = \left(\frac{ph}{\mu_0 U} \right), \quad \hat{x} = \left(\frac{x}{h} \right) = \overline{x} \left(\frac{B}{h} \right)$$

$$e^{\alpha(T-T_0)} = 1 + \theta = \left(\frac{1}{\overline{\mu}}\right) = \left(\frac{\mu_0}{\mu}\right)$$

$$\overline{U} = \left(\frac{\mu_0 \alpha}{k}\right) U^2$$

the energy and Reynolds equations become

$$Pe\,\overline{u}\frac{\partial \theta}{\partial \hat{x}} = \overline{U}\left(\frac{\partial \overline{u}}{\partial \overline{y}}\right)^2 + \frac{\partial^2 \theta}{\partial \overline{y}} - \frac{(\partial \theta / \partial \overline{y})^2}{(1+\theta)} \tag{3.21}$$

$$\frac{\partial}{\partial \overline{y}}\left(\overline{\mu}\frac{\partial \overline{u}}{\partial \overline{y}}\right) = \frac{\partial \hat{p}}{\partial \hat{x}} \tag{3.22}$$

Integrating the Reynolds equation twice with respect to y we obtain

$$\overline{u} = \left(\frac{d\hat{p}}{d\hat{x}}\right)\int_0^{\overline{y}} \frac{\overline{y}d\overline{y}}{\mu} + C_1 \int_0^{\overline{y}} \frac{d\overline{y}}{\mu} + C_2$$

Using the boundary conditions of

$$\overline{u} = 1 \quad \text{at} \quad \overline{y} = 0$$
$$\overline{u} = 0 \quad \text{at} \quad \overline{y} = 1$$

we obtain

$$\overline{u} = \left(\frac{d\hat{p}}{d\hat{x}}\right)\int_0^{\overline{y}} \frac{\overline{y}d\overline{y}}{\mu} - \left(\frac{1}{I_{00}} + I_{33}\frac{d\hat{p}}{d\hat{x}}\right)\int_0^{\overline{y}} \frac{d\overline{y}}{\mu} + 1 \tag{3.23}$$

where I_{00} and I_{33} are the familiar integrals from section 3.1.2 but now non-dimensionalized

$$I_{00} = \int_0^1 \frac{d\overline{y}}{\overline{\mu}} = \int_0^1 (1+\theta)d\overline{y}$$

$$I_{33} = \left[\int_0^1 \frac{\overline{y}d\overline{y}}{\overline{\mu}} \bigg/ \int_0^1 \frac{d\overline{y}}{\overline{\mu}}\right] = \frac{\int_0^1 (1+\theta)d\overline{y}}{\int_0^1 (1+\theta)\overline{y}d\overline{y}}$$

Integrating across the fluid film we must have

$$\rho \int_0^h u\,dy = \text{constant}$$

or, in normalized form,

$$\int_0^1 \overline{u}d\overline{y} = m = \text{constant}$$

Integrating the above by parts and using

$$\frac{\partial \overline{u}}{\partial \overline{y}} = \frac{1}{\mu}\left\{\overline{y}\left(\frac{d\hat{p}}{d\hat{x}}\right) - \left[\frac{1}{I_{00}} + I_{33}\left(\frac{d\hat{p}}{d\hat{x}}\right)\right]\right\}$$

we obtain

$$\frac{d\hat{p}}{d\hat{x}} = \frac{I_{33} - m}{I_{00}(I_{44} - I_{33}^2)} \tag{3.24}$$

where

$$I_{44} = \left[\int_0^1 \frac{\overline{y}^2 \, d\overline{y}}{\mu} \Big/ \int_0^1 \frac{d\overline{y}}{\mu}\right] = \frac{\int_0^1 (1 + \theta)\overline{y}^2 \, d\overline{y}}{\int_0^1 (1 + \theta) \, d\overline{y}}$$

Equ. (3.24) is essentially the one-dimensional Reynolds equation. When integrated with respect to x it will contain two arbitrary constants, m and a constant of integration. These can be evaluated from the two boundary conditions

$$p = 0 \quad \text{at} \quad x = 0 \quad \text{and} \quad x = B$$

The resulting expression $p(x)$ is, however, not fully defined since it contains the unknown function μ on which the three integrals I_{00}, I_{33}, and I_{44} depend. This function must come from the energy equation (3.20).

The solution of this problem has been obtained by Zienkiewicz (1957) using a finite difference "marching" technique in the x direction. The results are all for a particular geometry and set of operating conditions detailed in Table 3.2. The examination of this solution is of additional importance in that it automatically deals with the question whether thermal effects can be instrumental in generating hydrodynamic pressures in parallel films. At the beginning of the present volume it has been said that these theories have largely been discounted, but this was based mostly on results of $T = T(x, z)$ only. It will be of interest to examine what transverse temperature variations may add to this subject.

Two sets of solutions were obtained—one for equal surface temperatures, and the other for the moving surface being 40°F higher than the stationary part. Fig. 3.8 shows the pressures resulting from a transverse variation in temperature. The case of equal surface temperature generates a physically impossible profile yielding, as it does, negative absolute pressures. For the case of the moving surface hotter than the stationary surface, a perfectly regular pressure profile is developed, except perhaps that its peak is closer to the leading rather than to the trailing edge. For the specific parameters of Table 3.2 the ordinate $(ph/\mu_0 U)$ yields the equivalent of $(p/6.7)$ psi. Thus the unit loading in case (b) is some 250-300 psi. These results are contrary to experience on two accounts. Normally, parallel plate thrust bearings can support loads of no more than about 150 psi at film thicknesses much lower than the 3.6 mils used in the present example. More important, however, is the fact that in reality the stationary surface is usually hotter than the runner. By extension, then, should the pad be

TABLE 3.2

Input data for the Zienkiewicz solution

$h = 0.0036$ in.	$\rho_g = 0.85$
$B = 7.2$ in.	$c_p = 0.48 \dfrac{\text{Btu}}{\text{lb-}^\circ\text{F}}$
$U = 100$ ft/sec	$k = 0.89 \dfrac{\text{Btu-in.}}{\text{ft}^2\text{-hr-}^\circ\text{F}}$
$\alpha = 2.5 \times 10^{-2} \dfrac{1}{^\circ\text{F}}$	$Pe = 38,300$
$\mu_0 = 2 \times 10^{-5} \dfrac{\text{lb-sec}}{\text{in.}^2}$	$\bar{U} = 45.3$

kept at a temperature higher than the runner, the pressure over most of the pad would tend to be negative producing a suction instead of a load capacity. Furthermore, if negative pressures tend to form, then the boundary conditions to be used is not $p = 0$ at $x = 0$, but some other suitable condition at $x_1 > 0$, so as to eliminate the negative pressures. Whether such a new boundary condition would yield positive pressures at least over some portions of the pad for cases where $T_S > T_R$ is still to be resolved.

Fig. 3.9 gives the temperature profiles resulting from the present solution. Unlike in the solution for infinite plates the maximum temperatures tend to occur near the stationary surface for both cases, with T_{max}, as expected, rising with distance x. With equal surface temperatures heat flows to both surfaces; with unequal temperatures heat flows from the moving surface over the initial half of the slider, and into the moving surface over the trailing half. At the stationary part heat always flows out of the fluid film.

Finally, Fig. 3.10 shows the errors involved when certain simplifications are introduced into Equ. (3.20). The simplifications used in certain previous analyses were:

- omission of the conduction term
- approximation of $u(y)$ by a linear velocity profile

As seen, each of these simplifications adds its cumulative error to the film temperatures, producing discrepancies of the order of 10-20% from the present solution.

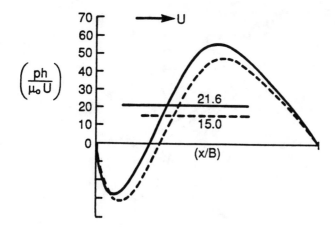

a) Equal Surface Temperatures

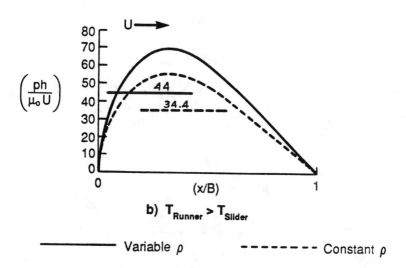

b) $T_{Runner} > T_{Slider}$

————————— Variable ρ – – – – – – – – Constant ρ

Fig. 3.8 Pressures in infinitely long parallel plates (Zienkiewicz, 1957)

3.4.2 Tapered Films

There are no essential differences in the treatment of wedge-shaped sliders from that applicable to parallel plates of finite extent. All the expressions derived in the previous section remain valid except that here h is a variable. Even so its expression is extremely simple and the particular

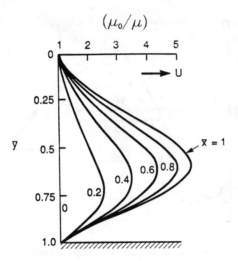

a) Equal Surface Temperatures

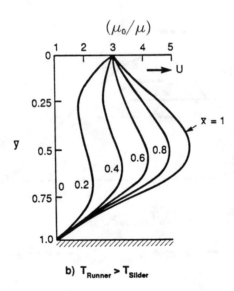

b) $T_{Runner} > T_{Slider}$

Fig. 3.9 Temperature profiles in infinitely long sliders (Zienkiewicz, 1957)

geometry treated here is given by

$$\overline{h} = (2 - \overline{x})$$

that is a slider with an inlet film thickness twice that of the outlet which is kept here at 0.0036 in. All the other parameters are nearly the same as in Table 3.2.

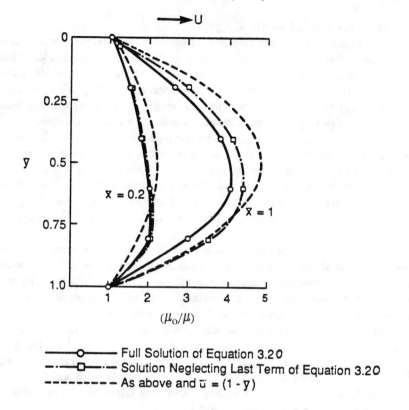

Full Solution of Equation 3.20

Solution Neglecting Last Term of Equation 3.20

As above and $\bar{u} = (1 - \bar{y})$

Fig. 3.10 Comparison of several thermal solutions for an infinitely long slider

Two sets of solutions will be presented, one an adiabatic case in which the boundary conditions are

$$\frac{\partial T}{\partial y}\bigg|_{0,h} = 0$$

and the second in which both surfaces are kept at a temperature equal to the inlet temperature of the lubricant, namely

$$T = T_0 \quad \text{at} \quad y = 0 \quad \text{and} \quad y = h$$

The main interest of these results worked out by Hunter and Zienkiewicz (1960) is in the comparison of these $\mu = \mu(x, y)$ solutions based on equations (3.21) and (3.24), and shown as case (d) in Fig. (3.11), with several other sets of data, namely

a) solutions based on the average viscosity at the exit of the slider i.e., $(\mu_2)_{av}$ at $\bar{x} = 1$

b) solutions in which $\mu = \mu(x)$ only which neglect all transverse effects, i.e., the kind of solutions treated in Chapter 2.

c) solutions with adiabatic $\mu(x, y)$.

The comparison of the two $\mu(x, y)$ solutions is straightforward. However, a solution in which $\mu = \mu(x)$ only is somewhat problematical. Here a $\mu(x)$ solution will be constructed as follows. At each x location a temperature will be extracted from a solution where $T = T(y)$. The obtained variable $T(y)$ at each x location will be averaged, and a $\mu(x)$ solution based on that average temperature, constant across y, will be generated.

Fig. 3.11 gives a mapping of the temperature profiles for the various solutions. Again we notice that, as remarked in the previous section, the high temperatures are closer to the stationary wall; in the adiabatic case they, in fact, reach a maximum at the slider surface. Clearly the temperature levels are much higher for the adiabatic case; the value of T_{max} at the coordinate (B, h) is more than double than the T_{max} of the isothermal wall case.

Fig. 3.12 compares the four solutions in terms of the pressures developed over the slider. The $\mu(x, y)$ solution with constant wall temperatures produces the lowest load capacity, the adiabatic $\mu(x)$ results yielding pressures nearly twice as large. The solution based on the average outlet temperature is somewhat artificial, as it is based not on some effective or bulk temperature but, on the temperature profile at the exit, i.e., T_{max}. The average temperature approach proposed in Section 2.1, on the other hand, is based on side leakage which yields an aggregate of temperatures ranging from T_0 at the inlet to T_{max} at the outlet. Clearly with an $L/D = \infty$ configuration there is no side leakage, and it is farfetched to consider the temperature at the exit as equivalent to T_{av}.

In the adiabatic case all the energy dissipated in the fluid film is convected away by the lubricant. In the other cases, however, it is of some interest to note the relative portions of heat that are carried away by convection and by conduction. The work done on the fluid is

$$\frac{H}{L} = U \int_0^B \mu \left(\frac{\partial u}{\partial y}\right)_0 dx = \frac{\overline{U} k}{\alpha} \int_0^1 \frac{(d\overline{u}/d\overline{y})_0}{(1 + \theta)} d\overline{x} \qquad (3.25)$$

The heat convected by the lubricant is

$$Q_c = \int_0^h u c_p \rho (T - T_0)_B dy = \frac{Pe \cdot k}{\alpha} \int_0^1 \overline{u} \ln(1 + \theta) d\overline{y} \qquad (3.26)$$

The heat conducted away to each of the two boundaries is

$$Q_{S,R} = \int_0^B k \frac{\partial (T - T_0)}{\partial y}\bigg|_{S,R} dx = \frac{k}{\alpha} \int_0^1 \frac{(\partial \theta / \partial \overline{y})_{S,R}}{(1 + \theta)} d\overline{x} \qquad (3.27)$$

the subscripts denoting temperature gradients corresponding to each of the surfaces. These integrals yield the energy distribution cited in Table 3.3

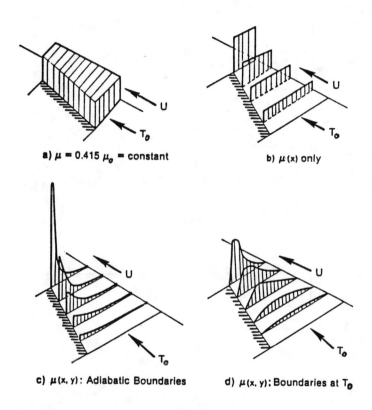

B = 18.3 cm	U = 30.5 m/s
h_2 = 0.0625 mm	μ_0 = 138 × 10^{-3} P-s
h_1/h_2 = 2	

a) μ = 0.415 μ_0 = constant

b) μ(x) only

c) μ(x, y): Adiabatic Boundaries

d) μ(x, y): Boundaries at T_0

Fig. 3.11 The (x,y) temperature distribution in infinitely long slider according to several thermal approaches

showing the stationary surface to be absorbing a larger share of heat than the moving surface.

The results presented above are based, as we have noted earlier, on an equation that omits the convection term in the transverse direction. When the term is included, instead of Equ. (3.20) we obtain

$$c_p \rho \left[u \frac{\partial T}{\partial x} + v \frac{\partial T}{\partial y} \right] = \mu \left(\frac{\partial u}{\partial y} \right)^2 + k \left(\frac{\partial^2 T}{\partial y^2} \right) \qquad (3.28)$$

To show the effect of neglecting the transverse convection, Dowson and Hudson (1964) solved the equation for the same slider treated previously

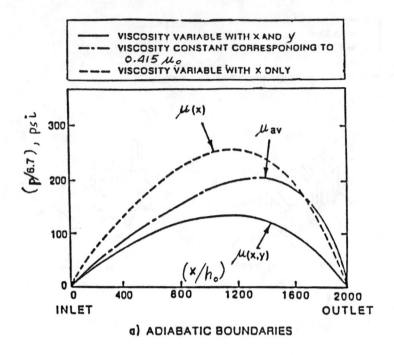

a) ADIABATIC BOUNDARIES

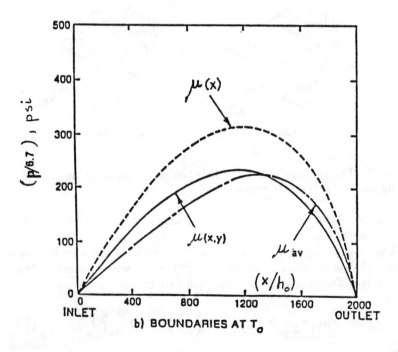

b) BOUNDARIES AT T_0

Fig. 3.12 Comparison of pressure profiles for various thermal solutions (Hunter and Zienkiewicz, 1960)

TABLE 3.3

Energy distribution in a slider

	H	Q_C^+	Q_S	Q_R
$E \times \frac{\alpha}{k} \times 10^{-3}$	36.4	20.6	12.2	3.6
% of Total	100	57	33	10

+Convected by the fluid

using the normalizations

$$\widehat{T} = \left(\frac{T}{T_0}\right) \quad ; \quad \bar{u} = \frac{u}{U} \quad ; \quad \widehat{U} = \frac{\mu_0 U^2}{kT_0} \quad ; \quad \widehat{x} = \frac{x}{h_2} \quad ; \quad \widehat{y} = \left(\frac{y}{h_2}\right)$$

resulting in

$$Pe\left[\bar{u}\frac{\partial \widehat{T}}{\partial \widehat{x}} + \bar{v}\frac{\partial \widehat{T}}{\partial \widehat{y}}\right] = \left(\frac{\partial^2 \widehat{T}}{\partial \widehat{y}^2}\right) + \widehat{U}\left(\frac{\partial \bar{u}}{\partial \widehat{y}}\right)^2 \tag{3.29}$$

First an expression for v had to be obtained. Using the continuity equation and the one-dimensional Reynolds equation this velocity becomes

$$\bar{v} = \left(\frac{\bar{h} - \widehat{y}}{\bar{h}}\right)\left(\frac{\widehat{y}}{\bar{h}}\right)^2\left(\frac{3\bar{h}_2}{\bar{h}} - 2\right)\left(\frac{d\bar{h}}{d\widehat{x}}\right) \tag{3.30}$$

Since v is proportional to (dh/dx), the slope of the slider, it can be seen that transverse convection does not appear in an analysis of parallel surfaces but does so when there is a variable film thickness. The direction of this transverse velocity is toward the moving surface for values of $h < (3/2)h_0$ and toward the stationary surface for $h > (3/2)h_0$.* For the present slider with $\bar{h}_1 = 2$ all the transverse velocities are in the direction of the moving surface, as shown in Fig. 3.13.

The problem with transverse convection was solved for three cases, all for the same slider specified in Table 3.2. The three cases are:

a.) Isoviscous with $\mu = \mu_0$ and $v\left(\partial T/\partial y\right) \neq 0$
b.) $\mu = \mu(x, y)$ and $v(\partial T/\partial y) \neq 0$
c.) $\mu = \mu(x, y)$ and $v(\partial T/\partial y) = 0$

The effects of the transverse convection term can be gleaned from Figs. 3.13c and 3.14. There is practically no effect on the load capacity and a small influence on the transverse temperatures, primarily at the trailing

---

* h_0 is the value of h where $p = p_{max}$.

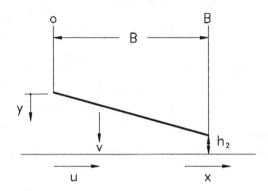

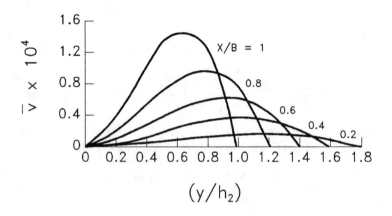

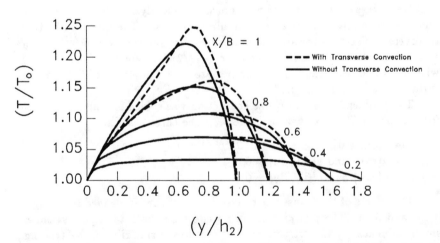

Fig. 3.13 Effect of transverse convection (Discussion by Dowson and Hudson to Hunter and Zienkiewicz, 1960)

edge. Neglect of the transverse effects reduces the peak temperatures and shifts them toward the center of the film. There is a similar equalizing tendency in the redistribution of energies absorbed, as shown in Table 3.4. The portion of heat absorbed by the stationary surface decreased by 15% from that obtained for the case of no transverse convection.

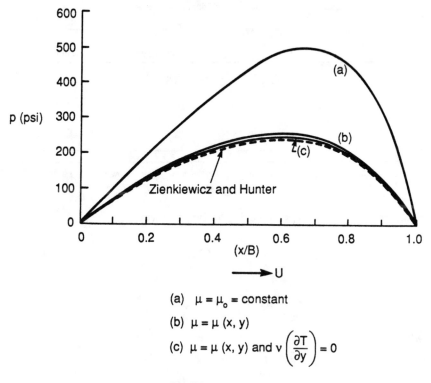

(a) $\mu = \mu_o$ = constant

(b) $\mu = \mu (x, y)$

(c) $\mu = \mu (x, y)$ and $v \left(\dfrac{\partial T}{\partial y} \right) = 0$

For Dimensions, See Table 3-1

Fig. 3.14 Pressure profiles for three thermal approaches (Dowson and Hudson, 1964)

As seen from Fig. 3.14 the reduction in pressure due to an x,y temperature variation is nearly to one half that of $\mu = \mu_0$ = constant. The differences in temperature distribution are shown in Fig. 3.15, the isoviscous plot in part (a) representing, of course, an upper limit on temperature levels. The maximum temperature is seen to occur at the end of the film, about one quarter of the film height above the stationary surface.

A general solution of the slider which does include transverse convection and which offers normalized solutions beyond the single numerical example given before was obtained by Rodkiweicz et al. (1973, 1975). Their work is of additional interest in that they handled the problem by means

TABLE 3.4

Energy distribution in a slider

Case	v	μ	Percentage heat to:		
			Oil	Stationary component	Moving component
a	$\neq 0$	μ_0	63·3	20·1	16·6
b	$\neq 0$	$\mu(x,y)$	65·4	19·6	15·0
c	0	$\mu(x,y)$	62·3	23·2	14·5

Dowson and Hudson, 1964

of a stream function Ψ, defined in the familiar way

$$\bar{u} = \frac{\overline{\Psi}_c}{\bar{h}} \frac{\partial \overline{\Psi}}{\partial \bar{y}} \tag{3.31a}$$

$$\bar{v} = -\overline{\Psi}_c \left[\frac{\partial \overline{\Psi}}{\partial \bar{x}} + \frac{(\bar{h}_1 - 1)}{\bar{h}} \bar{y} \frac{\partial \overline{\Psi}}{\partial \bar{y}} \right] \tag{3.31b}$$

where

$$\overline{\Psi} = \frac{\Psi}{\Psi_c}, \quad \overline{\Psi} = \int_0^1 \rho u \, dy, \quad \overline{\Psi}_c = \frac{\Psi_c}{\rho U h_2}, \quad \overline{U} = \frac{uB}{U h_2}$$

In terms of the stream function the Reynolds and energy equations become

$$\frac{\bar{h}^3}{\overline{\Psi}_c} \frac{d\bar{p}'}{d\bar{x}} = \frac{\partial}{\partial \bar{y}} \left(\bar{\mu} \frac{\partial^2 \overline{\Psi}}{\partial \bar{y}^2} \right) \tag{3.32}$$

$$\bar{h} \overline{\Psi}_c Pr \cdot Re^* \left(\frac{\partial \overline{\Psi}}{\partial \bar{y}} \frac{\partial \overline{T}}{\partial \bar{x}} - \frac{\partial \overline{\Psi}}{\partial \bar{x}} \frac{\partial \overline{T}}{\partial \bar{y}} \right)$$

$$= \frac{\partial^2 \overline{T}}{\partial \bar{y}^2} + Pr \cdot Ec \cdot \bar{\mu} \left(\frac{\overline{\Psi}_c}{\bar{h}} \frac{\partial^2 \overline{\Psi}}{\partial \bar{y}^2} \right)^2 \tag{3.33}$$

These equations are functions of five independent parameters:

$$\bar{h}_1; \quad Pr \cdot Re^* \equiv \frac{c_p \rho U h_2}{k}; \quad Pe; \quad \left(\frac{T_R}{T_S} \right); \quad Pr \cdot Ec \equiv \frac{\mu_0 U^2}{k} \bigg/ \left(\frac{T_R + T_S}{2} \right)$$

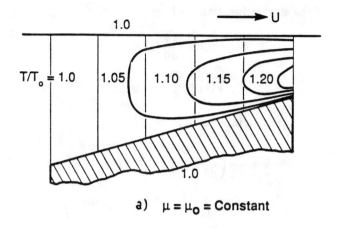

a) $\mu = \mu_o$ = Constant

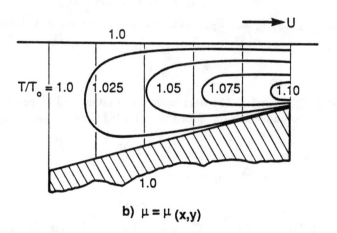

b) $\mu = \mu (x,y)$

Fig. 3.15 Transverse temperature maps in infinitely long sliders (Dowson and Hudson, 1964)

The boundary conditions on the differential equations are the usual ones of zero pressures and velocities on the boundaries except for $\bar{u} = 1$ at $\bar{y} = 1$. The temperatures are T_R and T_S on the runner and stationary boundaries respectively. At the inlet, a linear temperature profile, consistent with T_R and T_S, is imposed in the form of

$$\bar{\bar{T}}(y) = \bar{\bar{T}}(0) + \left[\bar{\bar{T}}(h) - \bar{\bar{T}}(0) \right] \bar{y}$$

where

$$\bar{\bar{T}} = T \Big/ \left(\frac{T_R + T_S}{2} \right) = \frac{T}{T_{av}}$$

In terms of the stream function we have

$$\overline{\Psi}(0) = \frac{\partial \overline{\Psi}}{\partial \overline{y}}(1) = 0$$

$$\overline{\Psi}(1) = \frac{\partial \overline{\Psi}}{\partial \overline{y}}(0) = \frac{\overline{h}}{\overline{\Psi}_c}$$

(3.34)

The Reynolds equation integrates into

$$\overline{\Psi} = \frac{\overline{h}^3}{\overline{\Psi}_c} \left(\frac{d\overline{p}}{d\overline{x}} \int_0^{\overline{y}} \int_0^{\overline{y}} \frac{\overline{y}d\overline{y}}{\mu} d\overline{y} \right.$$
$$\left. + C_1 \int_0^{\overline{y}} \int_0^{\overline{y}} \frac{d\overline{y}}{\mu} d\overline{y} + C_2\overline{y} + C_3 \right)$$

(3.35a)

$$\frac{d\overline{p}}{d\overline{x}} = -\frac{1}{\overline{h}^2 (A_2 - B_2)} + \frac{B_1 \overline{\Psi}_c}{\overline{h}^3 (A_2 - B_2)}$$

(3.35b)

where C_1, C_2, C_3, A_2 and B_2, all functions of x, are given in Table 3.5 and

$$\overline{p} = \frac{pB}{\mu_0 U} \left(\frac{h_2}{B} \right)^2$$

Equations (3.32) and (3.33) were solved and integrated to provide dimensionless performance solutions for the following range of parameters:

$$0.1 \leq Pe^* \leq 20$$

$$0.25 \leq (T_R/T_S) \leq 4.5$$

$$1.25 \leq \overline{h}_1 \leq 3.0$$

all evaluated at $Pr \cdot Ec = 1$. The particular viscosity function used is that for a typical industrial oil with 100 csks at 100°F and 10 csks at 200°F.

Figs. 3.16 through 3.19 give the dimensionless performance quantities for infinitely long sliders ($L/D = \infty$). The performance parameters have the following normalizations:

$$\overline{F}_r = (F_r/\mu_{av}UL)(h_2/B))$$
$$\overline{Q} = (Q/kT_{av}L)(h_2/B)$$
$$\overline{W} = (W/\mu_{av}UL)(h_2/B)^2$$

with $T_{av} = \frac{1}{2}(T_R + T_S)$ and μ_{av} corresponding to T_{av}. Some of the points worth emphasizing are

- The hotter the runner the lower the pressures and load capacity. When the fact that $\overline{p}$ and $\overline{W}$ are both normalized by μ_{av} is taken into account, a hot runner would produce load capacities even lower than indicated

TABLE 3.5

Values of constants in equation (3.29)

Constant	Value
$C_1(\bar{x})$	$\left[\dfrac{1}{A_3 - A_4}\right]\left[1 - \dfrac{\bar{h}^3}{\bar{\psi}_c}\left[\dfrac{d\bar{p}}{d\bar{x}}\right]\left[A_1 - A_2\right]\right]$
$C_2(\bar{x})$	$-\left[\left[\bar{h}^3/\bar{\psi}_c\right]\left[\dfrac{d\bar{p}}{d\bar{x}}\right]A_2 + C_1 A_4\right]$
$C_3(\bar{x})$	0
A_1	$\displaystyle\int_0^1 \int_0^{\bar{y}} (\bar{y}/\bar{\mu})\; d\bar{y}\; d\bar{y}$
A_2	I_1 Table 3.1
A_3	$\displaystyle\int_0^1 \int_0^{\bar{y}} (d\bar{y}/\bar{\mu})\; d\bar{y}$
A_4	I_0 Table 3.1
B_1	$A_4 / (A_3 - A_4)$
B_2	$A_4 (A_1 - A_2)/(A_3 - A_4)$
ψ_c	$-\displaystyle\int_0^1 \dfrac{d\bar{x}}{\bar{h}^2 (A_2 - B_2)} \;\Big/\; \int_0^1 \dfrac{B_1 \cdot}{h^{-3} (A_2 - B_2)}$

by the normalized coefficients $\bar{p}$ and $\overline{W}$. This is contrary to the parallel plate results of Zienkiewicz where a hotter runner produced higher positive pressures.

- The ratio (T_R/T_S) is important in determining performance but the actual value of the inlet temperature T_{av} has only a minor effect.
- Friction, and thus power loss, natually goes down with a rise in (T_R/T_S). This is so in face of the fact that a higher runner temperature produces steeper velocity gradients at the moving surface. The impact of lower viscosity due to a higher T_R predominates.
- Maximum load capacities are obtained in the region of $2 < \bar{h}_1 < 2.5$, which is about the same as with isothermal solutions.

Two concrete cases were run to re-emphasize the previous observation that a runner hotter than the bearing does not yield higher load capacities. The operational data are given in Table 3.6. In both cases the sum (T_R+T_S) was the same, yielding thus identical average inlet temperatures. As seen when the temperature levels were reversed to make the bearing hotter than the runner, a doubling of the load capacity was obtained.

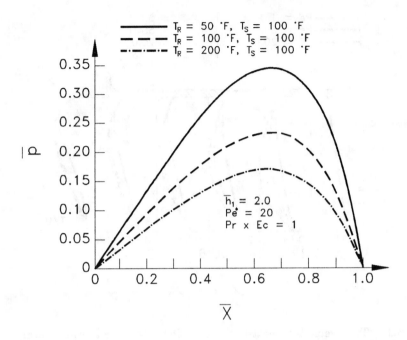

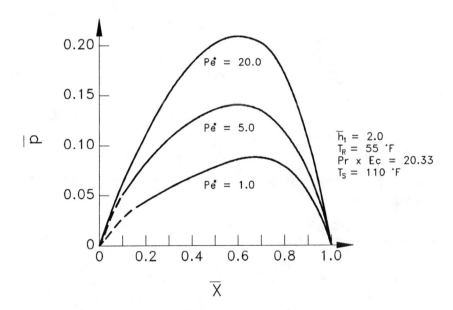

Fig. 3.16 Pressure profiles in infinitely long slider as function of thermal parameters (Rodkiewicz *et al.*, 1974)

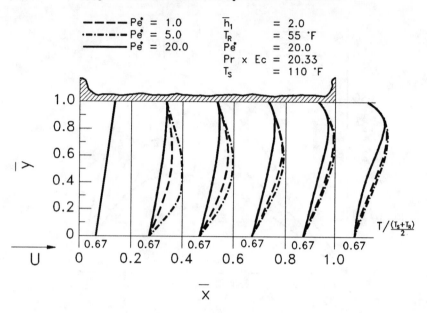

Fig. 3.17 Transverse temperatures as function of various thermal parameters (Rodkiewicz *et al.*, 1974)

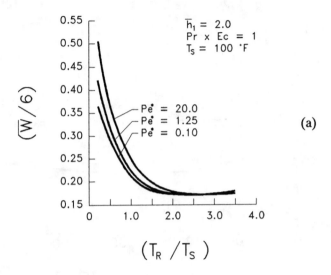

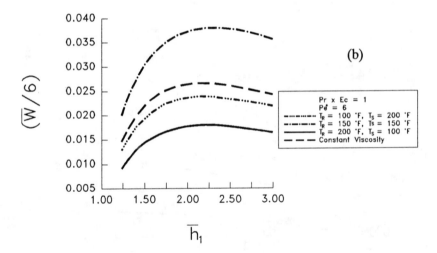

Fig. 3.18 Load capacity of infinitely long sliders as function of various parameters (Rodkiewicz *et al.*, 1974), (a) effect of surface temperature, (b) effect of taper, (c) effect of inlet temperature, (d) effect of Peclet number

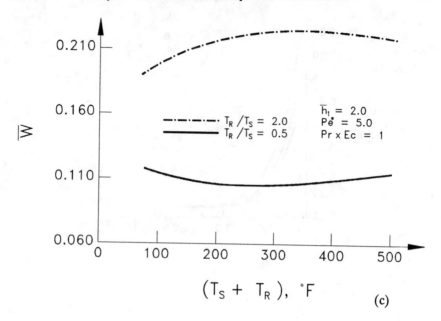

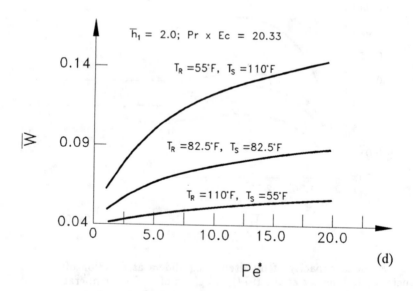

Fig. 3.18 Load capacity of infinitely long sliders as function of various parameters (Rodkiewicz *et al.*, 1974), (a) effect of surface temperature, (b) effect of taper, (c) effect of inlet temperature, (d) effect of Peclet number

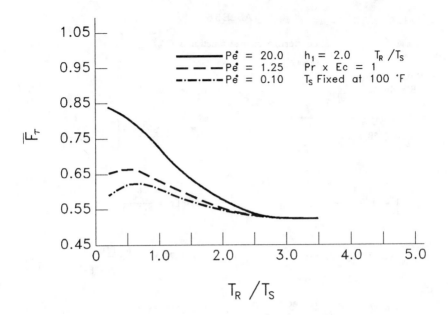

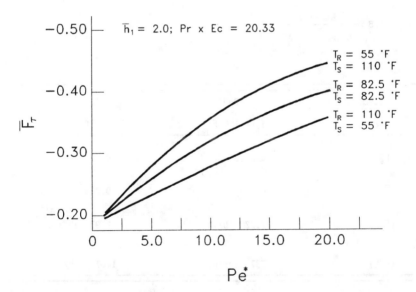

Fig. 3.19 Friction in infinitely long sliders as function of various thermal parameters (Rodkiewicz *et al.*, 1974)

TABLE 3.6

Slider performance as function of (T_R/T_S)

B = 4 in. $\bar{h}_1$ = 2.0

h_2 = 2.28 x 10^{-3} in. $Pe^* = \dfrac{\rho c_p U h_2}{k}\left(\dfrac{h_2}{B}\right)$ = 1.25

$\left(T_{av}\right)_0 = \dfrac{T_R + T_S}{2}$ = 175°F $Pr \cdot Ec = \dfrac{\mu_0 U^2}{k(T_{av})_0}$ = 1.0

μ_{av} at 175°F = 75 csks k = 1.75 x 10^{-6} $\dfrac{Btu-in.}{sec-°F-in.^2}$

ρ = 8.19 x 10^{-5} $\dfrac{lb-sec^2}{in.^4}$ U = 1150 in./sec

c_p = 178 $\dfrac{Btu-in.}{lb-sec-in.^2-°F}$

Item	T_R = 250°F T_S = 100°F	T_R = 100°F T_S = 250°F
$\bar{F}_\tau$	0.5315	0.671
$\bar{W}$	0.103	0.219
(F_τ/L)	2.32 lb/in.	2.93 lb/in.
(W/L)	787 lb/in.	1,675 lb/in.

Chapter 4

HEAT TRANSFER
SOLUTIONS

In Chapter 2 the thermal problem has been looked at from a planar perspective, that is in the (x, z) domain alone. In Chapter 3 the third dimension, the vital transverse variation, has been added. With that the three-dimensional implications of THD analysis have been brought into the open. However, this does not exhaust the complexity of the THD problem, for we are about to link the three-dimensional fluid film to the goings-on at the walls of the film, that is, the heat transfer processes occurring there. This calls for appropriate heat transfer equations in addition to the energy and Reynolds equations that have been so far the only ones to affect the thermal picture. Since, as was seen, there are numerous permutations on how the thermal fluid film can be handled and since additional demands are about to encroach on what is already a dense analytical problem, we may first summarize the THD picture as it has emerged so far in the course of looking at the problem. This is done in Table 4.1, and it will serve as a guide to the methods and shortcuts employed later on in some of the solutions.

4.1 THE HEAT TRANSFER EQUATIONS

By the laws of heat transmission the amounts of heat entering and leaving the one-dimensional element of Fig. 4.1 are, per unit thickness,

$$\Delta Q = Q_0 - Q_i = k\Delta y \left[\frac{\partial T}{\partial x}\bigg|_{x+\Delta x} - \frac{\partial T}{\partial x}\bigg|_{x} \right]$$

$$= k\Delta x \Delta y \left[\frac{\partial T}{\partial x}\bigg|_{x+\Delta x} - \frac{\partial T}{\partial x}\bigg|_{x} \right] \Big/ \Delta x$$

or

$$\Delta Q = k\Delta x \Delta y \left(\frac{\partial^2 T}{\partial x^2} \right)$$

TABLE 4.1

Spectrum of possible THD solutions

Mode	Solution method	Effects on relevant equations
Fluid convection	• $f(x)$	$v\,\dfrac{\partial T}{\partial y} = w\,\dfrac{\partial T}{\partial z} = 0$
	• $f(x,z)$	$v\,\dfrac{\partial T}{\partial y} = 0$
	• $f(x,y,z)$	v required
	• u & w based on $\mu(x,y,z)$	Form of Reynolds equation affected
Fluid conduction	None	$\nabla^2 T = 0$
	$f(y)$	$\dfrac{\partial^2 T}{\partial x^2} = \dfrac{\partial^2 T}{\partial z^2} = 0$
	$f(x,y,z)$	T at $z = \pm(L/2)$
Heat transfer	• Adiabatic	$\left.\dfrac{\partial T}{\partial y}\right|_{S,R} = 0$
	• To bearing	$(\partial T/\partial y)_R = 0$
	• To bearing & runner	$\left.\dfrac{\partial T}{\partial y}\right|_{S,R} \neq 0$
	• Heat sources and sinks	$\left.\pm\,Q\,(x,z)\right|_{0,1}$
	• Radiation	$Q_r \sim T^4$
	• Free Convection	$h_c \neq 0$
Energy dissipation	• Viscous • Viscous & dilatation	$p \cdot \nabla V = 0$ $\nabla V \neq 0$
Viscosity	• Isoviscous	$\mu = $ const; $T \neq$ const
	• $\mu(T)$	Form of Reynolds equation affected
	• $\mu(p,T)$	Non-linear Reynolds equation

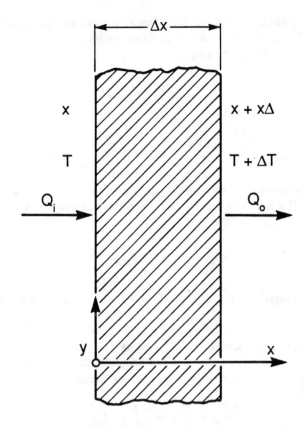

Fig. 4.1 Differential element for heat transmission

The difference between the heat fluxes entering and leaving the element is

$$\Delta Q = c_p \rho \Delta x \Delta y \left(\frac{\partial T}{\partial t} \right)$$

Equating the last two expressions, we obtain

$$\frac{k}{\rho c_p} \left(\frac{\partial^2 T}{\partial x^2} \right) = \frac{\partial T}{\partial t}$$

A similar heat balance in the y and z directions produces identical expressions for the other heat flows; for steady state conditions there is no storage or loss of heat in the element so that $\frac{\partial T}{\partial t} = 0$, and we end up with the Laplace equation representing the heat flow, or

$$\Delta^2 T = \frac{\partial^2 T}{\partial x^2} + \frac{\partial^2 T}{\partial y^2} + \frac{\partial^2 T}{\partial z^2} = 0 \tag{4.1}$$

In cylindrical coordinates the Laplace equation becomes

$$r^2 \frac{\partial^2 T}{\partial r^2} + r \frac{\partial T}{\partial r} + \frac{\partial^2 T}{\partial \theta^2} + r^2 \frac{\partial^2 T}{\partial z^2} = 0 \tag{4.2}$$

The boundary conditions on the heat flow equations are that at the interface of the fluid film and the bounding surfaces we have

• continuity in temperature, or

$$T_{0,h} = T_{S,R} \tag{4.3a}$$

• continuity of heat flux

$$k\frac{\partial T}{\partial y}\bigg|_{0,h} = k_{S,R}\frac{\partial T}{\partial y}\bigg|_{S,R} \tag{4.3b}$$

For complete generality we must also require that the heat flow into the solids equal the heat convected away into the atmosphere or any cooling medium. This requires that

$$-k\frac{\partial T}{\partial y}\bigg|_{S,R} = h_a(T_{S,R} - T_a) \tag{4.3c}$$

where h_a is some heat transfer coefficient and T_a the temperature of the environment.

For the moving boundary one ought to include in the heat balance a term representing the convection of heat by the surface translating in the x direction or a term given by $\rho c U(dT/dx)$. With this term included the heat transfer equation for the moving surface becomes

$$\frac{\partial^2 T}{\partial x^2} + \frac{\partial^2 T}{\partial y^2} - \frac{U}{\sigma}\left(\frac{\partial T}{\partial x}\right) = 0 \tag{4.3d}$$

These heat transfer equations will apply to system components such as the runners, collars, housings, and pedestals that are an inevitable part of seal and bearing assemblies. Equations 4.1 and 4.2 apply to the heat flow within a given homogeneous body. When heat crosses from one medium into another, as from a liquid or a gas into a solid, characteristic heat transfer coefficients determine the rate of heat transmission across such boundaries. These are mostly empirical quantities and they, together with conductivity k determine the rate of heat transmission. Tribological system components will have rather complicated geometries but most of them will be composed of some basic elements such as plates or annuli in series, shown in Fig. 4.2. For future reference a few basic solutions for such simple geometries are given below.

Fig. 4.3a shows the basic element of heat transmission across walls with h_i and h_0 as the inner and outer surface coefficients. The overall heat transmission is given by

$$Q = \frac{1}{\dfrac{1}{h_i} + \dfrac{B}{k} + \dfrac{1}{h_0}}$$

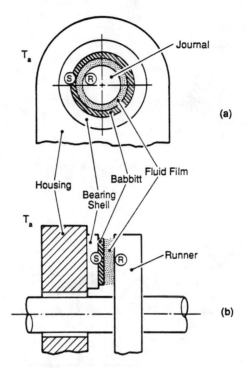

a) **Journal Bearing Assembly**

b) **Thrust Bearing Shell**

Fig. 4.2 Generic bearing assemblies

for a single wall. For two walls the heat transmission is

$$Q = \frac{1}{\dfrac{1}{h_i} + \dfrac{B_1}{k_1} + \dfrac{B_2}{k_2} + \dfrac{1}{h_0}} \qquad (4.4)$$

Given a series of concentric rings as shown in Fig. 4.3b the total heat transmitted across them is

$$Q = \frac{2\pi(T_1 - T_{n+1})}{\dfrac{1}{k_{m1}}\ln\left(\dfrac{R_2}{R_1}\right) + \dfrac{1}{k_{m2}}\ln\left(\dfrac{R_3}{R_2}\right) + \ldots + \dfrac{1}{k_{mn}}\ln\left(\dfrac{R_{n+1}}{R_n}\right)} \qquad (4.5)$$

The temperatures between the layers are given by

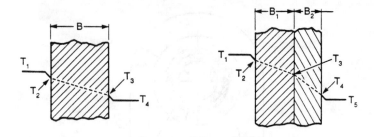

a) Transmission Across Parallel Walls

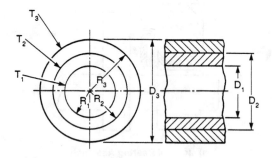

b) Transmission Across Concentric Annuli

Fig. 4.3 Heat transmission across simple bodies

$$T_1 - T_2 = \frac{Q}{2\pi k_{m1}} \ln(R_2/R_1)$$

$$T_2 - T_3 = \frac{Q}{2\pi k_{m2}} \ln(R_3/R_2) \qquad (4.6)$$

$$\cdots\cdots\cdots\cdots\cdots\cdots\cdots\cdots\cdots$$

$$T_n - T_{n+1} = \frac{Q}{2\pi k_{mn}} \ln(R_{n+1}/R_n)$$

4.2 SLIDERS

4.2.1 Infinitely Long Sliders

The general geometry and nomenclature for slider assemblies is given in Fig. 4.4. The first slider configuration treated here is the one from the previous chapter except that instead of assigning to it arbitrary boundary conditions, the fluid film is now linked to the thermal processes transpiring in the runner and pad. While it seems clear that the stationary surface temperature must be obtained from a proper THD solution, there is considerable uncertainty about the runner. The temptation is to assert that since the moving surface, be it a journal or the disc of a thrust bearing, is traversing various heat zones and is undergoing cyclic variations of heat flow at a high rotational frequency, it may, in fact, be considered as an isothermal surface.

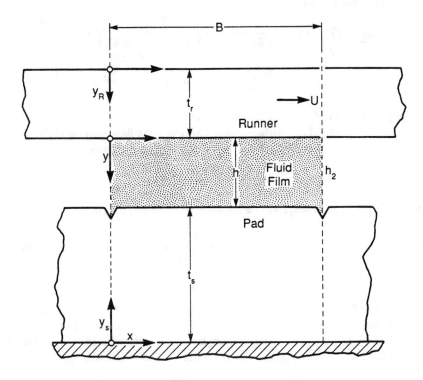

Fig. 4.4 Coordinate system for heat transfer analyses

In the previously cited work by Dowson and Hudson (1964) this aspect of the difficulty was approached by assigning to the runner a quadratic heat flux and then proceeding to estimate the resulting temperature map. This heat flux profile is given by

$$Q = Q_0[1 + C_0(2\bar{x}') + C_2(2\bar{x}')^2]$$

The surface temperature of such a moving semi-infinite solid was constructed from the response to an instantaneous point source developed by Jaeger (1943):

$$\frac{\pi k U(T - T_0)}{2\sigma} = \int_{x-b}^{x+b} Q e^{-u} K_0 |u| du \qquad (4.7)$$

where

$$u = U(x' - s)/2\sigma$$
$$x = (x'U/2\sigma)$$
$$b = (UB/4\sigma)$$

and $K_0|u|$ is the modified Bessel function of 2nd kind and zero order. For large values of b which is here of the order of 10^5, the authors show the above integral to reduce to

$$\frac{\pi k U (T - T_0)}{2Q_0\sigma} = \sqrt{2\pi(b - x)} \left\{ 1 + C_1\left(\frac{x}{b}\right) + C_2\left(\frac{x}{b}\right)^2 \right.$$
$$\left. + \frac{b - x}{3b}\left[C_1 + 2C_2\left(\frac{x}{b}\right)\right] + \frac{C_2(b - x)}{5b^2} \right\}$$

thus yielding the $T = T(x)$ of the runner.

For the pad it is postulated that there is heat flow only in the y direction so that from $(\partial^2 T/\partial y^2) = 0$ we obtain a linear temperature profile

$$T_S = C_3\bar{y} + C_4$$

where the two constants are to be obtained from applying

$$k\frac{\partial T}{\partial y}\bigg|_h = k_s\frac{\partial T}{\partial y}\bigg|_s = 0 \tag{4.8b}$$

$$y = y_t \quad , \quad T = T_0 \tag{4.8b}$$

The second boundary condition presupposes that at some distance y_t in the pad the prevailing temperature is known, in this case being T_0.

The physical properties required here in addition to the data given in Table 3.2 are those pertaining to the solid bodies which are

$$k_{S,R} = 26.2 \quad \frac{\text{BTU}}{\text{hr-ft-°F}}$$

$$\sigma_{S,R} \equiv (k/\rho c_p)_{S,R} = 1.24 \cdot 10^{-4} \quad \text{ft}^2/\text{sec}$$

The pad was considered part of a semi infinite slab with insulated surfaces except for the extent B over the fluid film. The temperature maps of the film, runner, and pad are shown in Fig. 4.5. The thermal effects due to the viscous dissipation in the fluid film extend in the pad to a height of some $2{,}600h_2$ (~ 10 in.) and downstream of the trailing edge of the slider to a distance equal to the extent of the pad (~ 7 in.). It is also seen that with heat transfer present T_{max} is located at the tip of the pad, whereas without heat transfer T_{max} is located well within the film. We also see that the moving surface indeed turned out to be nearly an isothermal surface, at

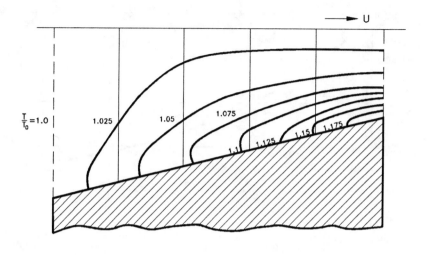

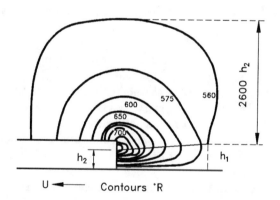

Fig. 4.5 Isotherms in infinitely long slider: T's in °R, Dowson and Hudson, 1964

least for the kind of quadratic heat flux postulated here. Such an isothermal runner surface would of course produce high transverse temperature gradients near the stationary wall, accounting for the high heat flow to the pad.

The variation of load capacity with the Peclet number in Fig. 4.6 (essentially a velocity parameter) shows that due to HT effects the load capacity increases less than linearly with a rise in velocity, which is the case with isothermal solutions. It is also found that heat transfer causes an additional drop in load capacity as compared with an isothermal boundary, due, clearly, to higher film temperatures.

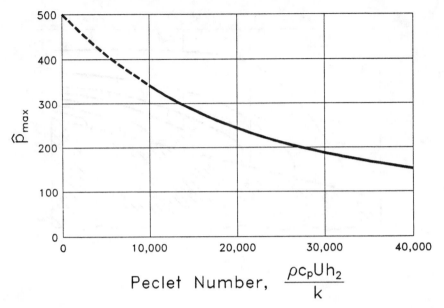

Fig. 4.6 Effect of Peclet number on load capacity of sliders, Dowson and Hudson, 1964

Additional data for an $(L/D) = \infty$ slider having the dimensions given in Table 4.2 are shown in Figs. 1.8 and 4.7a & b. They confirm, in general, the previous results including the point that consideration of heat transfer effects further lowers the load capacity vis-a-vis cases with constant surface temperatures. The understanding, of course, is that in both approaches the pad outer surface is kept at the same temperatures as would an isothermal stationary surface adjacent to the fluid film. An additional point worthy of note is that the analysis on which the results of Fig. 4.7 are based also yields the conclusion that the runner temperature is essentially constant.

A complete set of solutions of sliders with various conditions of taper, heat transfer, and thermal coefficients as obtained by Strömberg (1971) is given in Table 4.3. In this table

$$h = h_2[1 + K(1 - \bar{x})]$$
$$\overline{W} = (W/\mu_0 U L)(h_2/B)^2$$
$$\overline{H} = (H/\mu_0 U^2 L)(h_2/B)$$
$$\bar{\alpha} = \alpha\mu_0 U B/\rho c_p h_2^2$$
$$\bar{k} = kB/\rho c_p U h_2^2$$
$$\tilde{T} = \rho c_p (T - T_0) h_2^2/\mu_0 U B$$

Figs. 4.8 and 4.9 give an overview of the HT solutions versus the adiabatic case and, as was noted earlier, the HT solutions yield lower load capacities

TABLE 4.2

Data for slider of Fig. 4.7

$B = 3.6$ in.	$\mu_0 = 2.0 \dfrac{\text{lb-sec}}{\text{u}^2}$
$h_2 = 2.5 \times 10^{-3}$ in.	$c_p = 0.48 \dfrac{\text{Btu}}{\text{lb-}°\text{F}}$
$(h_1/h_2) = 2.0$	$k = 0.0742 \dfrac{\text{Btu}}{\text{hr-ft-}°\text{F}}$
$U = 100$ ft/sec	$k_R = k_s = 26.2 \dfrac{\text{Btu}}{\text{hr-ft-}°\text{F}}$
$t_R = 1.44$ in.	$T_0 = 100°\text{F}$
$t_S = 7.2$ in.	$\sigma_R = 1.24 \times 10^{-4} \dfrac{\text{ft}^2}{\text{sec}}$
$\alpha = 0.025 \dfrac{1}{°\text{F}}$	

$$Re^* \equiv \left(\frac{\rho U h_2}{\mu_0}\right)\left(\frac{h_2}{B}\right) = 8.76 \times 10^{-3}$$

$$Re^* \cdot Pr = 19.1$$

$$Pr \cdot Ec = 18.1$$

and higher temperatures than solutions which ignore temperature variations across the film. Some of the reasons for this phenomenon will be discussed in the subsequent section.

4.2.2 Finite Sliders

When finite configurations are to be considered one is faced again with a reconsideration of the form of the energy equation to be used. It was

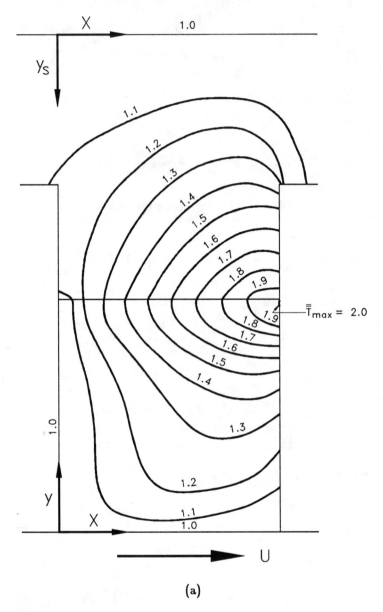

(a)

Fig. 4.7 HT solution for infinitely long sliders, (a) constant T_R, (b) variable $T_R(x)$, Hahn and Kettleborough, 1967

pointed out in Section 3.1 that compared to the transverse conduction term $(\partial^2 T/\partial y^2)$ all other conduction can be ignored. This was quite straightforward when only the (x, y) field was treated. When the three-dimensional film (x,y,z) is to be analyzed, the three conductance terms assume a some-

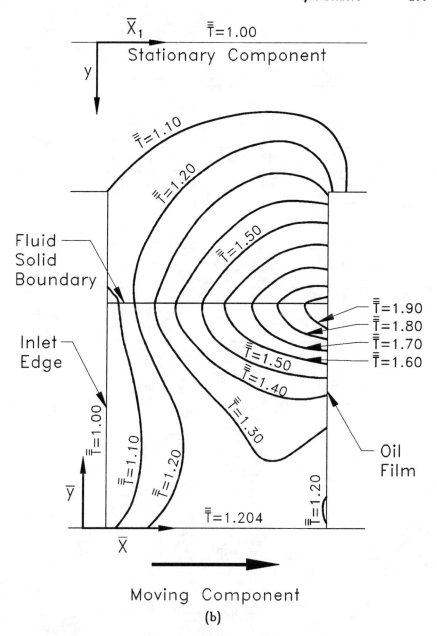

Fig. 4.7 HT solution for infinitely long sliders, (a) constant T_R, (b) variable $T_R(x)$, Hahn and Kettleborough, 1967

what different hierarchy in the energy balance. The new element that arises is that while in the x direction there is a high rate of heat convection induced by the translatory velocity U, in the axial direction this convection,

TABLE 4.3

Performance of infinitely long sliders

$$\bar{\alpha} = 0$$

K	$\bar{W}$	$\bar{H}$	$\dfrac{Q}{LUh_2}$	$\bar{T}$ at $\bar{k} =$				
				0.005	0.1	0.3	1.0	∞
1.0	0.15888	0.7726	0.6667	3.654	2.900	2.023	1.506	1.158
1.5	0.15773	0.7292	0.7143	3.621	2.868	1.982	1.435	1.019
2.0	0.14793	0.6972	0.7499	3.592	2.840	1.950	1.390	0.928

$\bar{\alpha}$	$\bar{k}$	K	$\bar{W}$	$\dfrac{Q}{LUH_2}$	$\bar{H}$	$\bar{T}_{max}$
0.5	0.05	1.0	0.09898	0.7416	0.5970	2.474
		1.5	0.10992	0.7925	0.5826	2.514
		2.0	0.10993	0.8319	0.5685	2.548
	0.3	1.0	0.11922	0.7078	0.6295	1.479
		1.5	0.12459	0.7613	0.6095	1.479
		2.0	0.12080	0.8012	0.5915	1.475
	1.0	1.0	0.12799	0.6936	0.6440	1.153
		2.0	0.12604	0.7865	0.6036	1.096
	4.0	1.0	0.13199	0.6869	0.6506	1.023
		1.5	0.13512	0.7390	0.6300	0.953
		2.0	0.12900	0.7777	0.6105	0.905
	∞	1.0	0.13378	0.6839	0.6536	0.955
		1.5	0.13696	0.7350	0.6338	0.861
		2.0	0.13069	0.7727	0.6147	0.794
1.0	0.05	1.0	0.07216	0.7964	0.4996	1.953
		1.5	0.08545	0.8496	0.4980	2.009
		2.0	0.08891	0.8913	0.4928	2.058
	0.3	1.0	0.09807	0.7370	0.5445	1.197
		1.5	0.10537	0.7952	0.5359	1.210
		2.0	0.10434	0.8387	0.5255	1.216
	1.0	1.0	0.10951	0.7126	0.5638	0.959
		1.5	0.11488	0.7706	0.5542	0.938
		2.0	0.11175	0.8135	0.5421	0.924
	4.0	1.0	0.11476	0.7015	0.5730	0.863
		1.5	0.11988	0.7577	0.5642	0.814
		2.0	0.11596	0.7990	0.5520	0.778
	∞	1.0	0.11660	0.6971	0.5743	0.823
		1.5	0.12208	0.7517	0.5675	0.754
		2.0	0.11809	0.7913	0.5563	0.701
2.0	0.05	1.0	0.04852	0.8778	0.3907	1.437
		1.5	0.06106	0.9370	0.3986	1.501
		2.0	0.06589	0.9823	0.4012	1.556
	0.3	1.0	0.07522	0.7793	0.4429	0.858
		1.5	0.08309	0.8445	0.4449	0.917
		2.0	0.08430	0.8940	0.4425	0.930
	1.0	1.0	0.08376	0.7402	0.4639	0.737
		1.5	0.09390	0.8060	0.4660	0.730
		2.0	0.09319	0.8547	0.4621	0.724
	4.0	1.0	0.09297	0.7224	0.4738	0.674
		1.5	0.09971	0.7857	0.4775	0.647
		2.0	0.09830	0.8319	0.4738	0.624
		1.0	0.09400	0.7166	0.4715	0.657
		1.5	0.10158	0.7775	0.4783	0.614
		2.0	0.10031	0.8212	0.4762	0.579

Stromberg, 1971

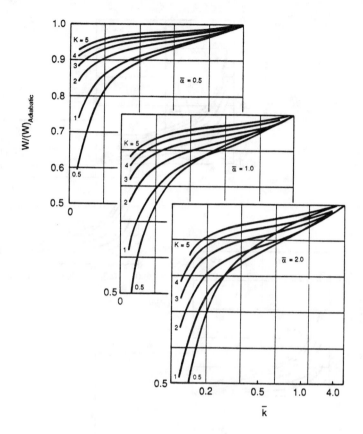

Fig. 4.8 Comparison of adiabatic and HT load capacity for sliders, Strömberg, 1970

determined by w, will be a much lower quantity. Thus, while ignoring conduction in the x direction will not have much effect on the amount of energy transported in the direction of U, conduction vis-a-vis convection in the z direction will have a more telling effect.

While the above paragraph constitutes a physical argument for dropping $(\partial^2 T/\partial x^2)$ from the equation, there is also a solid mathematical reason for doing so. The omission of the conduction term in the x direction yields the equivalent of an initial value problem in the direction of motion with the temperature at the start of the film known (T_0 or the mixing temperature T_1). If $(\partial^2 T/\partial x^2)$ were retained, the energy equation would be elliptical which would pose much greater difficulties than the parabolic kind treated here.

Consequently, the form of energy equation of a finite slider assumes

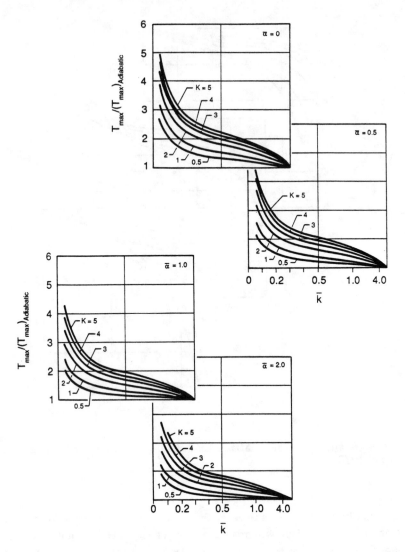

Fig. 4.9 Comparison of temperatures from adiabatic and HT solutions, Strömberg, 1970

the form

$$\rho c_p \left[u \frac{\partial T}{\partial x} + v \frac{\partial T}{\partial y} + w \frac{\partial T}{\partial z} \right]$$
$$= k \left[\frac{\partial^2 T}{\partial y^2} + \frac{\partial^2 T}{\partial z^2} \right] + \mu \left[\left(\frac{\partial u}{\partial y} \right)^2 + \left(\frac{\partial w}{\partial y} \right)^2 \right] \qquad (4.9)$$

Ezzat and Rohde (1973) also included the term

$$\delta T \left[u \frac{\partial p}{\partial x} + w \frac{\partial p}{\partial z} \right]$$

accounting for the thermal compressibility of the lubricant. Since $\delta = 0.36 \cdot 10^{-3}$ $1/°F$ and since density variation has been shown to have a negligible effect by a number of other investigators, this term was here left out of Equ. (4.9).

The boundary conditions employed for the finite slider were

$$\text{at} \quad x = 0 \quad \text{and} \quad y = 0, \quad T = T_0 \tag{a}$$

This, as seen, assumes a constant runner temperature, a conclusion based on the same Jaeger integral used in the previous section, Equ. (4.7). Next, note the condition

$$\left(\frac{\partial T}{\partial z} \right)_{\pm L/2} = 0 \tag{b}$$

This condition implies that at the axial edges of the film (and only there) the temperature gradients are zero and all the heat removed is by axial convection.

For the pad the Laplace equation

$$\frac{\partial^2 T}{\partial x^2} + \frac{\partial^2 T}{\partial y^2} = 0$$

is used subject to the following boundary conditions:

$$k \frac{\partial T}{\partial y} \bigg|_h = k_s \frac{\partial T}{\partial y} \bigg|_{s,0} \tag{c}$$

$$-k_s \frac{\partial T}{\partial y} \bigg|_{s,t} = h_s \left[T(y_s = t) - T_a \right] \tag{d}$$

$$k_s \frac{\partial T}{\partial y} \bigg|_0 = h_s \left[T(x = 0) - T_a \right] \tag{e}$$

$$k_s \frac{\partial T}{\partial y} \bigg|_B = h_s \left[T(x = B) - T_a \right] \tag{f}$$

From the two-dimensional form of the Laplace equation it is clear that in the pad itself all z variations were ignored. This approach was taken in light of the subsequent solutions which showed that the axial temperature profile along the pad $(y = h)$ was constant except for a slight decrease at the very ends of the bearing $(z = \pm L/2)$. It should also be noted that whereas in the previous section the pad was taken to be part of a semi-infinite solid, here it is considered to be an isolated element, facing on one side the fluid film, and on the three other sides the ambient air at a temperature T_a.

On the Reynolds equation which is given by its form of Equ. (3.10) the boundary conditions are simply

$$p = 0 \quad \text{at} \quad x = 0 \quad \text{and} \quad B$$
$$p = 0 \quad \text{at} \quad z = \pm(L/2)$$

For completeness, since several new terms appear in the energy equation, its non-dimensionalized form is

$$
\begin{aligned}
A_1 \Bigg\{ \overline{u}\left[\frac{\partial \overline{\overline{T}}}{\partial \overline{x}} - \left(\frac{\overline{y}}{\overline{h}}\right)\left(\frac{\partial \overline{h}}{\partial \overline{x}}\right)\frac{\partial \overline{\overline{T}}}{\partial \overline{y}}\right] + \left(\frac{\overline{v}}{\overline{h}}\right)\left(\frac{B}{h_2}\right)\frac{\partial \overline{\overline{T}}}{\partial \overline{y}} \\
+ \overline{w}\frac{\partial \overline{\overline{T}}}{\partial \overline{z}} \Bigg\} \quad = \left[\frac{\partial^2 \overline{\overline{T}}}{\partial \overline{z}^2} + \left(\frac{B}{h_2}\right)^2 \frac{1}{\overline{h}^2}\frac{\partial^2 \overline{\overline{T}}}{\partial \overline{y}^2}\right] \\
+ A_2 \frac{\overline{\mu}}{\overline{h}^2}\left[\left(\frac{\partial \overline{u}}{\partial \overline{y}}\right)^2 + \left(\frac{\partial \overline{w}}{\partial \overline{y}}\right)^2\right]
\end{aligned}
\tag{4.10}
$$

where

$$A_1 = \frac{\rho c_p U B}{k} \quad , \quad A_2 = \frac{\mu_0 U}{k T_0}\left(\frac{B}{h_2}\right)^2 \quad , \quad \overline{z} = \frac{z}{B}$$

and all velocities are normalized by U. Ten cases with various geometric and thermal characteristics were considered, all centered around a standard configuration with $(L/B) = 1$, $\overline{h}_1 = 2$ and using an SAE 30 oil at $T_0 = 100°F$. The ten permutations are given in Table 4.4. Fig. 4.10, showing essentially the load capacity, confirms the previously noted point that load does not rise linearly with speed in THD bearings, this being due to the drop in lubricant viscosity with speed. Figs. 4.11 through 4.14 provide the thermal characteristics of the finite slider solution. First we again note the considerable decrease in load capacity from an isothermal analysis— a drop of 54%—but also from a solution with pad surface temperature a constant—a 30% drop. The data on load capacity and friction given in Fig. 4.12 also show the effect of the inlet temperature T_0. Perhaps more significant is the comparison between the present results and the average or effective viscosity approaches taken to evaluate thermal effects. It is seen that a given method cannot be stated to be good or bad because it depends in which thermal regime it is being applied. Thus when $T_0 = 100°F$, which means that there are steep variations of viscosity, the solution based on the exit temperature T_2 almost offers correct results. However, when $T_0 = 200°F$ and the viscosity variation is much milder then the one based on an average viscosity, $T_{av} = T_0 + \Delta T/2$ brackets the correct solution.

One series of runs consisted in ascertaining the effect of varying the thermal characteristics of the pad. It turned out that, as shown in Table 4.5, orders of magnitude changes in the values of h_S and k_S did not affect slider performance to any serious degree. Likewise varying pad thickness from 1/2 in. to 2 in., or the ambient temperature T_a from 0°F to 300°F

TABLE 4.4

Slider characteristics for Figs. 4.10-4.14

Case no.	h_1 in.	h_2 (in.)	B (in.)	U (ft/sec)	Oil type	T_0 (°F)
1^a	0.005128	0.002564	3.6	100	(d)	100
2	0.004	0.002	3	Variable	SAE-30	′′
3	0.003	′′	3	′′	′′	′′
4	0.004	′′	6	′′	′′	′′
5	′′	′′	1.5	′′	′′	′′
6	0.002	0.001	3	′′	′′	′′
7	0.004	0.002	′′	′′	Variable	Variable
8	′′	′′	′′	′′	SAE-50	′′
9^b	′′	′′	′′	100	SAE-30	100
10^c	′′	′′	′′	′′	′′	′′

$$k = 0.075 \ \frac{Btu}{hr\text{-}°F\text{-}ft} \qquad\qquad T_a = 80°F$$

$$k_s = 30 \ \frac{Btu}{hr\text{-}°F\text{-}ft} \qquad\qquad L = 3 \text{ in.}$$

$$t = 1 \text{ in.} \qquad\qquad h_s = 100 \ \frac{Btu}{hr\text{-}°F\text{-}ft^2}$$

a) $L = 3.6$ in.   b) $T_a = 0°F$

c) $h_s = 10 \ \dfrac{Btu}{hr\text{-}°F\text{-}ft^2}$ d) $\mu = 20.1 \times 10^{-7} \ \dfrac{lb\text{-}sec}{in.^2}$ at 100°F

did not affect W, F, or even T_{max} beyond a few percentage points. The effect of the latter, however, on the temperature distribution as given in Figs. 4.13 and 4.14 is of some interest. While with the colder ambient T_a there was, as expected, a drop in the T_{max} value, there also resulted a shift in the locus of the maximum temperature line, from that at the pad surface to inside the fluid film $(\overline{y} = 0.8)$, yielding higher heat transfer rates. Likewise of importance are the plots of the axial temperature distribution $T(x, z)$ shown on the two figures. As seen even with the present rigorous solution, one would conclude that $T(z) \simeq$ constant.

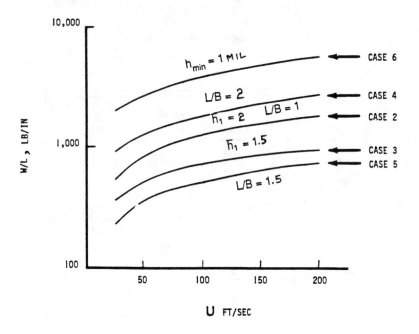

U FT/SEC

Fig. 4.10 Three-dimensional solution for load capacity of finite sliders: for input data see Table 4.4, Ezzat and Rohde, 1973

We next consider a slider whose pad is cooled on the underside by a continuous flow of water maintained at a suitably low temperature T_{W1}, an analysis due to Tahara (1967). This is an instance of a THD solution with a heat sink as listed in Table 4.1. The convective term in the y direction is here ignored and the following set of boundary conditions are specified:

• At the moving surface, $y = 0$, there are two imposed conditions

$$T = T_R = \text{constant} \qquad (a)$$

$$\int_0^{(L/2)} dz \int_0^B \left.\frac{\partial T}{\partial y}\right|_0 dx = 0 \qquad (b)$$

the latter implying that the heat flow to and from the runner integrated over the fluid film must add up to zero—a form of bulk adiabacity in place of the usual $\partial T/\partial y|_{0,h} = 0$.

The coolant remains at a constant temperature and it absorbs all the heat flowing into the pad from the film, or

$$\left.\frac{\partial \overline{T}}{\partial \overline{y}}\right|_{\overline{y}=1} = \Pi(\overline{T}_s - \overline{T}_{sw}) \qquad (c)$$

where

$$\Pi = \frac{h_{sw} h_2}{k}, \qquad h_{sw} = \frac{1}{\left(\dfrac{h}{k_s}\right) + \dfrac{1}{h_w}}$$

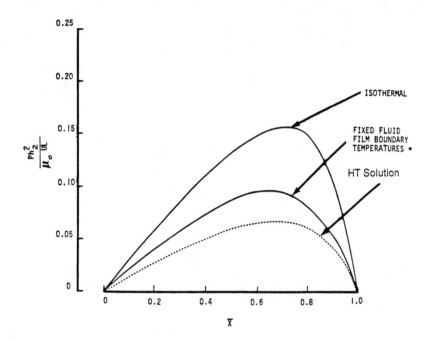

Fig. 4.11 Midplane pressure as obtained from three solutions for finite sliders: for slider data see Table 4.4 - Case 1. Ezzat and Rohde, 1973

h_{sw} being the combined transmissivity of the pad and water.

- The inlet temperature consists of a linear profile between T_R and T_0, or

$$T(y)\Big|_{x=0} = T_R - (T_R - T_0)\bar{y} \qquad (d)$$

It tells something of the nature of the heat flux at the runner surface that in the solution of the problem there was no need to rigorously apply boundary condition (b). A simple trial and error selection of T_R provided a fairly quick convergence on the runner temperature profile that satisfies the involved integral. Only very mild variations in $T(x,z)$ in the vicinity of runner could have made this approach possible.

The solution for the cooled slider had the following geometrical characteristics:

B=30 cm   t=0.5 cm

$k = 0.0134 \frac{\text{kg·cm}}{\text{cm·sec·°C}}$ $k_s = 5.5 \frac{\text{kg·cm}}{\text{cm·sec·°C}}$

$\alpha = 0.04°\text{C}^{-1}$ $T_w = 20°\text{C}$

$\mu_0 = 2.6 \cdot 10^{-6} \frac{\text{kg·sec}}{\text{cm}^2}$ $k_w = 7.5 \frac{\text{kg·cm}}{\text{cm·sec·°C}}$

$\rho c_p = 17 \frac{\text{kg}}{\text{cm}^2 \cdot °\text{C}}$ $\Pi = (h_{sw} h_2 / k) = 1.0$

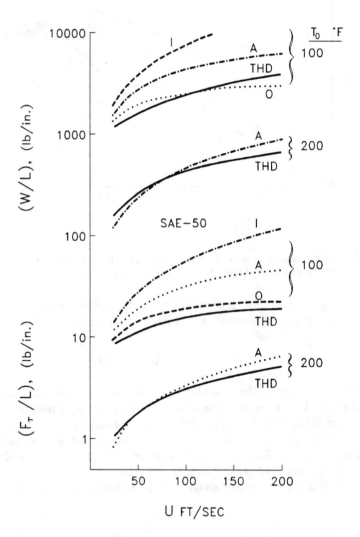

THD —   HT Solution
I —   Based on Inlet Viscosity μ_0
A —   Based on Average Viscosity μ_{av}
O —   Based on Outlet Viscosity μ_2

Fig. 4.12 Load capacity and friction in finite sliders for Case 8 of Table 4.4. Ezzat and Rohde, 1973

$h_2 = 3 \cdot 10^{-3}\text{cm}$ $(\mu_0 B U^2/k) = 7.7$

$\overline{h}_1 = 2$ $(\rho c_p U h_2^2/kB) = Pe(h_2/B) = 0.38$

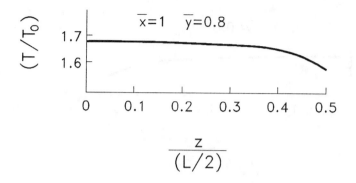

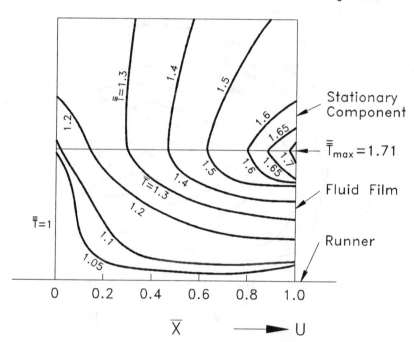

Fig. 4.13 Isotherms in finite sliders: Case 10 of Table 4.4. $T_a = 80°F$, Ezzat and Rohde, 1973

$U = 10^3 \, \text{cm/sec}$ $\qquad\qquad$ $T_0 = 40°C$

The transverse temperatures resulting from the underside cooling of the pad are shown in Fig. 4.15 at the inlet, mid-section and end of the slider along the axial center line. With a coolant at 20°C and inlet lubricant at 40°C the runner had a temperature of about 55°C. The ΔT of the pad relative to the runner was −20°C at the inlet and −8°C at the outlet. According to the author these results were nearly identical to those obtained

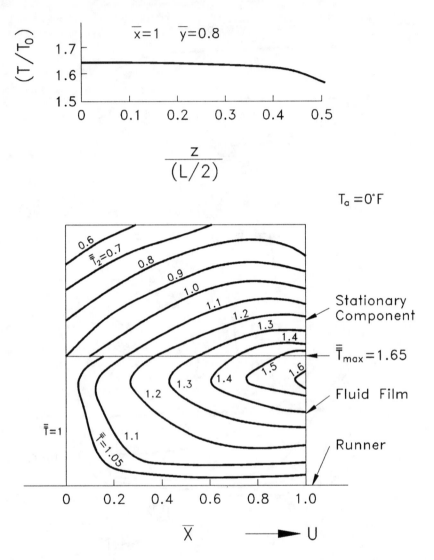

Fig. 4.14 Isotherms in finite sliders: Case 9 of Table 4.4. $T_a = 0°F$, Ezzat and Rohde, 1973

for a parallel analysis conducted for an $(L/B) = \infty$. The small effect of the (L/B) parameter on the thermal characteristics of a cooled slider is also shown in Fig. 4.16 which provides an index of cooling effectiveness in terms of the parameter $h_{SW} B / \rho c_p U h_2$. Large values of these parameters, when the amount of heat carried away by the pad approaches 100%, is achieved by either a large heat transfer coefficient h_{SW}, or small film thicknesses h_2. Thus effective cooling is not likely to be achieved by a low (L/B) ratio or high side leakage but by a reduction in h. This is, in a way, fortunate

TABLE 4.5

Effect of pad parameters on performance

Heat transfer parameters				Percentage change in performance			
T_a	h_s	k_s	t	$\bar{W}$	$\bar{F}_T$	$\bar{\bar{T}}_{max}$	$\bar{Q}_s$
80	10	30	1	-2	-1	+2	+0.2
80	100	30	1	0*	0	0	0
80	10,000	30	1	+10	+3	-2	-2
80	100	6	1	-5	-3	+2	+4.5
80	100	30	1	0	0	0	0
80	100	118	1	+10	+2.5	-1.5	-7.5
0	100	30	1	+2	+2.2	-3.5	-2
80	100	30	1	0	0	0	0
150	100	30	1	-1.5	-1.2	+4	+4
300	100	30	1	-1.5	-1.2	+13	+5
80	100	30	0.5	+1.5	0	-1.2	0
80	100	30	1	0	0	0	0
80	100	30	2	0	-0.8	0	0

Ezzat and Rohde, 1973

$$\bar{W} = \frac{W}{\mu_0 U} \left(\frac{h_2}{B} \right)$$

$$\bar{F}_T = \frac{F_2}{\mu_0 U B} \left(\frac{h_2}{B} \right)$$

$$\bar{Q}_s = \frac{Q_2}{U B^2} \left(\frac{h_2}{B} \right)$$

* Rows showing zeros correspond to the reference

because this is when cooling is most needed. On the other hand, not much will be achieved with forced cooling when the film thickness is large.

The degree of cooling, of course, affects the temperature differential between the two surfaces and this relationship is shown in Fig. 4.17. This can be expressed by means of an equation first developed by Hagg (1944) for infinite parallel surfaces

$$\alpha(T_R - T) = \ln\left[1 + \frac{\alpha\mu U^2}{Ak}\right] \tag{4.11}$$

with the value of the constant A given in Fig. 4.18 for various values of $\bar{h}_1$ in terms of the parameter $(\alpha\mu U^2/k)$. For $(L/B) \to \infty$ and parallel surfaces

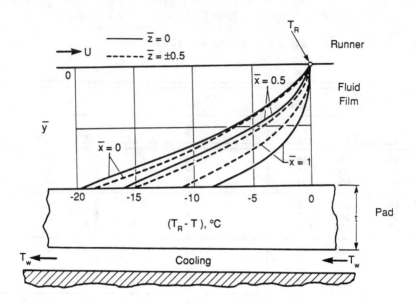

Fig. 4.15 Transverse temperatures in cooled finite slider, Tahara, 1967

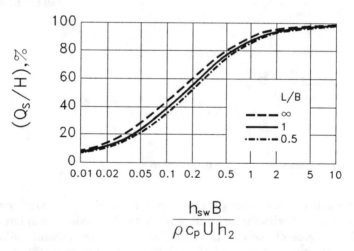

$$\frac{h_{sw}B}{\rho\, c_p\, U\, h_2}$$

Fig. 4.16 Cooling effectiveness, Q_s, in finite sliders, Tahara, 1967

the value of A is 2.0.

Finally, when an average viscosity was sought which would provide a solution close to the present one via an isoviscous analysis, the value obtained was

$$T_{av} = \left(\frac{3T_R + T_s}{4}\right)$$

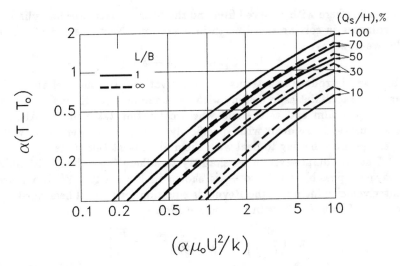

Fig. 4.17 Relation between degree of cooling, Q_s, and surface temperatures, Tahara, 1967

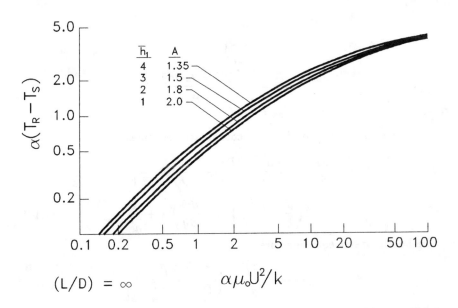

Fig. 4.18 Surface temperatures in fully cooled infinitely wide slider, Hagg, 1944

4.3 JOURNAL BEARINGS

In dealing with journal bearings there are a few new details in the setting up of the coordinate system of the journal bearing assembly. First, we are dealing here with a curved film and the bearing liner calls for cylindrical coordinates in solving the differential equation. Next, instead of a linear film we now have

$$h = C[1 + \epsilon \cos(\theta - \phi_0)] \tag{4.12}$$

which will introduce new terms in the derivatives. But most of all we shall run into uncertainties in dealing with the boundary conditions both at the start of the film and at the end where cavitation takes place. Although these questions are dealt with specifically in a later chapter, there will be no escape from having to deal with them already at this stage.

The coordinate system we shall try to adhere to — though it will not always be possible to do so — is shown in Fig. 4.19. With a journal center velocity included, the Reynolds equation for journal bearings taking account of Equ. (4.12) is in normalized form

$$\frac{\partial}{\partial \theta}\left[\overline{h}^3 I_{RR}\left(\frac{\partial \overline{p}}{\partial \theta}\right)\right] + \left(\frac{R}{L}\right)^2 \frac{\partial}{\partial \overline{z}}\left[\overline{h}^3 I_{RR}\frac{\partial \overline{p}}{\partial \overline{z}}\right]$$

$$= \frac{\partial}{\partial \theta}\left[\overline{h} - \overline{h}I_{33}\right] + \dot{\epsilon}_x \cos(\theta + \phi_0) + \dot{\epsilon}_y \sin(\theta + \phi_0) \tag{4.13a}$$

where

$$\tau = \omega t$$

$$\dot{\epsilon} = \left(\frac{d\epsilon}{d\tau}\right)$$

$$I_{RR} = \int_0^1 \frac{\overline{y}(\overline{y} - I_{33})d\overline{y}}{\overline{\mu}}$$

$$I_{33} = I_1/I_0$$

The energy equation in (θ, z) coordinates becomes

$$Pe\left\{\overline{u}\left[\frac{\partial \overline{\overline{T}}}{\partial \theta} - \frac{\overline{y}}{\overline{h}}\left(\frac{\partial \overline{h}}{\partial \theta}\right)\frac{\partial \overline{\overline{T}}}{\partial \overline{y}}\right] + \left(\frac{L}{D}\right)\overline{w}\frac{\partial \overline{\overline{T}}}{\partial \overline{z}}\right\}$$

$$= \Pi\left(\frac{\overline{\mu}}{\overline{h}}\right)\left[\left(\frac{\partial \overline{u}}{\partial \overline{y}}\right)^2 + \left(\frac{\partial \overline{w}}{\partial \overline{y}}\right)^2\right] + \frac{1}{\overline{h}^2}\left(\frac{\partial^2 \overline{\overline{T}}}{\partial \overline{y}^2}\right) \tag{4.13b}$$

The heat transfer equations are

$$\frac{\partial^2 \overline{\overline{T}}}{\partial \overline{r}^2} + \frac{1}{\overline{r}}\frac{\partial^2 \overline{\overline{T}}}{\partial \overline{\theta}^2} + \left(\frac{L}{D}\right)^2 \frac{\partial^2 \overline{\overline{T}}}{\partial \overline{z}^2} = 0 \tag{4.13c}$$

where

$$\Pi = \left(\frac{\mu_0 U^2}{kT_0}\right)$$

A complete solution has thus to treat and reconcile four differential equations: the Reynolds equation, the energy equation, and two Laplace equations, one for the bearing and one for the journal, all interlinked by common and complex boundary conditions. The following sections represent the various attempts to achieve this.

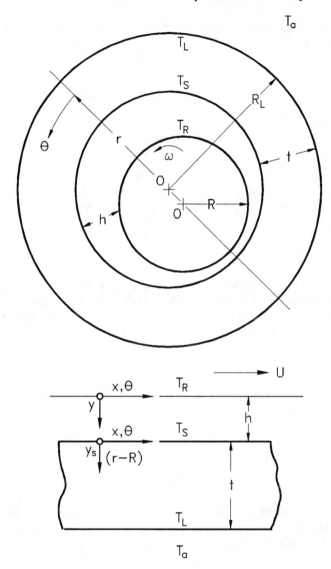

Fig. 4.19 Nomenclature for heat transfer problems of journal bearings

4.3.1 The Couette Approximation

It was seen in the study of sliders—and similar results are obtained in the treatment of journal bearings—that there is little temperature variation in the axial direction. This makes it particularly acceptable to apply here the Couette approximation, which in the treatment of thermal effects ignores the pressure gradients. This eliminates completely the axial velocity w from the energy equation. However, there is still a need for the transverse

velocity v and this in the present case will be shown to be small, too.
Starting with

$$\rho c_p \left[u \frac{\partial T}{\partial x} + v \frac{\partial T}{\partial y} \right] = k \frac{\partial^2 T}{\partial y^2} + \mu \left(\frac{\partial u}{\partial y} \right)^2$$

and writing

$$u = U\overline{y}$$

we obtain

$$\overline{h}^2 \left[\overline{u} \left(\frac{\partial \hat{T}}{\partial \overline{x}} \right) + \overline{uy} \frac{\epsilon \sin \theta}{\overline{h}} \left(\frac{\partial \hat{T}}{\partial \overline{y}} \right) + \frac{\overline{v}}{\overline{h}} \left(\frac{\partial \hat{T}}{\partial \overline{y}} \right) \right]$$
$$= \frac{1}{Pe^*} \frac{\partial^2 \hat{T}}{\partial \overline{y}^2} + \overline{\mu} \left(\frac{\partial \hat{u}}{\partial \overline{y}} \right)^2 \tag{4.14}$$

The v component is usually evaluated by integrating the continuity equation, but as continuity is violated when all the pressure terms are omitted, an estimation of v is made with the aid of Fig. 4.20. Assuming that the inclination to the horizontal of u varies linearly with y,

$$\delta(y) = \overline{y}\delta_0 = \overline{y} \left(\frac{\partial h}{\partial x} \right)$$

and

$$v = U \sin \delta \simeq u \cdot \delta = u\overline{y} \left(\frac{\partial h}{\partial x} \right)$$

$$\overline{v} = \frac{v}{U} \left(\frac{R}{C} \right) = \overline{uy} \left(\frac{\partial \overline{h}}{\partial \theta} \right) = -\overline{uy}\epsilon \sin \theta$$

This equation implies that the second and third terms of the left side of Equ. (4.14) are equal in magnitude and of opposite signs, and cancel each other, resulting in the following energy equation:

$$\overline{hu} \left(\frac{\partial \hat{T}}{\partial \theta} \right) = \frac{1}{Pe^*} \frac{\partial^2 \hat{T}}{\partial \overline{y}^2} + \overline{\mu} \left(\frac{\partial \hat{T}}{\partial \theta} \right)^2 \tag{4.15}$$

The result is that the problem can be solved without reference to the Reynolds equation, independent as it is of $p(x, z)$. Equ. 4.15 is then the equivalent of an $(L/D) = \infty$ case with the additional simplification of $(dp/d\theta) \simeq 0$.

In using Equ. (4.15) McCallion et al. (1970) also assumed that due to its cylindrical geometry, the flow of heat in the bearing shell is presumed to be in the circumferencial direction only with the journal temperatures kept constant. The operating characteristics of the bearing are given in Table

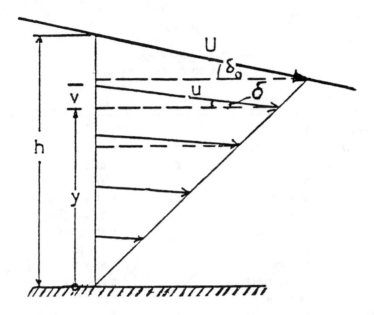

Fig. 4.20 Approximate values of transverse velocity in journal bearings

4.6. A number of revealing aspects emerge from the present simplified approach many of which reappear later in the course of more rigorous analyses. Perhaps the most interesting point emerges in Fig. 4.21, which compares the present results to those of adiabatic and isothermal approaches. Instead of being bracketed by these two extreme cases, the HT load capacity falls below that of the adiabatic solution. From an overall energy balance this would seem difficult to accept because with all the energy stored in the fluid film the adiabatic case should yield the highest average temperature and lowest average viscosity.

But this only proves once again the weakness of the average T or average μ approach. One reason why it should be possible for a non-adiabatic film to yield lower viscosities emerges from an examination of Figs. 4.22 and 4.23. In the first figure we have the isotherms for an inlet oil temperature of 37°C. Near the h_{min} region the hot fluid film will be pumping heat into the bearing whereas around h_{max} heat will flow from the bearing shell into the fluid film. Thus, at the start of the film the viscosity for the non-adiabatic case will due to the inflow of heat be lower than for the adiabatic case. From here on, even though more energy is stored in the film under adiabatic conditions, its corresponding viscosities may never fall below the HT case, the curve marked 1 in Fig. 4.23. But even if it did catch up, as in curve 2, it may not do so sufficiently to reverse the initial loss in viscosity. All the uncertainty is due to the non-linear nature of the $\mu - T$ relationship and consequently the adiabatic curve may not constitute the lower limit on

TABLE 4.6

Journal bearing operating data
for Couette Approximation Solution

Journal diameter	D	4 in.
External bearing diameter	D_0	9 in.
Radial clearance	C	0.0025 in.
Bearing length	L	3 in.
Bearing thermal conductivity	k_s	28.8 Btu/hr-ft-°F
Lubricant at inlet temperature	μ_0	4.35×10^{-6} Reyns
Lubricant specific heat	c_p	0.48 Btu/lb-°F
Lubricant specific weight	w	53.84 lb/ft^3
Lubricant thermal conductivity	k	7.52×10^{-2} Btu/hr-ft-°F
Lubricant viscosity coefficient	α	0.023 1/°F
Oil inlet temperature	T_0	98.24°F
Ambient temperature	T_a	76°F
Heat transfer coefficient	h_s	$14.08 \dfrac{Btu}{hr\text{-}ft^2\text{-}°F}$

Phosphorous bronze bearing shell-4 in.long; shaft-mild steel

load capacity.

The next point of interest is that the temperature profile decreases past the point of h_{min}. This, in the first place, would seem to be due to the purely circumferential heat flow lines in the bearing shell which more or less makes the line of centers OO′ a line of symmetry, the only perturbation being the admission of cold oil at $\theta = 0$ which shifts the location of T_{min} only some 10° or so. A temperature profile symmetrical about h_{min} is contrary to experience and the analysis would be even more in conflict with reality were it not for the fact that in actuality temperatures are lowered due to fluid film cavitation. This touches upon the question of boundary conditions, a subject reserved for later treatment.

It has been apparent before, and the present analysis corroborates the observation, that the HT results seem insensitive to parametric variations of a number of seemingly pertinent variables. Following is a tabulation of some of the parametric changes introduced and the resulting effect of load capacity:

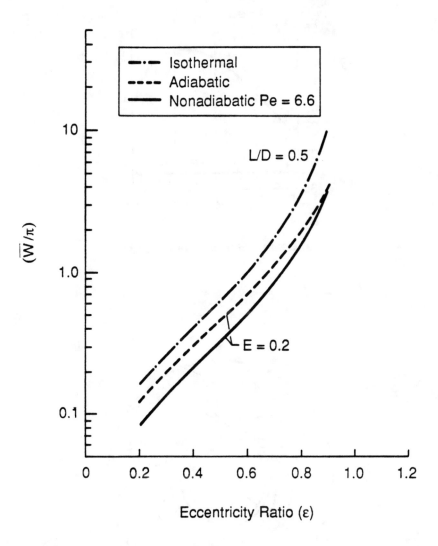

Fig. 4.21 Couette approximation solutions for journal bearings, McCallion, 1970

Item	Change	Effect on W
T_a	40°C–0°C	3%
Pe	7:1	20%
(k/k_s)	4:1	3%
Insulated bearing	—	17%

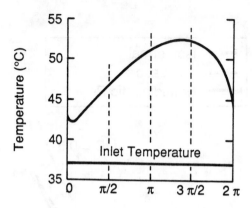

a) Bearing Surface Temperatures

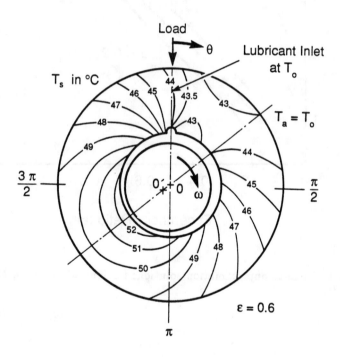

b) Bearing Shell Isotherms

Fig. 4.22 Bearing temperatures by the Couette approximation, Mccallion, 1970

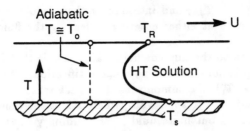

a) Temperature Profile at Start of Film

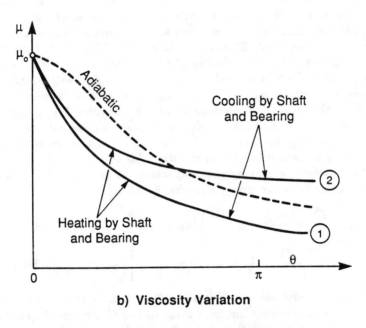

b) Viscosity Variation

Fig. 4.23 Viscosity in adiabatic and HT solutions

4.3.2 The Integral Method

It is the nature of HT analysis that the thermohydrodynamics of the fluid film can be obtained only in conjunction with a thermal analysis of the bounding solids. It would simplify the problem if one could detach the energy equation from the heat transfer equations. This can be done by incorporating into the energy equation terms that would in some fashion reflect the heat transfer process in the solids. En gross, this can be done by writing for the heat flux exchanged with the runner some $h_R(\Delta T)_0$ and for the bearing $h_S(\Delta T)_h$, and introducing these into the energy balance to account for the heat either entering or leaving the fluid from the two bounding surfaces.

One objection to the film coefficient approach is that in a tribological setting these expressions have no physical justification. The type of expression $Q = h(T_2 - T_1)$ is commonly used in heat transfer across liquid or gaseous layers adjacent to solid walls, where ΔT is the temperature differential across a small but finite fluid layer. Ordinary boundary layers, for example, extend far enough out from the wall to permit the existence of a small thermal layer. In lubrication the entire film thickness constitutes the thermal layer and the lubricant in the immediate vicinity of the solids must take on the same temperatures. A not completely satisfactory answer in justification of using the film coefficient approach would be to say that the quantities h_S and h_R may be given values ranging from zero to some finite or infinite value. The zero case would correspond to adiabatic conditions; in the infinite case the temperatures at the interfaces would assume the same value, as required. Thus the violation, $T \neq T_{S,R}$, could be mitigated by a proper selection of the value of h. Another possibility is to replace $T_{S,R}$ by some T_M at a suitable distance from the film-wall boundary and consider the heat transfer to be occuring across a finite distance.

Since the main motivation for obtaining a detailed $T(y)$ mapping was to be able to solve the heat transfer equations, with the last simplification the need for $T(y)$ has much diminished. One can thus treat the energy equation in a manner similar to the Reynolds equation and integrate out all the y variations, leaving the temperature to be a function of x and z only. On this basis the energy equation becomes

$$\rho c_p \int_0^h \left[u\left(\frac{\partial p}{\partial x}\right) + w\left(\frac{\partial p}{\partial z}\right) \right] dy + h_s(T - T_s) + h_R(T - T_R)$$

$$- k \int_0^h \left[\frac{\partial^2 T}{\partial x^2} + \frac{\partial^2 T}{\partial z^2}\right] dy = \int_0^h \mu \left[\left(\frac{\partial u}{\partial y}\right)^2 + \left(\frac{\partial w}{\partial y}\right)^2\right] dy \quad (4.16)$$

This is the problem that Motosh (1964) solved for a full journal bearing. Having reduced the problem from three to two dimensions, he carried the work forward by a separation of variables so that

$$T(x, z) = X(x) \cdot Z(z)$$
$$\mu(x, z) = \Pi(x) \cdot \Omega(z)$$
$$p(x, z) = K(x) \cdot G(z)$$

These functions were solved by finite difference methods and applied to a full journal bearing having the characteristics

$D = 12 \,\text{cm}$ $N = 1,000 \,\text{rpm}$

$L = 6 \,\text{cm}$ $T_0 = 40°\text{C}$

$(R/C) = 10^3$ $\mu = 6 \cdot 10^{-6} e^{(975/T+95)} \,\text{kg} \cdot \text{sec/m}^2$

In addition to providing data for thermal behavior, the results here are of interest in context of some of the conclusions reached in the previous section for a similar 360° journal bearing with an $L/D = 1/2$. It was seen that with heat transfer effects the lowering of the initial lubricant viscosity was hypothesized to have contributed to a drop in the load capacity below that of adiabatic operation. Fig. 4.24 of the Motosh solution confirms that the higher the heat transfer between fluid and bearing, the higher will the initial temperature be (and the lower μ), as was shown qualitatively in Fig. 4.23. This picture is consistent for the entire range of heat transfer coefficients. However, in the Motosh solution this is not sufficient to produce a drop in load capacity of the HT solutions vis-a-vis the adiabatic case, and far from it as shown in Fig. 4.25. There we see that the adiabatic case yields the lowest load capacity and the higher the heat transmissivity of the bearing surface the higher the load capacity. This is as one would expect the bearing to behave because, although initially the HT solutions have lower viscosities, they gain in viscosity over the adiabatic case in the region of h_{min} where the high pressures are generated, which account for most of the total bearing force. And indeed Fig. 4.25 shows that the load capacity, except for a slight initial kink, rises steadily with a rise in the value of h_s, and so does the value of the frictional force due to large shear gradients at the moving surface. The friction coefficient, however, decreases slightly with a rise in h_s.

As shown in Fig. 4.26 the present solution confirms the previous results that temperatures begin to drop beyond h_{min}. This is so even though here one cannot attribute this behavior to restrictive assumptions used in the heat analysis of the bearing shell. This behavior is seen to prevail for all values of temperature T_S assigned to the bearing.

4.3.3 The (x,y) Solution

The following solution adheres pretty much to the demands of a full heat transfer treatment of journal bearings, except that it ignores the thermal variations in the z direction. It does, however, take into account mixing inlet temperature effects and in the heat transfer solutions of the bearing shell it considers variations in both the radial and circumferencial directions. The boundary conditions then can be stated as follows:

- For the bearing shell equation these are

$$(a) \quad \text{at} \quad r = R_L; \quad k_s \frac{\partial T}{\partial r}\bigg|_{R_L} = h_s (T - T_a)$$

$$(b) \quad \text{at} \quad r = R; \quad k_s \frac{\partial T}{\partial r}\bigg|_{R} = k \frac{\partial T}{\partial y}\bigg|_{h}$$

$$(c) \quad T_h = T_s$$

- At the journal it is required that no net heat flow take place into the

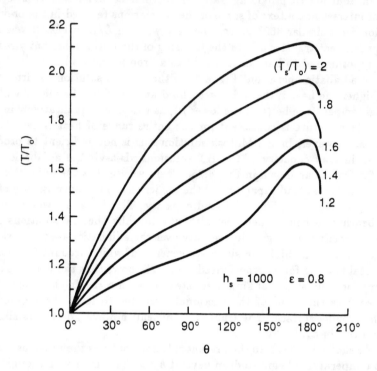

Fig. 4.24 Effect of heat transfer rate on journal bearing temperature profile, Motosh, 1964

shaft, so that

$$\int_0^{2\pi} \left.\frac{\partial T}{\partial y}\right|_0 dx = 0$$

- At the inlet to the film we must have

$$T_1 = \frac{Q_2 T_2 + Q_0 T_0}{Q_1}$$

which is Equ. (2.9) derived in Chapter 2.

It is this T_1 that will here constitute the reference temperature and we have, therefore,

$$\overline{\overline{T}} = T/T_1$$

The problem with the above set of boundary conditions has been solved by Khonsari and Beaman (1986) for the same bearing analyzed by the Couette approximation in Section 4.3.1. One of the major issues raised by the aforementioned analysis was whether an HT solution is likely to yield a smaller load capacity than what would seem to be a lower limit set by an

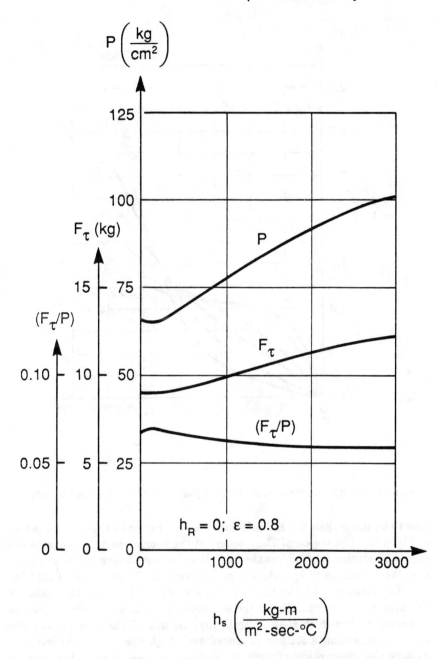

Fig. 4.25 Effect of heat transer rate on load capacity and friction, Motosh, 1964

adiabatic solution. The Motosh solution did not support such a conclusion; nor do the present results. Fig. 4.27 shows the load capacity of the adiabatic

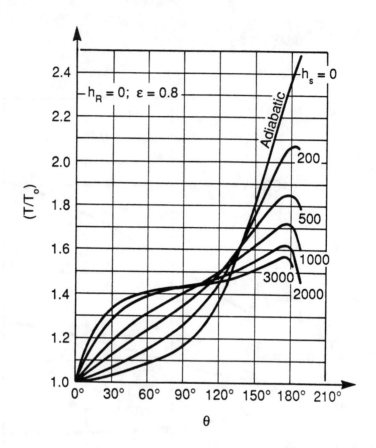

Fig. 4.26   Temperature distribution in journal bearings, Motosh, 1964

case to be lower than the HT solution. The isotherms in the film are shown in Fig. 4.28. The region of T_{max} occurs at the bearing surface and as in the previous solution it straddles the h_{min} line; by comparison in the adiabatic case the maximum temperature region extends to the end of the fluid film.

The isotherms in the bearing, given in Fig. 4.29, show that most of the heat flows out radially to the environment with only a small portion returning to the inlet zone. Herein may lie one of the reasons why the solution in Section 4.3.1 has produced such high upstream temperatures, namely the assumption of pure circumferencial heat flow in the bearing shell. There is little effect on thermal behavior of a change in load, but as shown in Fig. 4.30 an increase in speed tends to raise the temperature levels in the entire system, from the mixing temperature to the film, journal, and bearing temperatures. In general, journal temperature was found to be close to midway between the inlet and maximum temperatures. This,

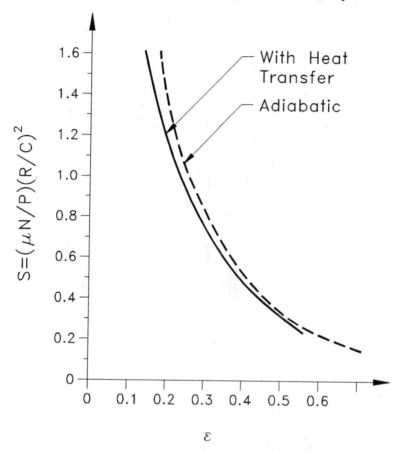

Fig. 4.27 Load capacity in adiabatic and HT solutions, Khonsari and Beaman, 1986

of course, is purely a consequence of the particular boundary condition imposed on the runner.

4.3.4 Three-Dimensional Solutions

The following results due to Boncompain and Frene (1980) are the product of a nearly rigorous treatment of the Reynolds and energy equations as they were formulated at the beginning of the chapter. The variations in the axial (z) direction have been retained and no terms were dropped. Moreover, in addition to the mixing inlet temperatures, the cavitation zone has been accounted for. By retaining the axial variable, there are then, of course, additional boundary conditions to specify and to fulfill. The conditions to be specified are as follows:

- Journal surface and edge conditions
- Fluid film inlet condition and the peculiarities of the cavitation zone

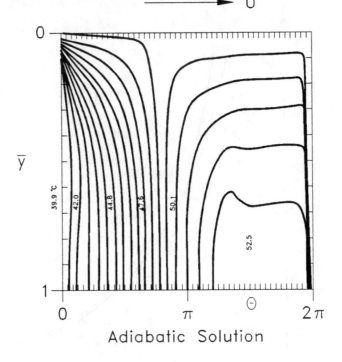

Adiabatic Solution

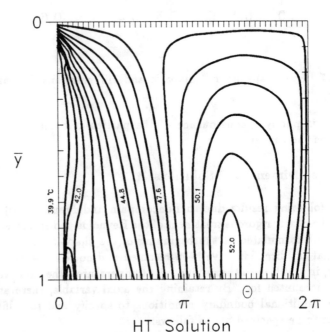

HT Solution

Fig. 4.28 Isotherms in fluid films of a journal bearing at $\epsilon = 0.58$, Khonsari and Beaman, 1986

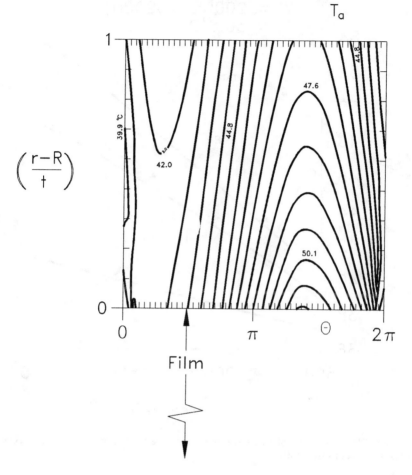

Fig. 4.29 Bearing shell isotherms at $\epsilon = 0.58$, Khonsari and Beaman, 1986

- The bearing surfaces on both the inside and outside and the conditions at the edges
- Ambient temperatures

There are nine conditions to be specified here, and these are given in Table 4.7 with the bearing characteristics listed in Table 4.8. Perhaps the most arbitrary statement is that of assigning a fixed temperature to the shaft at $z = \pm L/2$, a boundary condition necessitated by the fact that the shaft temperature is made to be a function of z (but not of θ). The ratio (L'/L) in the cavitation conditions reflects the fact that in this region the fluid film is incomplete and consists of streamlets whose total width $L' < L$

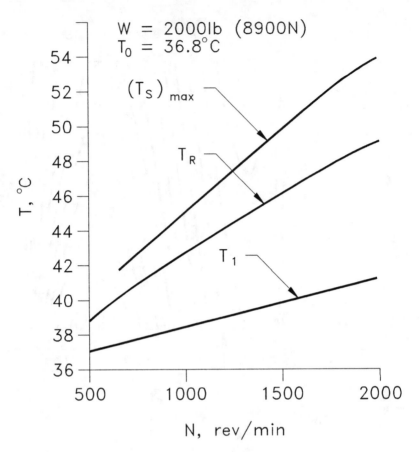

Fig. 4.30 Variation of temperature with speed in journal bearing, Khonsari and Beaman, 1986

is given by

$$\left(\frac{L'}{L}\right) = \frac{\overline{h}_2 \int_0^{\frac{1}{2}} \int_0^1 (\overline{u}_2 \, d\overline{y}) \, d\overline{z}}{\overline{h} \int_0^{\frac{1}{2}} \int_0^1 (\overline{u} \, d\overline{y}) \, d\overline{z}} \quad , \quad \theta_2 < \theta < 2\pi \qquad (4.17)$$

θ_2 denoting the start of the cavitating zone and h_2 the film height at that location.

Figs. 4.31 and 4.32 show the isotherms in the shaft-fluid film-bearing system. The maximum temperature region in the film is slightly away from the stationary surface and straddles the line of centers. The normals to the isotherms show that heat flows from the shaft to the film over an upstream arc of $0 < \theta < 110°$; and heat flows to the shaft over the remainder of the shaft surface. There is also a strong heat flux from the bearing shell outwards except over a small region past $\theta = 0$ where heat flows from the bearing to the fluid film. Thus again when both circumferencial and radial

TABLE 4.7

Boundary conditions for three-dimensional solution

Component		Location	Condition
Journal	1	Edges $z = \pm \frac{L}{2}$	$T = 23°C = $ constant
	2	Facing fluid film	$k_R \left.\frac{\partial T}{\partial r}\right|_R = -\frac{1}{2\pi} \int_0^{2\pi} \left(\frac{\partial T}{\partial \theta}\right)_h d\theta = f(z)$
Fluid film	3	Start $\Big\{$ Supply	$T_0 = 33°C = $ constant
	4	Inlet	$T_1 = \dfrac{Q_0 T_0 + Q_2 T_2}{Q_1}$
	5	Cavitation	$k = k_a - \left[\frac{L'}{L}\right](k_a - k)^*$
Bearing	6	Facing fluid film	$k_s \left.\frac{\partial T}{\partial \theta}\right|_R = k \left.\frac{\partial T}{\partial y}\right|_0$
	7	Edges at $z \pm \frac{L}{2}$	$\left.\frac{\partial T}{\partial r}\right|_{\pm L/2} = -Nu^+\left[T \pm (L/2) - T_a\right]$
	8	Outside surface	$\left.\frac{\partial T}{\partial z}\right|_{R_L} = -Nu^+(T_L - T_a)$
Ambient	9	$r > R_L$	$T_a = 23°C = $ constant

* To be used in Conditions 2 and 6. For (L'/L) see Equ. (4.17)

$^+Nu = (h_{s,R}/k)$

heat flows in the bearing shell are accounted for, there is not much upstream heating of the lubricant by the stationary surface.

The variation of temperature with ϵ is shown in Fig. 4.33 and as seen the relative amounts of heat absorbed by the fluid and by the bounding solids do not differ much with a variation in bearing operating conditions.

TABLE 4.8

Bearing characteristics for three-dimensional solution

Shaft radius	$R = 50$ mm
Bearing external radius	$R_L = 100$ mm
Bearing length	$L = 100$ mm
Clearance	$C = 0.1$ mm
Lubricant density	$\rho = 850$ kg/m^3
Lubricant viscosity at 33°C	$\mu_0 = 0.0323$ Pa-s
Lubricant specific heat	$C_p = 2000$ J/kg-°C
Thermo-viscosity coefficient	$\alpha = 0.03665/°C$
Lubricant thermal conductivity	$k = 0.13$ W/m-°C
Air thermal conductivity	$k_a = 0.025$ W/m-°C
Bearing conductivity	$k_s = 51.9$ W/m-°C
Shaft thermal conductivity	$k_R = 51.9$ W/m-°C
Convection heat transfer coefficient	$h_a = 56.8$ W/m^2-°C
Ambient temperature	$T_a = 23$°C
Inlet lubricant temperature	$T_0 = 33$°C
Angular rotational speed	$N = 2000$ rpm

Thus according to the present model the fluid film at any ϵ absorbs 70-80% of the dissipated energy. With a rise in ϵ the maximum temperature of the bearing interior increases faster than the fluid temperature. The mixing inlet film temperature decreases with an increase in the eccentricity because the total oil flow rate increases while the recirculating hot flow rate decreases drastically. The cumulative heat flow rate in the journal appears larger than in the bearing shell because of the boundary conditions imposed at the shaft surface. This condition is representative of a shaft which acts as a heat sink.

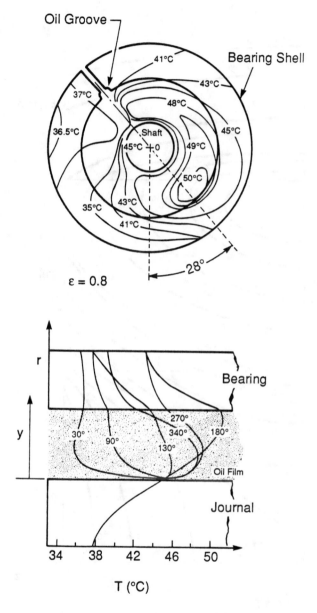

Fig. 4.31 Fluid film temperatures at midplane of journal bearing, Boncompain and Frene, 1980

The more significant result of the present full treatment is that contrary to previous conclusions there is a considerable variation in axial temperature, Fig. 4.32. It is, in fact, of the same order as the ΔT's across the film, both about 10°C, and not much less than the circumferencial temperature rise of some 13°C (relative to T_1). With this, certainly the issue is not

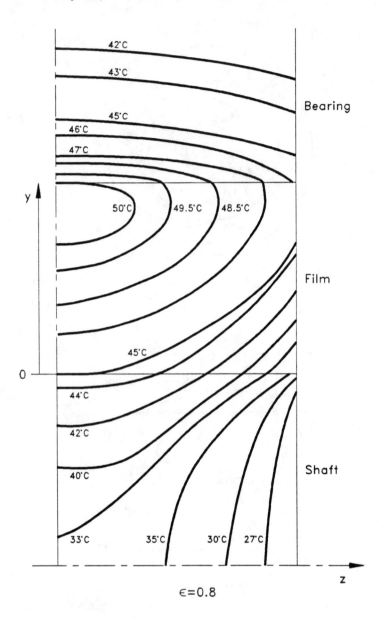

Fig. 4.32 Journal bearing isotherms in plane of maximum temperature $\theta = 28°C$, Boncompain and Frene, 1980

resolved because the present $T(z)$ variation could very well be attributed to the rather arbitrary boundary condition of $T = T_a$ at $z = \pm L/2$. It is quite unlikely that adjacent to the fluid film the shaft would have reached a low temperature of 23°C when the outflowing lubricant was according to the same figure leaking out at temperatures between 35°C and 48°C.

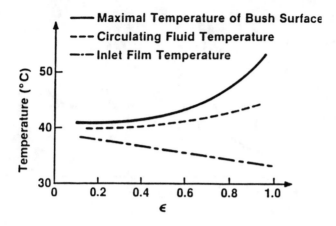

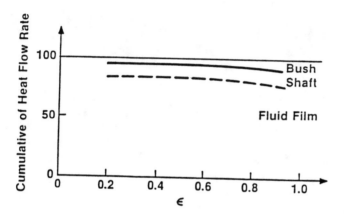

Fig. 4.33 Temperatures and heat flux in three-dimensional solution of journal bearing, Boncompain and Frene, 1980

Figs. 4.34 and 4.35 give the values of the spring and damping coefficients of the journal bearing. When a comparison is made on the basis of the same ϵ, there seems to be little difference between the isothermal and HT solutions. However, a comparison on the basis of the same ϵ does not convey the impact that variable viscosity has on the dynamic coefficients. For a given set of operating conditions the variable viscosity solution yields higher eccentricities and, therefore, the dynamic coefficients should, likewise, be obtained for different values of ϵ. When this is accounted for, the picture becomes even more blurred, even though the main colinear spring coefficient K_{xx} would certainly increase. This perhaps has to do with the conflicting effects a change in viscosity has on stability. While lower viscosity would yield softer spring and less damping, it would at the same

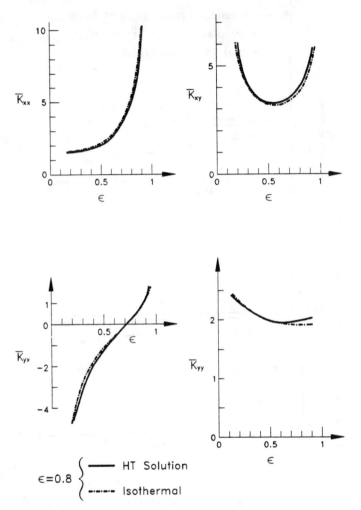

Fig. 4.34 Isothermal and HT solutions for spring coefficients, Boncompain and Frene, 1980

time increase journal eccentricity and decrease attitude angle, both known to contribute to bearing stability. Results obtained by Craighead *et al.* (1980), Fig. 1.2, confirm the small or straddling effect that thermal variation has on bearing stability by showing that the unstable regime of a shaft-bearing system is increased by thermal effects for low ϵ, while it is decreased for high ϵ. For a two-bearing system, no effect due to thermal variations was present at all.

A subsequent analysis also by Craighead *et al.* (1985) where more complex but more realistic boundary conditions were used, indeed showed that the maximum temperature was located not near h_{min} but midway between h_{min} and the end of the bearing pad. This result is shown in Fig.

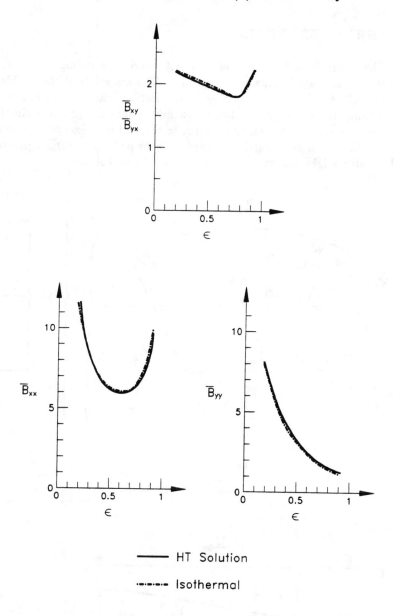

Fig. 4.35 Isothermal and HT solutions for damping coefficients, Boncompain and Frene, 1980

5.23 which is the same bearing as the one shown in Fig. 4.31. More on this subject will be said in the next chapter devoted specifically to the issue of boundary conditions.

4.4 THRUST BEARINGS

The general configuration of a thrust bearing including the system elements of interest is shown in Fig. 4.36. The essentials of the problem are not different from those arising in connection with journal bearings. If anything, the geometry of thrust bearings makes the problem simpler since in most cases there is no cavitation to contend with except, as will be seen, in the case of deliberately crowned pads, or pads (particularly pivoted pads) which might distort due to elastic and thermal stresses.

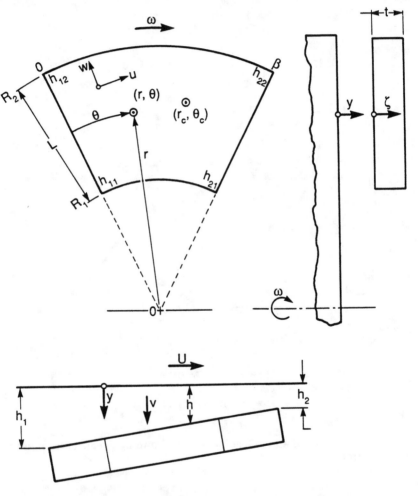

Fig. 4.36 Geometry of a thrust bearing

One important difference from a THD standpoint is that while in sliders and journal bearings constancy of temperature in the direction normal to motion could with some justification be asserted — for thrust bearings there is little doubt that such as assumption would be insupportable. There are at least two reasons for this. One is that the film here varies in both the θ and r directions; and more importantly the linear velocity is appreciably higher at the pad outer periphery and with it the dissipation and temperatures. Furthermore, it seems that due to the progressively higher temperatures toward the outer boundaries of the pad the assertion of constant runner temperature would also suffer. Since various r = constant portions of the runner surface are permanently exposed to rings of varying temperature, one would expect that at least in the radial direction the runner would exhibit an appreciable temperature gradient reflecting the rise in film temperatures.

4.4.1 The Governing Equations

In cylindrical coordinates the energy equation with $\mu = \mu(\theta, r, y)$ assumes the following form:

$$\rho c_p \left[\frac{u}{r}\left(\frac{\partial T}{\partial \theta} \right) + w\left(\frac{\partial T}{\partial r} \right) + v\left(\frac{\partial T}{\partial y} \right) \right] = k\left[\frac{1}{r^2}\frac{\partial^2 T}{\partial \theta^2} \right.$$
$$\left. + \frac{1}{r}\frac{\partial}{\partial r}\left(r\frac{\partial T}{\partial r} \right) + \frac{\partial^2 T}{\partial y^2} \right] + \mu\left[\left(\frac{\partial u}{\partial y} \right)^2 + \left(\frac{\partial w}{\partial y} \right)^2 \right] \qquad (4.18)$$

with the velocity components as defined in Fig. 4.36. In going from (θ, r, y) to $(\theta, r, \overline{y})$ we have

$$\frac{\partial}{\partial \theta} = \left[\frac{\partial}{\partial \theta} - \frac{\overline{y}}{\overline{h}}\left(\frac{\partial \overline{h}}{\partial \theta} \right)\frac{\partial}{\partial \overline{y}} \right]$$
$$\frac{\partial}{\partial r} = \left[\frac{\partial}{\partial r} - \frac{\overline{y}}{\overline{h}}\left(\frac{\partial \overline{h}}{\partial r} \right)\frac{\partial}{\partial \overline{y}} \right]$$
$$\frac{\partial}{\partial y} = \frac{1}{h_2 \overline{h}}\frac{\partial}{\partial \overline{y}}$$

yielding for the Reynolds and energy equations in normalized form the expressions

$$\frac{(L/R_1)^2}{\beta^2 \left[1 + (L/R_1)\overline{r} \right]^2}\frac{\partial}{\partial \theta}\left[\overline{h}^3 I_P\left(\frac{\partial \overline{p}}{\partial \theta} \right) \right] + \frac{\partial}{\partial \overline{r}}\left[\overline{h}^3 I_P\left(\frac{\partial \overline{p}}{\partial \overline{r}} \right) \right]$$
$$+ \frac{(L/R_1)}{\left[1 + (L/R_1)\overline{r} \right]}\overline{h}^3 I_P\left(\frac{\partial \overline{p}}{\partial \overline{r}} \right) = \frac{\partial}{\partial \theta}\left[\overline{h} I_V \right] \qquad (4.19)$$

$$\frac{(L/R_1)\overline{u}}{\beta\left[1 + (L/R_1)\overline{r} \right]}\left(\frac{\partial \overline{\overline{T}}}{\partial \overline{\theta}} \right) + \overline{w}\left(\frac{\partial \overline{\overline{T}}}{\partial \overline{r}} \right)$$

$$
+ \left[\frac{\bar{v}}{\bar{h}} - \bar{w}\left(\frac{\bar{y}}{\bar{h}}\right)\left(\frac{\partial \bar{h}}{\partial \bar{r}}\right) - \frac{(L/R_1)\bar{u}}{\beta\left[1 + (L/R_1)\bar{r}\right]} \frac{\partial \bar{\bar{T}}}{\partial \bar{y}} \right]
$$

$$
= \frac{1}{Pe}\left\{ \left(\frac{\bar{y}}{\bar{h}}\right)^2\left(\frac{\partial \bar{h}}{\partial \bar{r}}\right)^2 + \frac{1}{\beta^2}\left(\frac{\bar{y}}{\bar{h}}\right)^2 + \left(\frac{L}{h_0\bar{h}}\right)^2 \right\} \frac{\partial^2 \bar{\bar{T}}}{\partial \bar{y}^2}
$$

$$
+ \frac{\mu_0\omega(L/h_0)^2}{\rho c_p T_0}\left(\frac{\bar{\mu}}{\bar{h}}\right)\left[\left(\frac{\partial \bar{u}}{\partial \bar{y}}\right)^2 + \left(\frac{\partial \bar{w}}{\partial \bar{y}}\right)^2\right] \qquad (4.20)
$$

Above, only the conduction terms containing $(\partial^2 T/\partial y^2)$ were retained in the energy equation. Some of the non-standard normalization terms used above are as follows:

$$
\bar{r} = \left(\frac{r - R_1}{L}\right) \qquad\qquad \bar{p} = \frac{p\beta}{\mu_0 w}\left(\frac{h_0}{L}\right)^2
$$

$$
\bar{u}, \bar{v}, \bar{w} = \frac{u, v, w}{L} \qquad\qquad \overline{Pe} = \frac{\rho c_p \omega L^2}{k}
$$

and $\bar{h} = (h/h_0)$, where h_0 is h at the center of the pad. The terms I_P and I_V in the Reynolds equation are in terms of the integrals defined in Table 3.1

$$
I_P = \frac{1}{4}\left[1 - 2\frac{I_1(\bar{y})}{I_0(\bar{y})}\right]\left[\int_0^1 I_0(\bar{y})d\bar{y} + 2\int_0^1 I_1(\bar{y})d\bar{y}\right]
$$

$$
I_V = \left[1 - \int_0^1 I_0(\bar{y})d\bar{y} \Big/ I_0(\bar{y})\right]
$$

The velocity components appearing in the energy equation are, accounting for $\mu = \mu(\theta, r, y)$, given by

$$
\bar{u} = \frac{(L/R_1)}{\left[1 + (L/R_1)\bar{r}\right]}\left[\frac{I_0(\bar{y})}{I_0(1)} - 1\right]
$$

$$
- \left[\frac{I_1(1)}{I_0(1)}I_0(\bar{y}) - I_1(\bar{y})\right]\frac{\bar{h}^2(L/R_1)}{\beta^2\left[1 + (L/R_1)\bar{r}\right]}\left(\frac{\partial \bar{p}}{\partial \bar{\theta}}\right) \qquad (4.21a)
$$

$$
\bar{w} = -\left[\frac{I_1(1)}{I_0(1)}I_0(\bar{y}) - I_1(\bar{y})\right]\frac{\bar{h}^2}{\beta}\left(\frac{\partial \bar{p}}{\partial \bar{r}}\right) \qquad (4.21b)
$$

Solutions will be presented for three typical thrust bearing configurations. One is the tapered land bearing, essentially a planar surface inclined in one or two directions relative to the runner; one a crowned pad which may be a deliberate geometrical configuration or the unintended result of thermal and elastic distortion; and a pivoted pad.

4.4.2 Tapered Land Bearings

With the origin of an (x, y) coordinate system placed at the center of the pad and α_x, α_y representing tilt angles about the two axes, we have for the film thickness

$$\overline{h} = 1 - \left(\frac{L}{R_1}\right)\left\{ \left(\overline{r} + \frac{R_1}{L}\right)\left(\frac{R_1\alpha_y}{h_0}\right) \sin\beta \left(\overline{\theta} - 1/2\right) \right.$$

$$\left. + \left[\left(\overline{r} + \frac{R_1}{L}\right)\cos\beta \left(\overline{\theta} - 1/2\right) - \left(\frac{R_1 + R_2}{2L}\right)\right]\left(\frac{R_1\alpha_x}{h_0}\right)\right\}$$

where h_0 is the film thickness at the pad center.

The solution obtained by Jeng et al. (1986) is based on the following fluid film boundary conditions:

- $T_R = \text{const}$
- At $y = h$ in the film

$$k\frac{\partial T}{\partial y}\bigg|_h = -k_s\frac{\partial T}{\partial \varsigma}\bigg|_0$$

where ς is the transverse coordinate in the direction of pad thickness.
- At $\theta = 0$, $T = T_0 = \text{const}$.
- At all exit planes, $\theta = \beta$, $r = R_1, R_2$, no appropriate boundary conditions could be formulated. It was assumed that at these locations

$$\frac{\partial^2 T}{\partial r^2} = \frac{\partial^2 T}{\partial \theta^2} = 0$$

An interesting aspect of this boundary condition was that ignoring it made no difference in the solution as long as the energy balance was satisfied in general.

In the pad the only heat transmission was assumed to be in the transverse direction, along ς. Thus

$$\frac{\partial T}{\partial r}\bigg|_s = \frac{\partial T}{\partial \theta}\bigg|_s = 0$$

From the Laplace equation for the pad it then follows that

$$\frac{\partial^2 T}{\partial \varsigma^2} = 0$$

At the faces of the pad we have

$$-k\frac{\partial T}{\partial y}\bigg|_h = -k_s\frac{\partial T}{\partial \varsigma}\bigg|_0$$

$$-k_s\frac{\partial T}{\partial \varsigma}\bigg|_t = h_{sa}(T - T_a)$$

$$T_s = T_h$$

The above set of relationships then leads to an expression for the pad temperature

$$T = T_a + t\left(\frac{k}{k_s}\right)\left[\frac{\varsigma}{t} - \left(\frac{k_s}{h_{sa}t} + 1\right)\right]\left(\frac{\partial T}{\partial \varsigma}\right)$$

The operating conditions and geometry of the bearing for which the results of Figs. 4.37 and 4.38 are plotted appear in Table 4.9. As seen, this is a fairly large bearing with a tilt in the circumferential direction only, which nevertheless gives a film thickness variable in both the r and θ directions. There is little temperature variation across the film but unlike in journal bearings there is a steady rise in temperature with both r and θ, the peak occurring at the trailing upper corner, a good distance beyond p_{max}.

4.4.3 Crowned Pads

An important difference between crowned pads and the geometry considered in the previous section is that here cavitation occurs and we must satisfy the zero pressure gradient boundary condition, that is,

$$p = \frac{dp}{d\theta} = 0 \quad \text{at} \quad \theta = \theta_2(r)$$

With the reference film thickness h_0 in the center of the pad as shown in Fig. 4.39 the crown is ascribed a parabolic shape

$$\delta = R_c\left[1 - \left(\frac{z}{R_c}\right)^2\right] \sim \frac{x^2 + y^2}{2R_c}$$

with

$$x = r\sin\left(\theta - \frac{\beta}{2}\right)$$

$$y = r\cos\left(\theta - \frac{\beta}{2}\right) - \frac{R_1 + R_2}{2}$$

Writing $\overline{\delta} = (\delta/h_0), \overline{R}_2 = (R_2/L)$ we have

$$\overline{\delta} = \overline{\delta}_{max}\left\{4(\overline{r} + \overline{R}_2 - 1)^2 + 2(\overline{R}_2 - 1)^2\right.$$
$$\left. - 4(\overline{r} + \overline{R}_2 - 1)(2\overline{R}_2 - 1)\cos[\beta(\overline{\theta} - 1/2)]\right\}$$

Thus

$$\left(\frac{h}{h_0}\right) = 1 + \overline{\delta}(\overline{r}, \overline{\theta})$$

The same geometrical configuration as that given in Table 4.9 was analyzed here too except for somewhat different thermal parameters, namely

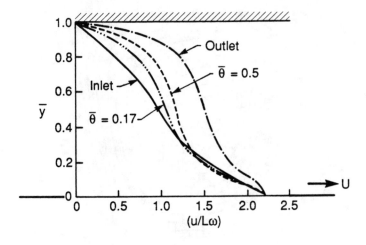

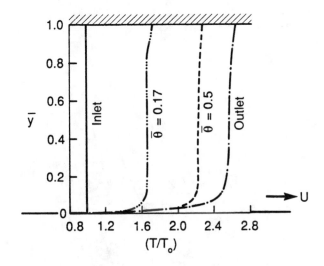

$(r - R_1)/L = 0.667$

Fig. 4.37 Transverse temperatures and velocities in tapered land thrust bearing, Jeng *et al.*, 1986

$Pr = 686$
$Pe = 1.2$
$Nu = 100$

$\overline{\delta}_{max} = 0.7$
$h_0 = 0.06\,\text{mm}$
$(\mu_0\omega/\rho c_p T_0)\,(L/h_0)^2 = 2.0$

As seen in Fig. 4.40 the maximum temperature occurred inside the film. This shift of T_{max} inwards may have been due to the fact that the authors ignored the dissipation losses in the cavitation region, a good 25%

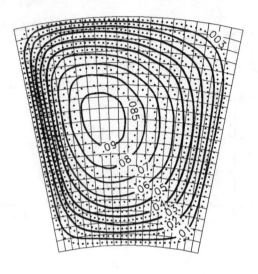

Isobars

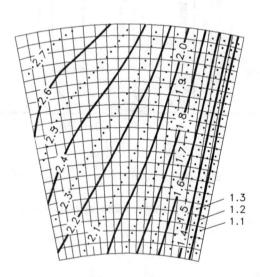

Isotherms

Fig. 4.38 Pressures, $\bar{p}$, and temperatures, $\bar{\bar{T}}$, in tapered land thrust bearing; for data see Table 4.9., Jeng *et al.*, 1986

TABLE 4.9

Tapered land bearing

$$T_0 = T_R = 38.5°C$$

$$\mu = 2.6006 \times 10^{-5} \exp \left[\frac{1234.9}{T(°C) + 125.44} \right] \frac{N \cdot s}{m^2}$$

$$R_2/L = 2.545, \; \beta = 24°, \; h_0 = 1.2856 \times 10^{-4} \text{ cm}$$

$$\alpha_x = 0; \; R_1 \frac{\alpha_y}{h_0} = 1.0, \; R_2 = 140.0 \text{ cm}$$

$$Re = 500, \; Pr = 714.69$$

$$Pe = 1.092 \times 10^9, \; Nu = 100$$

$$k/k_s = 0.0022480, \; h_0/t = 0.00044194$$

$$\frac{\mu_0 \omega}{\rho c_p T_0} \left(\frac{L}{h_0} \right)^2 = 1.5814$$

---

of the pad area. Had these losses been included, T_{max} may have occurred even here at the end of the pad.

Fig. 4.41 gives the variation of T_{max} and load capacity as a function of amount of bending, showing that over a range of $\frac{1}{4} < \bar{\delta}_{max} < \frac{1}{2}$ there is not much variation in the hydrodynamic pad force. The value of T_{max} seems to be declining with $\bar{\delta}$. Since the extent of cavitation rises with $\bar{\delta}$ this last effect, too, may be due to the fact that the power losses have been neglected in the cavitation region.

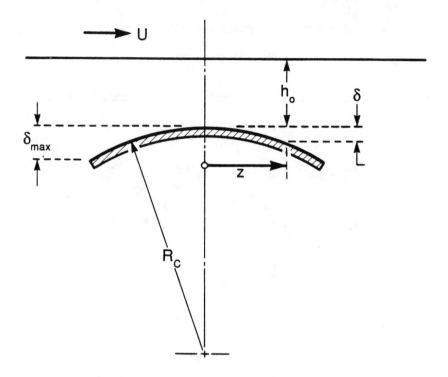

Fig. 4.39 Geometry of crowned pad

4.4.4 Pivoted Pads

With the pad tapered in both the radial and circumferential directions and supported at a point, the additional requirement in such a solution is that the resultant force pass through the pivot. Various r and θ slopes must then be tried until this condition is satisfied. Other than that, the boundary conditions in the solution developed by Kim et al. (1983) were the same as in the previous section, except for one notable difference. The r and θ fluid film heat conduction was eliminated and parabolic temperature gradients at $r = R_1$ and $r = R_2$ were introduced, a variation presumably based on experimental data.

The particular bearing solved for is specified in Table 4.10 with a reference prototype fixed at $h_{min} = 0.1\,\text{mm}$ at the coordinate (β, R_2) a speed of 1500 rpm, and a pivot position given by $\bar{r} = 0.51$ and $\bar{\theta} = 0.61$.

An important aspect of the present work lies in its comparison with adiabatic and isothermal solutions, shown in Fig. 4.42. As with the Mc-Callion solution (sec. 4.3.1), the HT solution here has lower load capacities and higher temperatures. A previously given rationale for this was that upstream the film is heated. An additional argument is offered here, namely that while an adiabatic solution "smears" or averages out the transverse

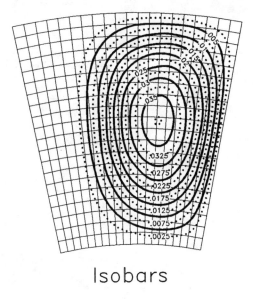

Isobars

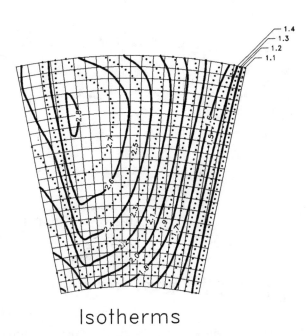

Isotherms

Fig. 4.40 Pressures, $\bar{p}$, and temperatures, $\bar{\bar{T}}$, in crowned pad thrust bearing, Jeng *et al.*, 1986

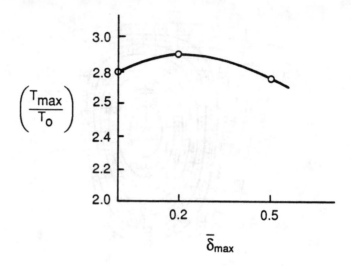

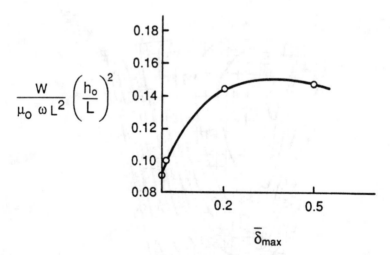

Fig. 4.41 Load capacity and maximum temperatures in crowned thrust bearing, Jeng *et al.*, 1986

temperatures, an HT solution has a variable transverse profile, its peak being higher than the adiabatic value. Conceivably these peak portions of the temperature profile with the accompanying reduction in viscosity may outweigh the fact that in the adiabatic case all the heat is stored in the film whereas in the HT case it is not.

That the HT solution does not always yield higher temperatures and

TABLE 4.10

Bearing characteristics for pivoted pad

R_1 = 16 cm, L = 7.7 cm, R_2 = 23.7 cm

t = 2.5 cm, $\theta_0 = \frac{\pi}{3}$ = 22.5°

c_p = 2.09 x 10^3 J/kg-°C , k = 0.214 W/m-°C

α = 7.34 x 10^{-4} $\frac{-1}{C}$, β = 4.91 x 10^{-2} $\frac{-1}{C}$

μ = 5.0 x 10^{-2} Pa-s at 27°C

ρ = 8.55 x 10^2 kg/m^3 at 27°C

k_s = 1.08 x 10^2 W/m-°C, h_{sa} = 5.81 x 10^2 W/m^2-°C

lower loads is indeed verified in Fig. 4.43. When plotted against the parameters of speed, h_{min}, and T_0 there is a crossover point at which the
HT solution yields higher load capacities and lower temperatures. These
cross-over points occur at different locations for the different performance
items, showing that here, too, there could not be a unique adiabatic approach that would simulate the results of an HT analysis. For the case
where $T_0 = T_a = T_R$ there is altogether very little difference between an
adiabatic and an HT solution.

Fig. 4.44 shows the isotherms in the fluid film-pad boundary as a function of the pad heat transfer coefficient. There is in all cases a slow temperature rise at the start of the pad and near the runner (kept at constant
T_R). The gradients grow rapidly near the outlet. The peak temperature
occurs at the end of the film, some 10– 15% away from the pad surface.
The main heat drain is in the axial (ς) direction, weakening the argument
that upstream heat flow in the pad is the cause for the lower load capacities
of the HT solutions.

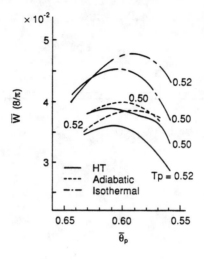

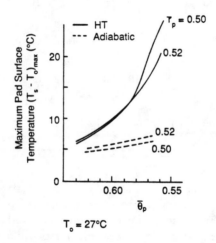

Fig. 4.42 Results of three methods of solution for tilting pad thrust bearing, Kim *et al.*, 1983

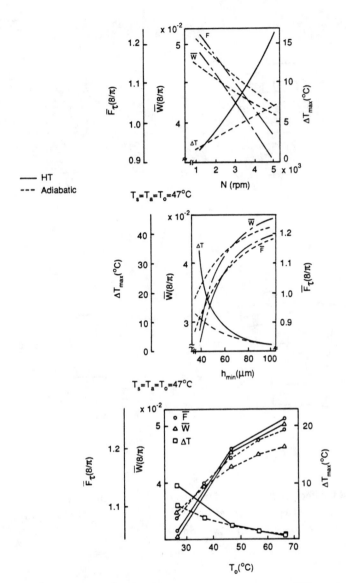

Fig. 4.43 Effect of various operating parameters on the THD performance of tilting pad thrust bearings, Kim *et al.*, 1983

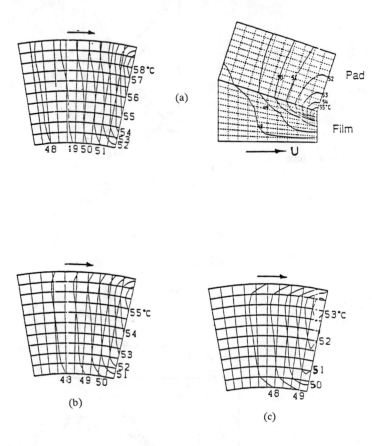

Fig. 4.44 Effect of heat transfer rate on thermal map of tilting pad thrust bearing at 3,000 rpm, Kim *et al.* 1983, (a) $h_{sa} = 0.581\text{kW/m}^2 \cdot^\circ \text{C}$, (b) $h_{sa} = 5.81\text{kW/m}^2 \cdot^\circ \text{C}$, (c) $h_{sa} = 58.1\text{kW/m}^2 \cdot^\circ \text{C}$.

Chapter 5

BOUNDARY CONDITIONS AND SYSTEM INTERACTION

Throughout the previous chapters, whatever has been the approach to a solution, the boundary conditions were for the most part stated without much concern as to their plausibility. This reflected the predilections of the individual investigators, as well as objective difficulties. Here, a more orderly attempt will be made to examine the nature of these boundary conditions, even if it will not always be possible to give them simple mathematical formulations. At this stage it should be clarified what in the present context is meant by boundary conditions. It was repeatedly stressed that the film's thermal map is co-determined by system components—runner, bearing shell, adjacent pads—and the subsequent discussion will show this interdependence to be even more intimate than has been suggested thus far. A rigorous treatment calls for a heat transfer analysis of a whole complex of machinery which is likely to overwhelm and obscure the core of our interest—the fluid film. A venture into a full-fledged analysis contains the paradoxical consequence that such information as the exact heat map in housings or shafts would be superfluous. A good case in point is the mixing inlet temperature. To obtain an exact solution one would have to conduct a three-dimensional flow and heat analysis in the lubricant supply groove. But were one to obtain such a solution, all the detail and rigor of the lubricant dynamics in the groove would be of little relevance to the fluid film. Similar arguments can be made with regard to the two-phase fluid flow and thermodynamics of the cavitation region. What is therefore being striven for in THD analysis is, while being cognizant of the relevance of the various system components, to yet so formulate their effects as to be able to use them as boundary conditions without resorting to a thermo-elastic analysis of a whole turbine or gear drive.

Considering the hydrodynamic film as a three-dimensional layer of fluid we have the following elements to consider:

(a) Mixing Inlet Temperature, T_1. This refers to the determination of the temperature of the lubricant entering a pad which consists of a mixture of cold supply fluid and hot lubricant carried over by the runner from an upstream pad.

(b) Back Flow. If the pressure gradient at the start of the pad is high there will be reverse flow and a number of complications arise affecting the thermal boundary condition.

(c) Runner Temperature, T_R.

(d) Stationary Surface Temperature, T_S.

(e) Edge Condition. This refers to the boundary condition at the outflow edges of the bearing. In a pad with a full film there will be three such surfaces, two at the sides and one at the end of the pad. In a cavitating film there will be only two such surfaces, the outlet edge falling into the category given below.

(f) Cavitation. This embraces not only the condition at θ_2 where the hydrodynamic film ends but the whole domain $\theta_2 \leq \theta \leq \theta_E$ where cavitation prevails.

(g) Starvation, T_1'. When the supply of lubricant is insufficient to fill the inlet at the start of a pad, a full film will not be formed before some $\theta_1 > \theta_S$ and the question arises of what is the starting temperature at θ_1.

As seen, items like cavitation or starvation, even by their terminology, do not suggest boundary conditions in the conventional sense, and even those that do, like the mixing inlet temperature T_1, represent actually a fluid dynamics problem in the lubricant supply groove. But, as explained above, the state of the fluid prevailing at the surfaces bounding the fluid film will have to be reduced to equivalents of conventional boundary conditions, so that they may facilitate a solution and not overwhelm the THD problem.

5.1 INLET CONDITIONS

5.1.1 The Nature of Inlet Mixing

It has been customary to consider the supply temperature T_0 as the temperature of the lubricant entering a bearing pad. This was so even after elaborate THD analyses were developed to account for the viscosity variation in the fluid film. However, considerable evidence has accumulated to indicate that this is in gross disagreement with reality. Of this evidence Fig. 5.1 is but one example. The data is from experiments performed by Ettles and Cameron (1968) on a set of parallel plate thrust bearings. The denominator in the ordinate of Fig. 5.1 represents the difference between runner and supply temperatures. If one considers runner temperature as an average of fluid film temperatures, then the results indicate that 70-90%

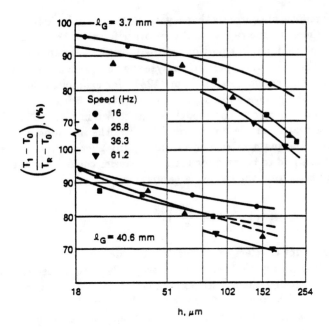

Fig. 5.1 Mixing inlet temperatures in parallel plate thrust bearing, Ettles and Cameron (1968)

of the overall ΔT occurs right at the beginning of the pad. In subsequent experiments the same authors found the level of carry-over heat energy to be close to 85%, regardless of the flow conditions in the groove. This percentage was little affected by a tenfold increase in the groove's angular extent, indicating that the hot layer manages to adhere to the runner regardless of exposure to the cold fluid. Increasing speed lowered the hot lubricant carry-over somewhat, while increasing film thickness reduced it. All this leads to the general postulate that the mechanism of mixing at the entrance conforms to that portrayed in Fig. 1.6 and a good start for representing this mixing mechanism is given by the simple equation

$$T_1 = \frac{Q_0 T_0 + Q_2 T_2}{Q_1} \tag{5.1}$$

Yet the same experiments cited above also emphasize the point that Equ. (5.1) cannot possibly account fully for the process of mixing in the groove. The experiments that produced the data of Fig. 5.1 were performed on parallel surfaces. Therefore the thickness of the inlet film was equal to the exit h of the upstream pad; and if the exit film adhered fully to the runner, there would be no room for cold lubricant infusion and we ought to have had $T_1 = T_R$, that is, a 100% heat carry-over. Yet the values obtained ranged as low as 70%.

The reality, of course, is that not all of the film in its journey across the groove adheres to the runner and that other forms of heat exchange occur between the cold and hot portions of the lubricant. A number of investigators have therefore attempted to analyze the goings-on in the groove. They have not been very successful. Vohr (1981), for example, resorted to a full analysis of the groove and coupled its thermal history to the overall heat balance in the bearing. The model used is shown in Fig. 5.2 in which convection of heat from the hot thermal layer to the incoming cold lubricant constitutes the main mixing mechanism. However, in the end empirical formulations had to be used for the heat transfer coefficients in the groove, as can be seen from the following expression used by Vohr:

$$Q_G = \left[\left(\frac{6}{\beta_G} \right) \cdot \left(\frac{N}{150} \right) \right]^{\frac{1}{2}} (T_R - T_0) A_G, \quad \text{watts}$$

where subscript G refers to groove, β_G is in degrees, and N in rpm.

Needless to say that when Vohr originally tried to use the results of Ettles to reconcile them with experimental data on a spring supported thrust bearing, he found little correlation between the two.

The reasons that most of these formal approaches do not yield useful engineering data should be clear from a mere inspection of the happenings in the supply groove, shown in Fig. 5.3. The mechanics and thermodynamics of the fluid in the configuration of a supply groove constitute a three-dimensional problem of utmost complexity subject to the following perturbations

(a) Loss of a portion of hot fluid Q_2 during transition of runner over the groove.

(b) Back-up of cold fluid into the upstream pad where cavitation occurs. This is particularly likely when the cold fluid is supplied under positive gauge pressure.

(c) Frictional heating in the groove, both in the thin layer adjacent to the runner and to a lesser extent in the circulating bulk of fluid in the groove. According to Vohr, the latter is some 20% of the friction in the film.

(d) Heat transfer along the walls of the groove.

(e) Three-dimensional effects. While the postulate here deals with a single exit temperature T_2, and a single inlet temperature T_1, in fact these temperatures are variable, both in the axial (z) direction and transversely across the film (y).

Various analytical examinations and most of all experimental programs have yielded some basic characteristics of flow in the groove, which can be restated as follows:

(a) The $\Delta T_1 = (T_1 - T_0)$ at the inlet to the pad is substantial, and in general exceeds 50% of the average temperature rise in the bearing.

(b) The thermal boundary layer in the groove is thicker than the velocity layer. Ettles and Cameron give an approximate expression for it in the

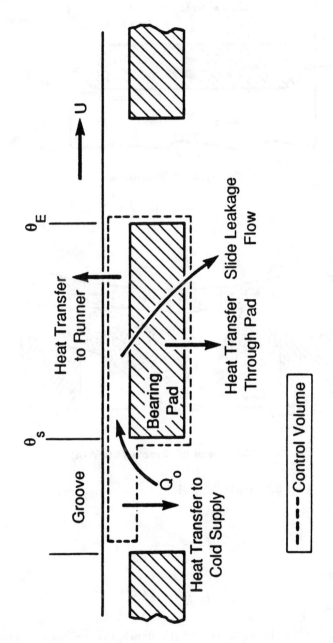

Fig. 5.2 Thermal model including supply groove

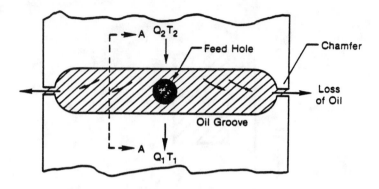

a. Top View of Oil Groove

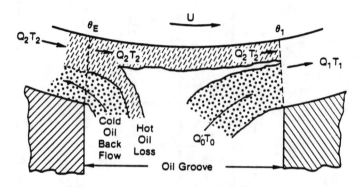

b. Side View of Oil Groove (Cut A-A)

Fig. 5.3 Perturbing elements in groove mixing

form

$$\bar{\delta}_T = \frac{(\bar{\delta} - 1)}{Pr^{1/2}} + 1 \tag{5.2}$$

where

$$\bar{\delta} = (\delta/h_2) = 5\left(\frac{x/h_2}{Re}\right)^{\frac{1}{2}}$$

(c) The geometry of the groove—length, depth, shape—has only a minimal effect on the mixing intensity. The most telling parameter aside from the variables appearing in Equ. (5.1) is the linear velocity.

(d) As shown in Fig. 5.4 the transverse temperature profile is a function of the level of mixing with a sharp transition to a thin hot layer next

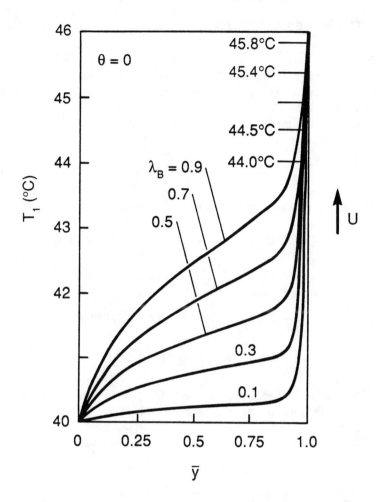

Fig. 5.4 Effect of mixing intensity on $T_1(y)$, Mitsui et al. (1983)

to the runner.

(e) The theoretical model given by Equ. (5.1) yields values of T_1 that are higher than what is usually measured.

In view of the intrinsic difficulties with a formal approach to the problem there remains an empirical approach to Equ. (5.1). This approach, while adhering to the basic model of mixing, as postulated by the existence of a thermal layer attached to the runner, would have to account for the multiple factors that cause a deviation from the theoretical model.

Three such empirical formulations of the mechanism of mixing will be cited here, and as will be seen they all essentially amount to the introduction of weighting functions or coefficients on the relative amounts of cold and hot lubricant entering the clearance space.

5.1.2 The Mixing Coefficient

The three investigations concerned with determining mixing coefficients are detailed in Table 5.1. The mixing coefficients in these investigations will be marked respectively with subscripts A, B, C denoting the results of each individual contribution.

The results of investigation A are essentially contained in Fig. 5.1. The mixing coefficient used is

$$\lambda_A = \frac{T_1 - T_0}{T_R - T_0} \tag{5.3}$$

T_R, of course, is usually not known but the authors conclude that a good representation of T_R is given by $T_R = (T_1 + T_2)/2$. When the above is introduced into Equ. (5.3), we obtain an expression for T_1 in the form

$$T_1 = \frac{2(1 - \lambda_A)T_0 + \lambda_A T_2}{(2 - \lambda_A)} \tag{5.4}$$

In investigation B a mixing coefficient λ_B was defined as

$$\lambda_B = \frac{(T_1 - T_0)Q_1}{(T_2 - T_0)Q_2} \tag{5.5}$$

While investigation A did not relate its coefficient to the prevailing flows, the above formula refers directly to Equ. (5.1) to which form it would revert if we set $\lambda_B = 1$. A value of $\lambda_B < 1$ would thus reduce the amount of hot lubricant Q_2 entering the film. The inlet temperature is here given by

$$T_1 = \frac{Q_0 T_0 + \lambda_B Q_2 T_2}{Q_1} \tag{5.6}$$

A plot of the coefficient λ_B is shown in Fig. 5.5. The values range from the theoretical $\lambda_B = 1$ to as low as $\lambda_B = 0.4$. Unfortunately, the authors of that figure plot the results against a ratio of (measured/calculated) flows, implying that it was the upstream cooling of the increased cold oil supply that caused a decrease in $\lambda_B = 1$. But it is more likely that a change in hydrodynamic conditions caused the variation in the value of λ_B.

While the results of both previous investigations are based on single test bearings, the program of investigation C employed at least eight different bearings and represents a more comprehensive evaluation of the phenomenon of mixing. Here a mixing coefficient is defined in the form of

$$\lambda_C = \frac{Q_0(T_0 - T_r) + Q_2(T_2 - T_r) - Q_1(T_1 - T_r)}{Q_1(T_1 - T_2)} \tag{5.7}$$

where T_r is a reference temperature.

TABLE 5.1

Investigations of inlet mixing coefficients

		Bearings tested		
ID	Source	Kind	Groove length x width	N
A	Ettles & Cameron (1968)	Parallel thrust R_2 x R_1 x h	1 in. $\Big\langle \begin{matrix} 0.145 \text{ in.} \\ 1.5 \text{ in.} \end{matrix}$	960-3670 rpm
			2-1/4 in. x 1-1/8 in. $\Big\langle \begin{matrix} 10^{-3} \\ 10^{-2} \end{matrix}$	
		6 & 3 pads		
B	Mitsui, Hori & Tanaka (1983)	Journal D x L x C = 100 mm x 70 mm x 78.10^{-3}mm	60 mm x 8.7 mm	0-3600 rpm
C	Heshmat & Pinkus (1986)	Journal 4 in. < D < 12 in. L/D = 0.93	$\dfrac{\ell}{L} \sim 0.9$	500-5000 rpm
		$1.5 \leq (C/R) \times 10^3 \leq 2$	$\beta_G = 30°$	
		Tapered thrust		

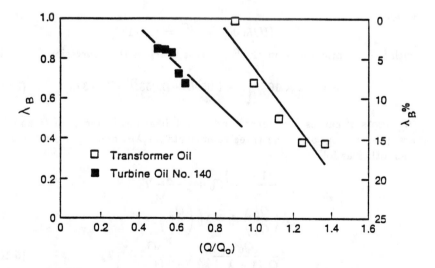

Fig. 5.5 Inlet mixing coefficient λ_B, Mitsui et al. (1983)

As seen the numerator in (5.7) represents Equ. (5.1) which, without
a mixing coefficient, adds up to zero. The departure of Equ. (5.7) from
its theoretical zero is thus here formulated as an indicator of the amount
of mixing. Any $\lambda_C > 0$ indicates that not all of the heat carried by Q_2
entered the fluid film. The inlet temperature is given then by

$$T_1 = \frac{Q_0 T_0 + Q_2 T_2 - \overline{\delta}\left(\frac{160}{9}\right)\lambda_C Q_1}{Q_1(1 + \lambda_C)} \tag{5.8}$$

where $\overline{\delta} = 0$ for °F and $\overline{\delta} = 1$ for °C.

A large number of tests spanning different kinds of bearings, loads,
lubricants, and linear speeds up to $13,000\,\text{ft/min}$, which is the upper limit
of laminar flow, were conducted. A sample plot of the kind of parametric
dependence shown by the tests, is given in Fig. 5.6. Based on these tests
the following characteristics of the coefficient λ_C were obtained.

- The main parameter affecting λ_C is linear speed. The value of λ_C at
  very low speed is slightly negative, rising with speed to level off at a
  value of U close to the limit of laminar operation.
- All curves seem to intercept the U axis at about the same location; also
  the value of U where the slope $(d\lambda_C/dU)$ approaches zero is similar
  for all curves.
- The parameter next in importance to U is the supply temperature T_0,
  the value of λ_C decreasing with a rise in T_0.
- The effect of loading on λ_C is small.

Based on a large body of data similar to Fig. 5.6 an empirical equation for
journal bearings was obtained in the form of

$$\lambda_C = -9 \times 10^{-3} \left[1 - 1.017\widetilde{U} + 0.017\widetilde{U}^2\right](5 - 3\widetilde{T}_0) \tag{5.9a}$$

where

$$\widetilde{U} = (U/5000), \qquad \widetilde{T}_0 = (T_0/120)$$

with U in ft/min and T in °F. For thrust bearings the expression for λ_C is

$$\lambda_C = -9 \times 10^{-3} \left[1 - 1.053\widetilde{U} + 0.053\widetilde{U}^2\right](5 - 3\widetilde{T}_0) \tag{5.9b}$$

Plots of curves for journal and thrust bearings in terms of $\widetilde{U}$ and $\widetilde{T}_0$
are given in Fig. 5.7. The three coefficients λ_A, λ_B, and λ_C are related to
each other as follows:

$$
\begin{aligned}
T_1 &= \frac{2(1 - \lambda_A)}{2 - \lambda_A}T_0 + \frac{\lambda_A}{2 - \lambda_A}T_2 \\
&= \left(\frac{Q_0}{Q_1}\right)T_0 + \lambda_B\left(\frac{Q_2}{Q_1}\right)T_2 \\
&= \frac{Q_0}{Q_1(1 + \lambda_C)}T_0 + \frac{Q_2}{Q_1(1 + \lambda_C)}T_2
\end{aligned} \tag{5.10}
$$

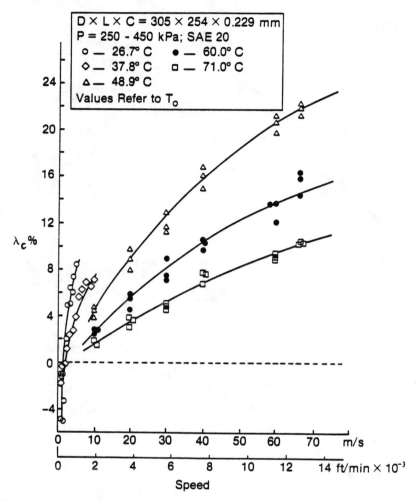

Fig. 5.6 Typical set of results for inlet mixing coefficient λ_C, Heshmat and Pinkus (1986)

5.1.3 Reverse Flow

When there are regions of reverse flow they present computational complications. The energy equation is in most cases treated as an initial value problem in which the coordinate x, imitating time, marches only forward. With a negative velocity u present, this is no longer possible and the ordinary approaches do not yield converging solutions.

The temperature distribution along a streamline which reverses direction is shown schematically in Fig. 5.8 as line IKO. During the flow forward, to point K, the temperature of a fluid particle rises due to heating and power dissipation. In the reverse flow, from K to O, the rate of

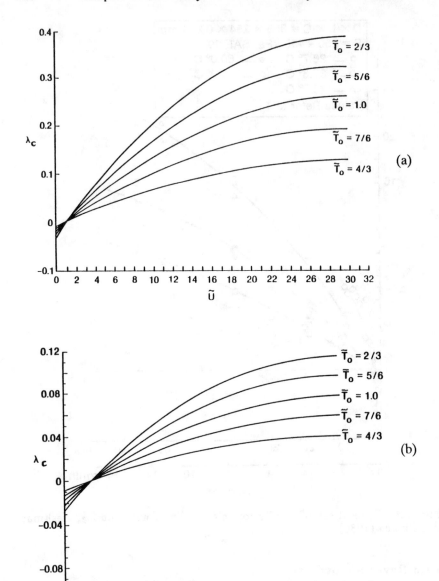

Fig. 5.7 Plot of coefficient λ_C for journal and thrust bearings, (a.) journal bearings, (b.) thrust bearings, Heshmat and Pinkus (1986)

dissipation is very low because of the low velocity gradients there, and the temperature can either decrease, as in Fig. 5.8b, or rise, depending on the level of heat conduction to the neighboring fluid layers. In general the

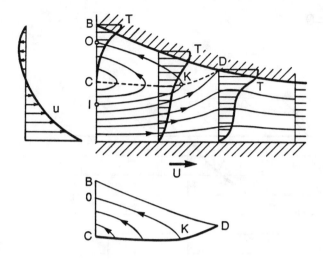

a) Reverse Flow in a Bearing Pad

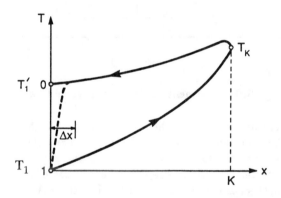

b) Temperature Distribution Along the Streamline IKO

Fig. 5.8 Effects of reverse flow on temperatures at inlet to pad, Ott and Paradissiadis (1988)

lubricant returns to the inlet at a T_1' higher than the inlet temperature T_1.

One way to deal with reverse flow is to first solve the thermal problem under the assumption of a uniform inlet temperature T_1 across the entire height $0 < y < h_1$. Such a solution will parallel closely the correct final solution except that at the very end of the reverse flow path, where the fluid instead of ending up at a temperature T_1', will be forced, by the imposed assumption, to return to T_1; this is shown by the dotted line in Fig. 5.8b. Thus the initially calculated distribution coincides with the correct one over almost the whole streamline except for a limited region, Δx in Fig.

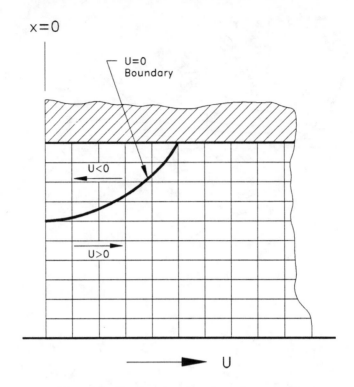

Fig. 5.9 Treatment of region of reverse flow

5.8b, where it decreases sharply to the prescribed T_1. This behavior results from the fact that the reverse flow in a bearing is convection dominated. The more convective the flow is, the shorter is the region Δx in which the solution is affected by the reverse flow. In the case of a finite element solution, the length of Δx is shorter than the length of the elements used for the discretization, causing oscillations in the obtained solutions.

When numerical methods are used to solve such problems the fluid film is divided into two zones, as shown in Fig. 5.9. In the first one which corresponds to the zone where u is positive the energy equation is solved using the standard techniques with initial temperatures assumed at the inlet boundary. In the zone of reverse flow the same iterative technique is used, but the computing process is started from the $u = 0$ boundary using some initial temperature at the boundary. The iterative process is then carried until convergence.

Fig. 5.10 shows results obtained for a plane slider bearing with reverse flow. As seen, in the region of reverse flow the temperatures at the inlet are higher than downstream caused by the shearing of the returning fluid. A similar result was obtained for the pumping ring back flow considered in Chapter 2.

A more complex example of reverse flow is given in Fig. 5.11 for a 90°

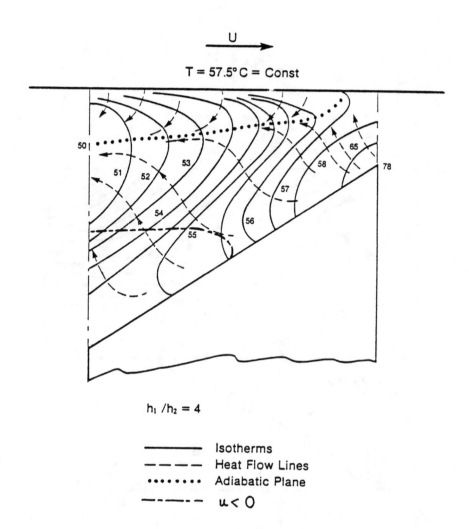

Fig. 5.10 Temperature distribution in slider with reverse flow, Boncompain and Frene (1980)

journal bearing pad. Although the reverse flow area extends over 40% of the pad length, the maximum temperature rise of the backflowing lubricant

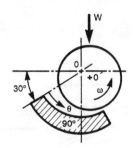

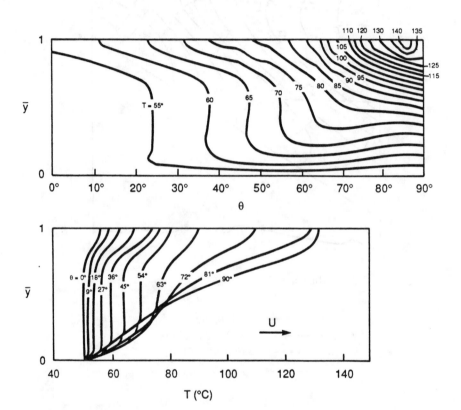

Fig. 5.11 Temperatures in 90° journal bearing pad with reverse flow, Ott and Paradissiadis (1988)

— at the pad surface — amounts to only 6°C. It should be noted that in this case the pad operates under extreme conditions of a unit loading of 114 atm. Generally, for usual operating conditions, the temperature rise at the film inlet due to reverse flow is small and has no significant influence on performance. The calculation of the reverse flow is, however, necessary if a solution of the energy equation free of oscillations is to be achieved.

5.1.4 Groove Parameters

In addition to mixing there are several other factors that affect the profile of the inlet temperatures, and consequently the average value of T_1. These are the geometry of the groove and the supply pressure of the lubricant. Both of these can, under certain conditions, lead to an incomplete fluid film at the entrance to a pad and thus to a variation of inlet temperature levels. A third factor leading to similar results is starvation—a supply of lubricant below the hydrodynamic requirements of the device. The latter is taken up in the next section while the geometrical and supply pressure effects are briefly discussed below.

The main geometrical feature that determines the shape of the inlet film is the axial extent of the supply groove. A proper supply groove extends nearly to the edges of the bearing, guaranteeing availability of fluid across its entire width. But often the groove extends only over a length ℓ which is considerably less than the bearing width; and at times the supply of lubricant is via a single hole, extending over a small fraction only of the bearing width. The effect on the value of the mixing temperature T_1 and the temperature map at the inlet can be seen in the three cases of Fig. 5.12 for which Table 5.2 provides the parametric data. As seen, the narrower the groove the greater the variation in the value of the inlet temperature, which can be summarized as follows:

	T_1 (°C)	
ℓ/L	Range	ΔT_1
0.9	50.4–51.0	0.6
0.5	59.8–61	2.2
0.1	56.8–62	5.2

The effect of supply pressure p_0 is shown in Fig. 5.13. While at 1.47×10^5 Pa there was a more or less uniform $T_1 = 46°C$ inlet temperature across the bearing, when the pressure was lowered to 1.24×10^5 Pa, T_1 ranged from 46°C to 49°C along the inlet.

Of course, the effect of the geometry and inlet pressure manifest themselves not only in the variation of T_1. They affect the entire temperature distribution, and thus performance as a whole. This can be seen from the two figures above as well as in Fig. 5.14 which plots the side leakage temperature T_S as it is affected by a variation in the inlet pressure and eccentricity.

5.1.5 Starvation

While geometry and low supply pressure are often the cause of insuf-

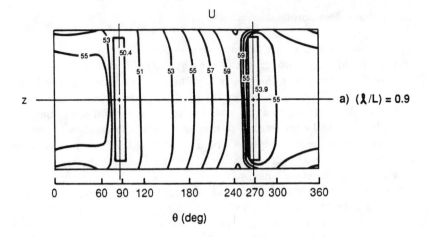

a) $(\ell/L) = 0.9$

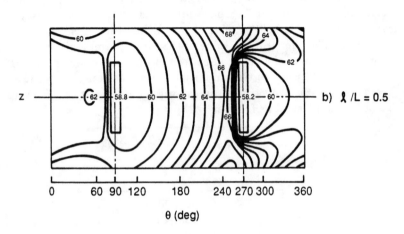

b) $\ell/L = 0.5$

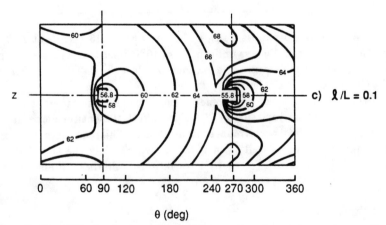

c) $\ell/L = 0.1$

Fig. 5.12 Effect of groove geometry on bearing temperatures, So and Shieh (1987)

TABLE 5.2

Bearing data for Fig. 5.12

Density of the oil, ρ, at 38°C and 0.1 MPa	881.6 kg/m^3
Thermal conductivity of the oil, k	0.1595 W/(m-K)
Specific heat	1840 J/(kg-K)
Temperature--viscosity coefficient	0.0315 K^{-1}
Oil viscosity at 37.8°C	0.0417 Pa-s
Radius of the journal, R	5 cm
Clearance ratio (C/R)	0.002
Angle of groove	15°
ϵ	0.5
P_0	4 X 10^5 Pa
L/D	1
N	2,000 rpm
T_0	38°C
54° < ϕ_0 < 59°	

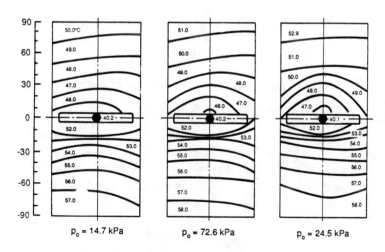

P_o = 14.7 kPa P_o = 72.6 kPa P_o = 24.5 kPa

W = 3.43 kN; N = 1750 rpm; T_o = 40°C

Fig. 5.13 Effect of supply pressure on inlet mixing temperatures, Mitsui *et al.* (1983)

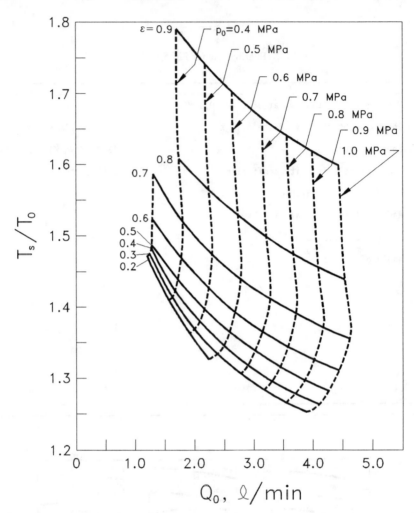

Fig. 5.14 Effect of initial conditions on average bearing temperatures, So and Shieh (1987)

ficient lubricant supply, actual starvation refers to cases when a bearing is supplied less than the hydrodynamically required quantity of lubricant. This is the quantity $Q_0 = Q_S$ to make up for the side leakage, or $Q_0 = Q_1$, if there is also leakage from the trailing end of the pad. An index of starvation can thus be defined by the ratio

$$\widehat{Q}_S = \frac{Q_{\text{supplied}}}{Q_{\text{required}}}$$

with $\widehat{Q}_S \geq 1$ denoting full film lubrication and $0 < \widehat{Q}_S < 1$ designating various levels of starvation. For conventional journal and thrust bearings this index is given by

$$\hat{Q}_S = (Q_0/Q_S)$$

where Q_S is the side leakage under full fluid film conditions, and Q_0 the actual amount of lubricant delivered.

The various ramifications of solving bearings under starved conditions will not be gone into here; they can be found in relevant works on the subject. What interests us here is the behavior of the temperature field and the effects of the variable temperature on starved bearing performance. Such results are given for journal bearings in Table 5.3 and for thrust bearings in Table 5.4. As can be seen, T_1, T_S, and T_{max} all increase with the severity of starvation. T_{max} increases much faster than the others, and therefore T_{max} is in starved bearings a more critical parameter than h_{min}. There are also some anomalies with starved bearings in that, for example, an increase in speed, instead of reducing, actually causes an increase in ϵ due to the severe rise in the value of T_1. Finally, as was broached earlier and as is shown in Fig. 5.15, the question of T_1 in starved bearings needs special treatment because $\theta_1 > \theta_S$ and before the full film forms at θ_1, the Q_0 and Q_2 streams that mix to form T_1, are heated by both the runner and bearing surfaces. Thus the actual T_1' would be higher than that calculated for full film lubrication.

5.2 RUNNER TEMPERATURE

There are at least two issues in considering the boundary condition represented by the runner surface. One is whether or not the runner can be considered to be at constant temperature; and, assuming it is an isothermal surface, what is its value and how is it to be determined.

It was seen in the previous chapters that the latter point was often resolved by writing for the runner heat flux

$$\int_0^{2\pi} k \left(\frac{\partial T}{\partial y_R} \right) d\theta = 0$$

that is, a condition of zero net heat exchange with the fluid. This means that the runner conveys no heat to its adjacent parts or to the atmosphere. Often the less stringent condition of

$$T_R = \frac{1}{2\pi} \int_0^{2\pi} T(\theta) \, d\theta$$

is used which assigns to the runner an average of the film temperatures. But this, too, comes close to imposing adiabatic conditions because the runner will then absorb heat over all $T_R < T$ and discharge it over the part of the bearing where $T_R > T$.

TABLE 5.3

Adiabatic performance of starved journal bearings

$\hat{Q}_s$	Q_s ml/s (GPM)	T_1 °C(°F)	ϵ	ϕ (deg)	M J × 10³ (in.-1b)	θ_2 (deg)	θ_1 (deg)	$(T_{av}-T_1)$ °C (°F)	$(T_{max}-T_1)$ °C (°F)
1.000*	72.7 (1.15)	56.7 (134.)	.589	44.8	10.9 (62.3)	250	97.0	1.11 (2.0)	8.89 (16.)
.774	56.3 (.892)	57.8 (136.)	.619	40.2	10.4 (59.6)	240	108	2.22 (4.0)	9.44 (17.)
.386	28.1 (.445)	65.0 (149.)	.714	30.0	9.02 (51.5)	230	129	1.67 (3.0)	10.6 (19.)
.304	22.5 (.357)	67.2 (153.)	.743	26.9	8.54 (48.8)	224	134	2.22 (4.0)	11.1 (20.)
.232	16.9 (.267)	72.2 (162.)	.781	24.0	7.87 (44.9)	220	139	2.22 (4.0)	11.7 (21.)
.155	11.3 (.178)	79.4 (175.)	.828	20.0	6.96 (39.7)	215	145	2.22 (4.0)	13.3 (24.)
.077	5.63 (.089)	96.1 (205.)	.889	15.0	5.43 (31.0)	205	153	2.78 (5.0)	16.1 (29.)
.039	2.81 (.045)	123. (253.)	.934	12.0	4.07 (23.2)	202	159	2.78 (5.0)	20.6 (37.)

*Full film   Heshmat and Pinkus, 1985

$W = 22,240$ N; $N = 1800$ rpm: $T_0 = 48.9$°C; $\beta = 150$°

$L/D = 0.93$; $C/R = 2 \times 10^{-3}$

5.2.1 Constancy of Runner Temperature

Using Jaeger's expression for a moving heat source as given by Equ. (4.7), Hahn and Kettleborough (1967) showed that the runner surface temperature must remain approximately constant. With the outside surface of the runner fixed at some constant ambient T_a and normalizing the runner surface temperature by T_a, $\overline{T}_R = (T_R/T_a)$, we have

$$\frac{\partial^2 \overline{T}_R}{\partial \overline{x}^2} + \frac{B^2}{t_R^2} \cdot \frac{\partial^2 \overline{T}_R}{\partial \overline{y}_R^2} - \frac{UB}{\sigma} \cdot \frac{\partial \overline{T}_R}{\partial \overline{x}} = 0 \qquad (5.11)$$

The boundary conditions are

$$\overline{y}_R = 0, \qquad \overline{T}_R(\overline{x}, 0) = 1$$

TABLE 5.3 (Continued)

Adiabatic performance of starved journal bearings

$\hat{Q}_s$	K_{xx}	K_{xy}	K_{yx}	K_{yy}	B_{xx}	B_{xy}	B_{yx}	B_{yy}	W_{cr}
	N/m x 10^{-6}		(1b/in.) x 10^{-6}		N-S/m x 10^{-6}		(1b-s/in.) x 10^{-6}		Kg (1b)
1.0*	262	65.8	-423	355	1.74	-1.42	-1.44	7.5	197
	(1.5)	(.376)	(-2.41)	(2.03)	(9.96)	(-8.10)	(-8.24)	(42.80)	435
.774	255	52.0	-400	414	1.54	-1.00	-1.31	8.99	359
	(1.45)	(.297)	(-2.28)	(2.36)	(8.78)	(-5.73)	(-7.46)	(51.30)	792
.385	228	38.1	-395	607	1.33	-5.05	-1.61	17.1	1780
	(1.30)	(.217)	(-2.36)	(3.46)	(7.61)	(-2.88)	(-9.18)	(97.70)	3930
.304	213	24.7	-392	708	1.20	-.942	-1.27	20.2	∞
	(1.22)	(.141)	(-2.24)	(4.04)	(6.84)	(-.538)	(-7.28)	(115.0)	
.232	205	22.8	-416	832	1.23	.162	-1.70	28.5	
	(1.17)	(.130)	(-2.37)	(4.75)	(7.01)	(.923)	(-9.73)	(163.0)	
.155	192	11.6	-436	1070	1.25	2.24	-1.14	15.5	
	(1.10)	(.066)	(-2.49)	(6.12)	(7.12)	(12.8)	(-6.53)	(88.60)	
.077	174	12.4	-483	1710	1.47	5.52)	.894	84.2	
	(.992)	(-0.71)	(-2.76)	(9.79)	(8.42)	(31.5)	(5.11)	(481.0)	
.039	171	-9.77	-638	2660	2.16	9.14	.634	49.5	
	(.990)	(-.056)	(-3.64)	(15.2)	(12.30)	(52.2)	(3.62)	(283.0)	

Heshmat and Pinkus, 1985

W = 22,240 N;   N = 1800 rpm: T_0 = 48.9°C;   β = 150°

L/D = 0.93;   C/R = 2 X 10^{-3}

* Full film

$$\overline{y}_R = 1, \qquad \frac{\partial \overline{T}_R}{\partial \overline{y}_R} = \phi(\overline{x}) = \frac{t_R}{h_2} \cdot \frac{k}{k_R} \cdot \left(\frac{\partial \overline{T}}{\partial \overline{y}} \right)_{\overline{y}=0}$$

$$\overline{T}_R(\overline{x}, \overline{y}_R) = \overline{T}_2(\overline{x}+1, \overline{y}_2)$$

the last condition representing a requirement of periodicity with each pad.
Since ϕ is periodic, we can write

$$\phi(\overline{x}) = \sum_{n=0}^{\infty} [a_n \sin(k_n \overline{x}) + b_n \cos(k_n \overline{x})], \quad k_n = 2\pi n$$

It can be shown that for such boundary conditions, the solution of

TABLE 5.4

Performance of starved thrust bearings

β	$\bar{W}_T$	$\bar{h}_2$	$\bar{\beta}_s$	$\bar{Q}_1$	$\bar{Q}_S$	$\hat{Q}$	$\bar{\bar{T}}$	$\bar{\bar{T}}_{max}$	$\bar{H}$
27°	7.807	0.125	1.00	0.43	0.29	1.00	0.61	1.69	1.07
	7.367	0.125	0.96	0.40	0.27	0.92	0.62	1.68	1.05
	6.373	0.125	0.75	0.30	0.17	0.58	0.68	1.62	0.91
	3.966	0.125	0.50	0.20	0.08	0.27	0.72	1.59	0.69
	4.325	0.250	1.00	0.54	0.30	1.00	0.45	1.17	0.83
	4.104	0.250	0.94	0.50	0.26	0.87	0.46	1.16	0.80
	4.104	0.250	0.72	0.40	0.16	0.53	0.49	1.11	0.67
	3.156	0.250	0.47	0.30	0.07	0.24	0.48	1.05	0.49
	0.0	0.250	0.20	0.02	0.01	0.04	0.11	0.92	0.21
	1.116	0.690	1.00	0.91	0.30	1.00	0.21	0.53	0.52
	0.825	0.690	0.81	0.80	0.20	0.66	0.21	0.50	0.45
	0.502	0.690	0.60	0.70	0.11	0.37	0.20	0.45	0.36
	0.173	0.690	0.36	0.60	0.04	0.14	0.15	0.37	0.25
	0.571	1.000	1.00	1.15	0.31	1.00	0.14	0.36	0.42
	0.233	1.000	0.72	1.00	0.16	0.53	0.13	0.31	0.33
	0.032	1.000	0.28	0.80	0.03	0.08	0.07	0.19	0.16
42°	7.113	0.125	1.00	0.55	0.42	1.00	0.75	1.89	1.17
	6.343	0.125	0.79	0.40	0.27	0.65	0.82	1.84	1.04
	5.284	0.125	0.62	0.30	0.18	0.42	0.86	1.81	0.91
	3.392	0.125	0.42	0.20	0.08	0.20	0.88	1.77	0.71
	3.925	0.250	1.00	0.65	0.42	1.00	0.57	1.36	0.96
	3.344	0.250	0.78	0.50	0.27	0.63	0.61	1.31	0.83
	2.668	0.250	0.61	0.40	0.17	0.40	0.63	1.27	0.71
	1.572	0.250	0.41	0.30	0.08	0.18	0.61	1.21	0.54
	0.00	0.250	0.20	0.20	0.01	0.03	0.16	1.10	0.26
	1.062	0.690	1.00	1.01	0.43	1.00	0.29	0.68	0.68
	0.704	0.690	0.70	0.80	0.22	0.51	0.29	0.62	0.53
	0.459	0.690	0.52	0.70	0.13	0.30	0.27	0.56	0.44
	0.179	0.690	0.33	0.60	0.05	0.12	0.20	0.47	0.32
	0.556	1.000	1.00	1.25	0.43	1.00	0.29	0.48	0.57
	0.298	1.000	0.64	1.00	0.18	0.43	0.19	0.30	0.42
	0.037	1.000	0.26	0.30	0.03	0.07	0.10	0.26	0.22
57°	5.862	0.125	1.00	0.63	0.51	1.00	0.86	2.03	1.19
	5.126	0.125	0.70	0.40	0.28	0.55	0.95	1.97	1.01
	4.378	0.125	0.55	0.30	0.18	0.36	0.98	1.93	0.90
	2.895	0.125	0.38	0.20	0.09	0.17	1.00	1.91	0.70
	3.256	0.250	1.00	0.73	0.51	1.00	0.67	1.49	1.02
	2.714	0.250	0.70	0.40	0.28	0.55	0.95	1.97	1.01
	2.234	0.250	0.54	0.40	0.18	0.35	0.73	1.38	0.73
	1.997	0.250	0.37	0.30	0.08	0.16	0.70	1.33	0.58
	0.0	0.250	0.20	0.20	0.01	0.02	0.21	1.23	0.29
	0.903	0.690	1.00	1.07	0.51	1.00	0.67	1.49	1.02
	0.594	0.690	0.64	0.80	0.23	0.46	0.35	0.71	0.60
	0.408	0.690	0.48	0.70	0.14	0.27	0.32	0.65	0.50
	0.174	0.690	0.31	0.60	0.06	0.11	0.25	0.55	0.38
	0.482	1.000	1.00	1.31	0.51	1.00	0.26	0.58	0.70
	0.348	1.000	0.73	1.10	0.30	0.58	0.25	0.52	0.57
	0.260	1.000	0.58	1.00	0.20	0.39	0.23	0.47	0.49

Heshmat, Artiles, and Pinkus, 1986

$$\bar{h}_2 = (h_2/\delta)$$

$$\bar{M} = \left[\frac{\mathbb{T}M}{\mu_0 \omega/R_2^2\, L} \right] \left[\frac{\delta}{L} \right]$$

$$\bar{W}_T = \left[\frac{P}{\mu_0 N} \right] \left[\frac{\delta}{L} \right]^2$$

$$\bar{Q} = \frac{Q}{\mathbb{T}N R_2 L \delta}$$

$$\bar{\beta}_s = \left[\frac{\beta - \theta_1}{\beta} \right]$$

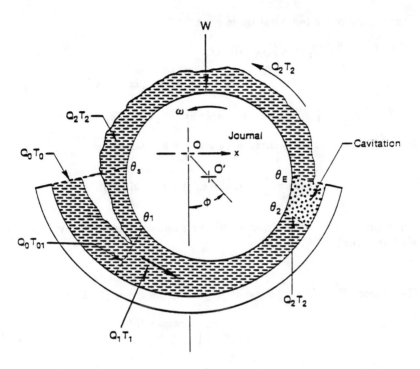

Fig. 5.15 Mixing inlet temperatures in starved bearing

Equ. (5.11) is

$$\overline{T}_R(\overline{x}, \overline{y}_R) = \overline{T}_2(\overline{x}, 0) + b_0 \overline{y}_R + \sum_{n=1}^{\infty} \{a_n[\eta_n \sin(k_n \overline{x})$$

$$+ \xi_n \cos(k_n \overline{x})] + b_n[\eta_n \cos(k_n \overline{x}) - \xi_n \sin(k_n \overline{x})]\}$$

(5.12)

where

$$\eta_n = \frac{u_n \cosh(u_n) \sinh(u_n \overline{y}_R) + v_n \cos(v_n) \sin(v_n \overline{y}_R)}{u_n^2 \cosh^2 u_n + v_n^2 \cos^2 v_n}$$

(5.13a)

$$\xi_n = \frac{u_n \cosh(u_n) \sin(v_n \overline{y}_R) - v_n \cos(v_n) \sinh(u_n \overline{y}_R)}{u_n^2 \cosh^2 u_n + v_n^2 \cos^2 v_n}$$

(5.13b)

$$u_n + w_n = r_n e^{i\theta_n}$$

(5.13c)

$$r_n = \frac{t_R k_n}{B}[1 + (UB/\sigma k_n)^2]^{1/4}$$

$$\theta_n = \frac{1}{2}\tan^{-1}(UB\sigma k_n)$$

(5.13d)

Hence the moving temperature is given by

$$\overline{T}_R(\overline{x}, 1) = \overline{T}_R(\overline{x}, 0) + b_0$$
$$+ \sum_{n=1}^{\infty} \{a_n[\eta_n \sin(k_n\overline{x}) + \xi_n \cos(k_n\overline{x})] \qquad (5.14)$$
$$+ b_n[\eta_n \cos(k_n\overline{x}) - \xi_n \sin(k_n\overline{x})]\}_{\overline{y}_R=1}$$

It is sufficient to approximate ϕ by its first five terms, or

$$\phi(\overline{x}) \doteq \sum_{n=0}^{5} [a_n \sin(k_n\overline{x}) + b_n \cos(k_n\overline{x})]$$

For typical slider values quoted in the numerical example by Hahn and Kettleborough (1967)

$$\frac{UB}{\sigma} = 2.4 \times 10^5$$

Hence from Equ. (5.13d) $\theta_n = 45$ deg and from equs. (5.13b-d)

$$u_n = v_n = \frac{t_R}{B}\left(\frac{\pi U B n}{\sigma}\right)^{1/2} \qquad (5.14)$$

Since for the parameters making up Equ. (5.14) we have approximately

$$u_n = 1600, \quad \xi_n = 0, \quad \eta_n = \frac{1}{u_n}$$

then to within 0.1% accuracy, the periodic terms in Equ. (5.12) may be neglected. Thus the solution is reduced to

$$\overline{T}_R(\overline{x}, 1) = \overline{T}_R(\overline{x}, 0) + \int_0^1 \left(\frac{\partial \overline{T}_R}{\partial \overline{x}}\right)_{\overline{y}_R=1} d\overline{x} \qquad (5.15)$$

and since $\overline{T}_R(\overline{x}, 0) = 1$, T_R is not a function of x and must be constant along the runner surface.

5.2.2 Heat Flow in Shaft

Of the major components of a bearing assembly the shaft seems the least likely body to operate under adiabatic conditions. A shaft situated in two or more bearings is usually long, exposed to the atmosphere, and, naturally, rotating at high speed — all of which conspire to make it lose heat. This heat transfer consists of free or rotational convection as well as radiation. This can be written as

$$Q = h_{Ra}A(T - T_a) + \eta\epsilon(T^4 - T_a^4) \qquad (5.16)$$

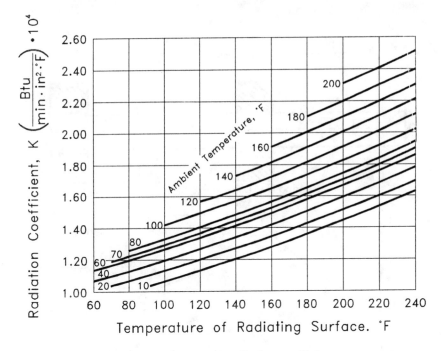

Fig. 5.16 Values of radiation coefficient

where $\eta = 0.2 \cdot 10^{-12}$ (BTU/min·in²·(°R)⁴ is the Stefan-Boltzman constant and ϵ is the shaft emissivity. Equ. (5.16) is often simplified by writing it in the form of

$$Q = h'_{Ra}(T - T_a) = (h_{Ra} + k\epsilon)(T - T_a) \qquad (5.17)$$

with the value of $k = \eta(T^4 - T_a^4)/(T - T_a)$ plotted in Fig. 5.16.

A more or less generic shaft configuration, one end of which is free and the other supported by a second journal bearing is sketched in Fig. 5.17. The task is to obtain the quantities of heat drained along portions c and b of the shaft — these energies naturally coming from the heat dissipated in the bearing under study.

The location of the adiabatic plane can be obtained as follows. From Fig. 5.18

$$-h_{Ra}(\pi D dz)(T - T_a) = dQ_c = d\left[-k\frac{\pi D^2}{4}\left(\frac{dT}{dz}\right)\right] \qquad (5.18)$$

whence

$$\frac{d^2T}{dz^2} - m^2 T = -m^2 T_a \qquad (5.19)$$

where

$$m^2 = \frac{4h'_{Ra}}{kD}$$

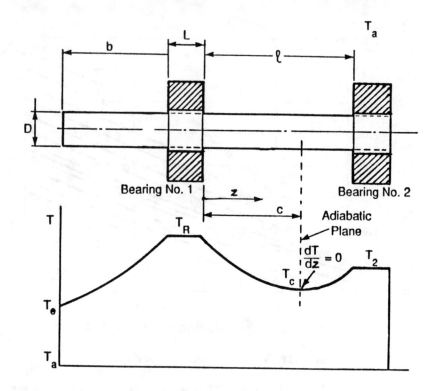

Fig. 5.17 Temperature profile in a shaft

The solution of Equ. (5.19) is

$$T = C_1 e^{mz} + C_2 e^{-mz} + T_a \tag{5.20}$$

and

$$\frac{dT}{dz} = m(C_1 e^{mz} - C_2 e^{-mz}) \tag{5,21}$$

The constants of integration C_1 and C_2 may be evaluated from the conditions that $T = T_R$ at $z = 0$ and at $T = T_2$ at $x = l$.

The position of the minimum value of T is the location of the adiabatic plane. Its position may be found by setting $\frac{dT}{dz}$ equal to zero and z equal to c in Equ. (5.21). Accordingly

$$c = \ln\left[\frac{(T_2 - T_a) - (T_R - T_a)e^{ml}}{(T_R - T_a)e^{ml} - (T_2 - T_a)}\right] \cdot \frac{1}{2m} \tag{5.22}$$

The analysis given by Burr (1959) arrives at the following results:

(a) The temperature and heat loss between the bearing and the adiabatic station c are

$$(T - T_a) = \frac{T_R - T_a}{2\cosh(mc)}\left[e^{-mc}e^{mz} + e^{mc}e^{-mz}\right] \tag{5.23}$$

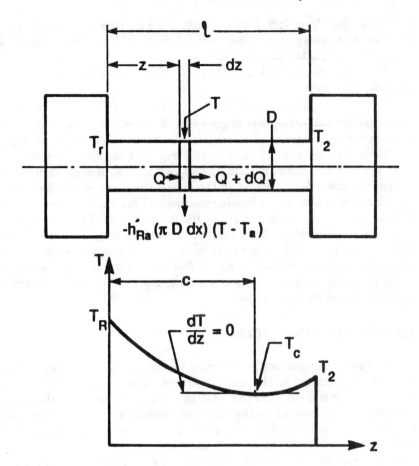

Fig. 5.18 Heat flow through a shaft heated at both ends

$$Q_c = mk_R \left(\frac{\pi D^2}{4}\right) \left[(T_R - T_a)\tanh(mc)\right] \qquad (5.24)$$

(b) The temperature and heat lost between the bearing and the free end, i.e. section b, are

$$(T - T_a) = \frac{(T_R - T_a)}{(h'_{Ra}/mk_R)\sinh(mb) + \cosh(mb)} \qquad (5.25)$$

$$Q_b = mk_R \left(\frac{\pi D^2}{4}\right) \frac{(T_R - T_a)\left[\left(\frac{h'_{Ra}}{mk_R}\right) + \tanh(mb)\right]}{1 + (h'_{Ra}/mk_R)\tanh(mb)} \qquad (5.26)$$

Above, $m = \sqrt{4h'_{Ra}/kD}$, a value which depends on the natural or rotational convection from the shaft, as well as on the level of radiation. Thus

$$h'_{Ra} = (h_{Ra} + k\epsilon) \qquad (5.27)$$

can be obtained from Figs. 5.16 and 5.19. For a slowly rotating shaft which may fall in the category of either part (a) or (b) of Fig. 5.19, an h_{Ra} is to be picked which is the higher of the two.

The total heat extraction by the runner from the fluid film is, of course, the sum of the two, or

$$Q_R = Q_c + Q_b$$

Fig. 5.20 shows an example of the position of the adiabatic plane in a shaft between two bearings one of which has varying temperatures, both higher and lower than the bearing under study. Fig. 5.21 represents the case of a shaft at $\epsilon = 0$, i.e., one of no side leakage and no fluid heat convection. For the specific conditions specified in the legend the figure shows the relative amounts of heat lost via the bearing housing and the shaft. At low speed — 800 rpm — the amounts are equal; at 1600 rpm the loss of heat through the shaft is about 40% higher than that lost via the bearing; this trend toward higher losses via the shaft increases with speed. Thus, the postulate that the runner surface has a net zero heat flow or that it is simply the average of the film temperatures is not supported by more careful considerations of the heat flow in rotating shafts.

5.3 THE BEARING SURFACE

The stationary surface is, according to most analytical and experimental evidence, hotter than the runner and this, in fact, may be due partly to the larger heat transfer from the rotating surface. By and large the boundary conditions presented by the stationary surface have been treated with greater circumspection than the runner in that it is rarely assumed to be isothermal or adiabatic (except in adiabatic solutions) and its temperature is usually arrived at by requiring a continuity of heat flux between the fluid film and the bearing shell. When heat transfer coefficients are postulated in the form of

$$Q = h_S (T - T_S)$$

the discontinuity in temperature would imply the presence of some sort of thermal barrier or boundary layer which is non-existent in the lubricant film. The flexibility in the approach is that by allowing the heat transfer coefficient to very from $h_S = \infty$ to $h_S = 0$ one can simulate actual conditions from $\Delta T = 0$ at the fluid-solid interface to adiabatic conditions.

The main shortcoming in the treatment of the bearing shell is that, as briefly illustrated in Section 4.1, the bearing pad is never an isolated, single component but part of an assembly that includes bearing liner, housing, pedestals, and seals, often made of different materials. It is thus a major simplification to consider simply the bearing pad as being exposed on one side to the fluid film and on the other to ambient conditions at some constant T_a. It would probably be more consistent with reality to consider the bulk of all of these parts as the stationary element with which the fluid film exchanges heat.

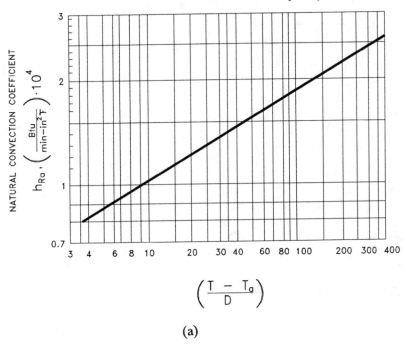

(a)

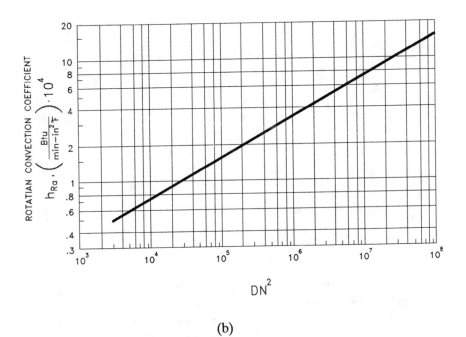

(b)

Fig. 5.19 Convection coefficients for shafts in air, (a.) non-rotating shafts (D in inches), (b.) rotating shaft (N, rpm; D, inches), Burr (1959)

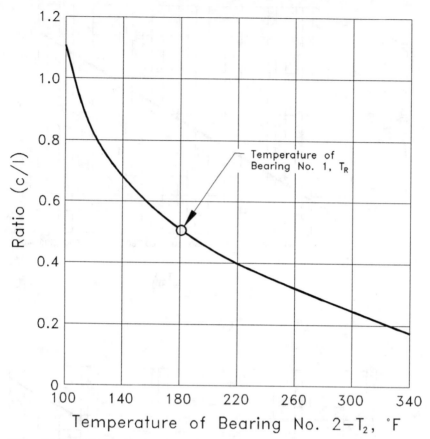

Fig. 5.20 Adiabatic plane position, $l = 10$ in.; $N = 1,800$ rpm; $T_R = 180°$F; $T_a = 80°$F; $D = 2$ in.; $\epsilon = 0.7$; $k_R = 36.2 \times 10^{-3}$ BTU/min·in·°F, Burr (1959)

5.4 THE EDGE CONDITIONS

As pointed out on a number of occasions the fluid conduction terms

$$\frac{\partial^2 T}{\partial x^2} \quad \text{and} \quad \frac{\partial^2 T}{\partial z^2}$$

are considered small as compared with the other terms in the energy equation and are usually ignored. This lowers the order of the differential equation and avoids the need to satisfy the unknown boundary conditions at $x = B$ and $z = \pm L/2$. The parabolic nature of the differential equation and the "marching technique" employed to solve it simply lead to the solution being halted at the bearing edges and whatever values have been obtained are accepted as the prevailing temperatures.

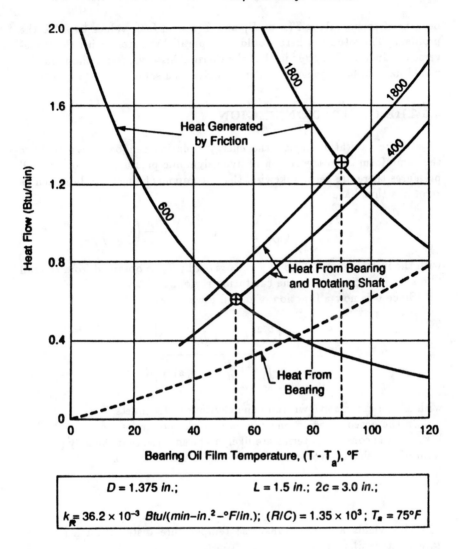

Fig. 5.21 Heat losses through shaft and bearing, Burr (1959)

It was seen that most solutions yield temperature profiles which are axially invariant, or even with a slight rise at the edges of the bearing. This is in clear disagreement with experimental evidence. Most measurements show a substantial drop in temperature at the edges of the bearing. In short bearings, $L/D < 1/2$, the profile is clearly of a parabolic shape.

It is not clear whether it is the neglect of the fluid conduction terms, or the inadequate formulation of the heat exchange with bearing and runner that is responsible for this discrepancy. At any rate, in terms of boundary conditions, should the x and z conduction terms be retained, there would be a need to postulate $T(x = B)$ and $T(z = \pm L/2)$ values to solve the problem. The situation here would then parallel the story with the transverse conduction $\partial^2 T/\partial y^2$ in that the thermal history outside the fluid film would have to be engaged in order to arrive at a solution.

5.5 THE CAVITATION REGION

In the cavitation region the energy equation can be shown to have the same form as in the region of hydrodynamic pressures except that all pressures there are zero. Likewise, the w component there can be taken as zero. We thus have

$$u\left(\frac{\partial T}{\partial x}\right) + v\left(\frac{\partial T}{\partial y}\right) = \frac{k}{\rho c}\left(\frac{\partial^2 T}{\partial y^2}\right) + \frac{\mu}{\rho c}\left(\frac{\partial^2 u}{\partial y^2}\right) \qquad (5.28)$$

with the initial temperature at $\theta = \theta_2$ given by T_2 as obtained from the full film solution upstream of the cavitation region.

Since u is now a function of μ and y we have

$$L' = L\frac{\int_0^{L/2}\left[h\int_0^h u(\theta, y, z)dy\right]dz}{h\int_0^{L/2}\left[\int_0^h u(\theta, y, z)dy\right]dz} \qquad (5.29)$$

with L' the cumulative width of the cavitation streamlets.

In the gaseous region one can assume the power loss to be negligible. Likewise the convection terms are likely to be small so that the only possible term of significance would be the transverse conduction or

$$\frac{\partial^2 T}{\partial y^2} = 0$$

This yields a linear transverse temperature distribution across the gaseous streamlets, or

$$T = (T_S - T_R)y + T_R \qquad (5.30)$$

5.5.1 Streamlets Cavitation

The basic feature of the cavitation region is that it is filled with a mixture of liquid, and gas-vapor streamlets. For the determination of pressure, flow, and most other performance quantities the exact configuration of this two- or three-phase flow region is of no great interest, but it is so when thermal effects are to be included. While in all cases the hydrodynamic pressures will be zero, the power dissipation and temperature rise do depend on the mode of cavitation.

The cavitation mode most commonly accepted in isothermal analysis is that given by A of Fig. 1.13. For such streamlets Ott and Paradissiadis (1988) obtained a solution of the cavitation temperatures assuming an isothermal runner and an adiabatic bearing surface. Each filmlet was treated as if it were an autonomous fluid film unaffected by the adjacent streamlets, the gases released from the lubricant, or any other peculiarities of this region.

Fig. 5.22 gives the temperatures along and across each filmlet. As seen, the temperature gradient across h is steepest in the cavitation region. However, the temperature profile along the cavitation path, is constant. Since the bearing surface is adiabatic presumably all the heat dissipation along the cavitation region is transferred to the runner at a constant rate.

A more rigorous solution was obtained by Boncompain and Frene (1980) in that the bearing-fluid film interface was required to satisfy continuity of heat flux, which was also required of the journal, though it was kept at a constant temperature. On the outside of the bearing shell the relationship

$$\frac{\partial T}{\partial y_S} = h_{sa}(T - T_a)$$

was used while, at the inlet, mixing according to Equ. (2.9) was accounted for with the inlet temperature parabolic across y. Table 5.5 gives the additional thermal parameter for this solution.

The temperature map at the midplane for all three components, journal, fluid film, and bearing shell are given in Fig. 5.23. The transverse temperatures in the fluid film are given in Fig. 5.24. As seen, in the fluid film the temperatures remain nearly constant from θ_2 to the end of the bearing. The temperatures of the bearing surface drop somewhat along the path of cavitation. All this is ascribed to the heat exchange between the fluid film and its two adjacent surfaces, as shown in Fig. 5.25. This simply shows that starting with θ_S there is heat input from both the moving and stationary surfaces into the fluid film over roughly 150°; a circulatory heat flow over the next 60°; and starting with h_{min} till θ_E heat outflow from the fluid film into both shaft and bearing. It is this latter heat drain that is responsible for the roughly constant temperature in the cavitation zone. This solution is thus consistent with the previous one, and several others, in which the heat exchange with runner and bearing is the main reason for a lack of rise in temperature in the cavitated region. These, of course, are theoretical predictions. Whether or not they are superior to other suppositions and, most of all, whether they conform to experimental evidence is considered next.

5.5.2 Possible Modes of Cavitation

The starting point of the scrutiny of the results of the previous section is the fact that while occasionally the temperatures in the cavitation region

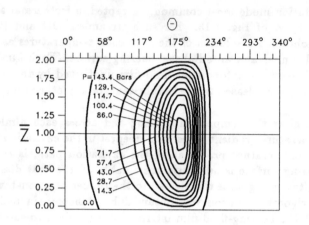

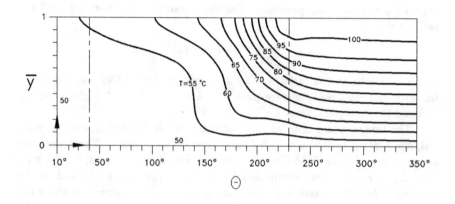

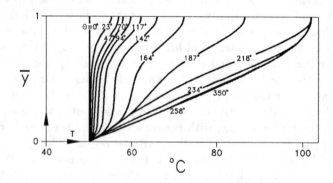

Fig. 5.22 Cavitation temperatures in journal bearings, $(L/D) = 0.3$, $(C/R) = 0.0015$, $N = 3000$ rpm, $D = 0.25$ m; $\epsilon = 0.75$; $\theta_0 = 314.5$ deg.; $\rho c_p = 1.75 \times 10^6$ J/m$^3 \cdot$°K; $h_R = 0.14$ W/m$^2 \cdot$°K; $\mu = 0.03$ kg/ms for $T_1 = 50$°C; $\alpha = 3.466 \times 10^{-2}$, Ott and Paradissiadis (1988)

TABLE 5.5

Operating conditions for Fig. 5.24 to 5.26

Journal radius	R = 50 mm
External bearing radius	R_2 = 100 mm
Bearing length	L = 80 mm
Radial clearance at 20°C	C = 145 μm
Rotational speed range	1000 < N < 4500 rpm
Load range	1000 < W < 10,000 N
Lubricant viscosity at 40°C	μ = 0.028 Pa·s
Lubricant density at 40°C	ρ = 860 kg/m^3
Lubricant specific heat	c_p = 2000 J/kg·°C
Lubricant thermal conductivity	k = 0.13 W/m·°C
Air thermal conductivity	k_a = 0.025 W/m·°C
Bearing thermal conductivity	k_s = 250 W/m·°C
Shaft thermal conductivity	k_R = 50 W/m·°C
Convection heat transfer coefficient of the bearing	50 < h_s < 500 W/m^2·°C
Convection heat transfer coefficient of the shaft	h_R = 100 W/m^2·°C
Inlet lubricant temperature	T_0 = 40°C
Ambient temperature	T_a = 45°C
Inlet lubricant pressure	P_0 = 70 X 10^3 Pa
Groove angle	18°
Bearing expansion coefficient	λ_s = 17 x 10^{-6}/°C
Shaft expansion coefficient	λ_R = 12 x 10^{-6}/°C

do stay more or less constant, in most practical cases there is a noticeable drop in temperature. In addition, there remains the skepticism about handling of the cavitation region in the same manner as the full film by simply reducing film width. In fact, cavitation represents a much more complex region and Booser and Wilcock (1987) attempted to look at some additional factors responsible for what they call "temperature fade" in the cavitation zone. The factors examined for possible causes of temperature drop in the region beyond θ_2 were:

- supply pressure of lubricant
- presence of subambient pressure at $\theta > \theta_2$
- heat transfer to the surfaces
- backflow from a supply groove into the cavitation zone

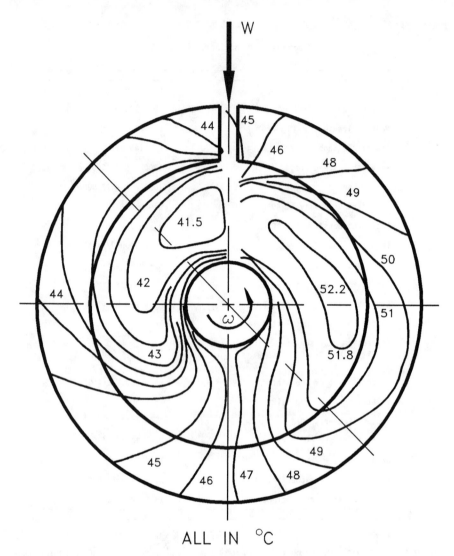

Fig. 5.23 Isotherms in journal bearing system, Boncompain and Frene (1980)

- drain of hot oil from supply groove (via the chamfers)
- level of power dissipation over the cavitation filmlets.

Of the above factors none had a noticeable effect on the level of temperatures except heat transfer and the power dissipation in the cavitation zone. The results of varying these two parameters are given in Table 5.6 and these can be summarized as follows:

(1) With zero heat transfer, T_{max} occurs at the circuferential end of the bearing.

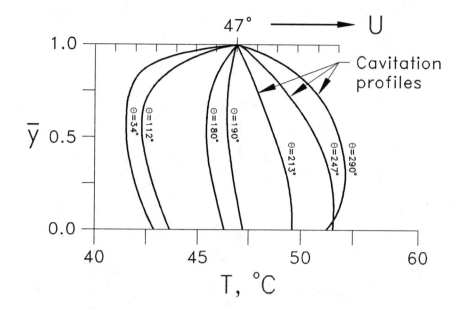

Fig. 5.24 Cavitation temperature profiles at the midplane of a journal bearing, $\epsilon = 0.74$; N=2000 rpm, Boncompain and Frene (1980)

(2) Heat loss to shaft with an adiabatic bearing reduces the level of all temperatures at all θ's beyond h_{min} and produces a slight decline in T along the cavitation path.

(3) Heat loss to both shaft and bearing reduces temperature levels throughout the fluid film and introduces a slight drop in temperature along the cavitation path. Still, the levels of temperature, here as well as in case (2) remained above experimental values measured during the same investigation.

(4) An assumed reduction in power dissipation over the cavitation zone lowers the overall temperatures and produces a noticeable decrease in temperature along the cavitation path. It required a significant reduction in (H/H_c) to bring theoretical results in line with experimental readings. A comparison of the theoretical values tabulated in column (c) of Table 5.6 with experimental values is given in Fig. 5.26.

Additional evidence that the streamlet pattern of striation may not be accurate is provided by the experimental data of Barrett and Branagan (1988). When shearing losses over the conventional streamlets of model A of Fig. 1.13 were used, the comparison with theoretical values was good over the loaded zone but very poor over the cavitation region, as shown in Fig. 5.27 (a); when model B of Fig. 1.13 was used, that is, shearing losses over an air film were used, the comparison was good over the entire zone, including a proper drop in temperature over the cavitated region. Still an intermediate model between those of A and B is possible as the one shown

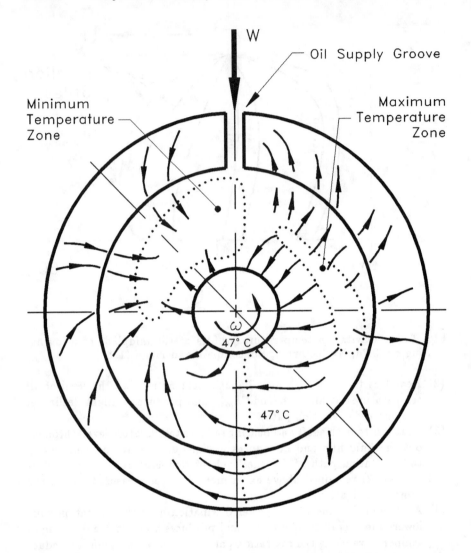

Fig. 5.25 Heat flux in journal bearing system, Boncompain and Frene (1980)

in Fig. 1.13c or in Fig. 5.28. The latter retains the pattern often observed in visualization experiments of the streamlet-like flow, yet because the area of contact between the liquid and the moving surface is considerably reduced, the losses in the cavitation region would likewise be much lower than in the conventional approach. The pattern portrayed in Fig. 5.28 has been verified by visualization experiments as detailed in Heshmat and Pinkus (1986).

TABLE 5.6

Effects of heat transfer and power dissipation
in the cavitation zone

Parameter	(a) Reference	Variation of parameter		
		(b) k	(c) k_S	(d) (H/H_C)*
$[k/\omega \ C^2 \ \rho \ c_p]$	0	0.091	0.091	0.091
$[2 \ k_S/\omega \ C \ R_0 \ \ln \ (R_0/R)]$	0	0	0.057	0.057
(H/H_C)*	1	1	1	0.25
ϵ	0.551	0.565	0.532	0.514
Fluid film temperature, $\bar{T}$				
Entrance	0.609	0.541	0.461	0.389
Maximum	1.291	1.086	0.892	0.813
Exit	1.291	1.058	0.818	0.661
Bearing surface temps, $\bar{T}_S$				
Entrance	0.454	0.541	0.383	0.324
Maximum	1.068	1.085	0.831	0.686
Exit	0.985	1.058	0.686	0.555
Shaft temperature, $\bar{T}_R$	0.94	0.92	0.75	0.65

*H_C - Power loss for conventional streamlet viscous shearing
    Booser and Wilcock, 1987

W = 2,000 lbs, N = 2,000 rpm, α = 0.037 1/°F

$L \times D \times C$ = 4 in. x 3 in. X 0.0025 in., T_0 = 38.6°C

$\bar{T} = \alpha \ (T - T_0)$, k = 0.13 J/m-s-K

E = 0.168, k_S = 0.057 J/m-s-K

R_0 = bearing shell outer radius

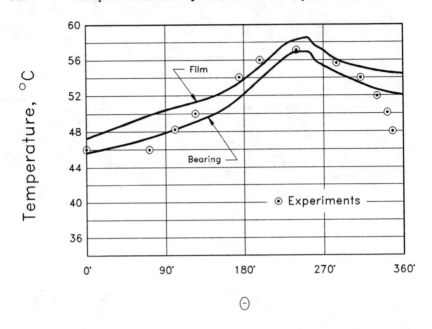

Fig. 5.26 Comparison of theoretical and experimental results, Booser and Wilcock (1987)

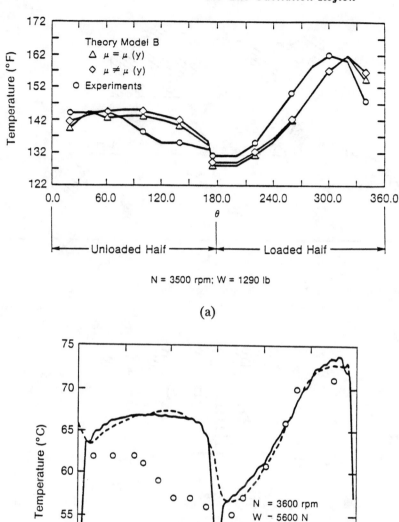

Fig. 5.27 Comparison of theoretical and experimental results in journal bearing, (a.) cavitation model A of Fig. 1.19, (b.) cavitation model B of Fig. 1.19, Barrett and Branagan (1988)

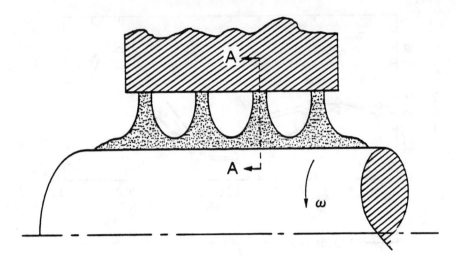

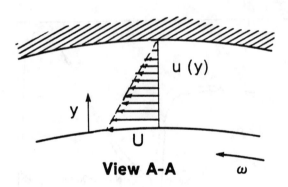

View A-A ω

Fig. 5.28 Alternate mode of cavitation

PART II

THE THERMAL WORKSHOP

The material considered in the second part of the book covers subjects which have been taken up by the 1988 workshop on thermal problems in tribology. However, in addition to the items presented at the Workshop (denoted by TW, 1988), the chapters contain extensive material from the general literature. The topics presented have several features in common. The most notable of these is that their formulation and technical status are not tribological concerns per se—but challenges that belong to the science of fluid mechanics in general. Thus, while the tribologist may have to grapple with the application of the concepts of turbulence, or of two-phase flow, to the specific circumstances of bearing and seal operation—the conceptualization of turbulent flow, or of two-phase fluid dynamics, is beyond this purview. Consequently, the state of knowledge of turbulent fluid films or of liquid-vapor lubricants can be no better than the models handed down by the science of fluid dynamics. Such problematics are present to a higher or lesser degree in most of the subjects treated in the following chapters. In addition to the turbulence and two phase flow areas mentioned above these include non-Newtonian fluids for which no generally valid models have as yet been constructed; the rheology of lubricants under extreme shear rates and high pressure; cavitation phenomena; the morphology of materials under conditions of near seizure; and several other topics discussed in the following chapters.

Chapter 6

TURBULENCE

6.1 TURBULENCE AND THE FLUID FILM

As noted in the opening remarks the treatment of turbulence in THD analysis is perforce subject to several limitations. One is the general status of turbulence as a branch of fluid dynamics which lacks a firm conceptual foundation, as well as consistency in the various empirical approaches employed in the field. In a comprehensive review of the status of turbulence as a scientific discipline Lumley (1983) has this to say: "In 1981-82 a competition was held between various methods for calculating (turbulent) isothermal flows of engineering interest... The conclusion of the judges was that the range of satisfactory performance of the models was rather narrow and that we should probably expect to use for some time a variety of models optimized for particular geometrical situations and parameter ranges. The judges were split on the likelihood of ultimately improving this situation...."

A second limitation in the present context is the necessity to link the relevant turbulence expressions to THD analysis in such a manner that they would not vitiate the basic thermal and hydrodynamic equations of lubrication. The latter constraint is eased by the first in the sense that whatever assumptions are introduced to safeguard the simplicity of the THD equations, these are not likely to violate any rigorous axioms of turbulence which, as was noted, are tentative to start with.

An aspect of turbulence which is perhaps more serious is that turbulence implies the presence of inertia effects. It would thus seem mandatory to consider fluid inertia along with turbulence. This would indeed carry the THD equations beyond the customary postulates of lubrication theory. However, it has been shown by numerous studies, for example in Kahlert, 1948, and Osterle and Saibel, 1955, that even when the inertia terms seem to be of the same order as the viscous terms, they yet do not contribute significantly to the resulting fluid film forces, whereas turbulence is known to have a profound effect on nearly all performance parameters. On the other

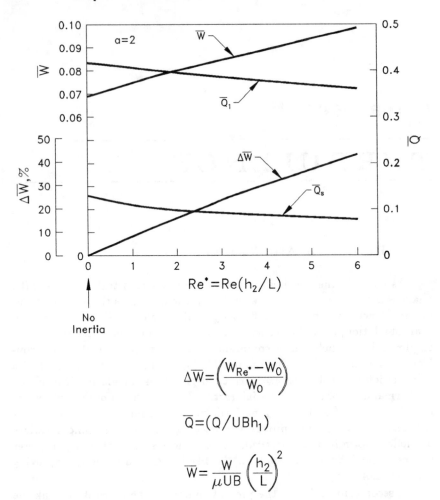

$$\Delta\overline{W}=\left(\frac{W_{Re^*}-W_0}{W_0}\right)$$

$$\overline{Q}=(Q/UBh_1)$$

$$\overline{W}=\frac{W}{\mu UB}\left(\frac{h_2}{L}\right)^2$$

Fig. 6.1 Effect of inertia on slider bearing performance, Launder and Leschziner, 1978

hand the work by Launder and Leschziner, 1978, shows a radical departure from all previous results. Using an integro-differential approach, they deduce that there is a considerable impact of the inertia forces at $Re^* > 1$. The authors ascribe this to the fact that in all analyses previous to their work the frictional terms were a priori decoupled from their dependence on inertia thereby weakening the impact of inertia on the resulting pressures. For both laminar and turbulent conditions the boost in the generated forces due to inertia can be seen in Fig. 6.1 to be as high as 50%. Until this issue is resolved we will here follow past practice in keeping turbulent effects decoupled from the possible effects of pure inertia.

The objective of the various analytical techniques with regard to turbulence is to be able to express the fluctuating components in terms of the mean values. For this purpose nearly all bearing analyses utilized the eddy diffusivity concept dating back to Boussinesq. This related the turbulent stresses of a 2-dimensional flow to the mean velocities via

$$-u_i u_j = \epsilon_m \left[\frac{\partial \overline{u}_i}{\partial x_j} + \frac{\partial \overline{u}_j}{\partial x_i} \right]$$

where ϵ_m is the eddy diffusivity, itself a function of the mean flow. This permits one to treat the fluctuating stresses in a manner analogous to the viscous stresses (with ϵ_m replacing μ) in laminar flow. In this approach the quantities affected by turbulent flow are multiplied by the factor $\left(1 + \frac{\epsilon_m}{\nu}\right)$. The expression for the eddy viscosity term $\left(\frac{\epsilon_m}{\nu}\right)$ is in most cases modeled on Reichhardt's formula constructed for turbulent flow adjacent to a wall, namely

$$\frac{\epsilon_m}{\nu} = k_0 \left[\frac{y_w}{\mu} \sqrt{\frac{|\tau_w|}{\rho}} - \delta_0 \tanh \left(\frac{y_w}{\mu} \frac{1}{\delta_0} \sqrt{\frac{|\tau_w|}{\rho}} \right) \right]$$

in which k_0 and δ_0 are empirical constants.

The added complication in tribological fluid films is that we have here two closely spaced walls whose mutual effect on the postulated turbulent model is not clear. The problem is usually handled by treating each wall independently of the other and providing for continuity of the eddy diffusivity function at the interface. Thus we have

$$\frac{y}{\nu} \cdot \sqrt{\frac{|\tau_w|_0}{\rho}} = \frac{h-y}{\nu} \cdot \sqrt{\frac{|\tau_w|_h}{\rho}}$$

which yields for the common coordinate y

$$y_c = h \left[\frac{\sqrt{|\tau_w|_h}}{\sqrt{|\tau_w|_0} + \sqrt{|\tau_w|_h}} \right]$$

Combined, we then obtain

$$\frac{\epsilon_m}{\nu} = k_0 \left[y - \delta_0 \tanh(\frac{y}{\delta_0}) \right]$$

with

$$y = \begin{cases} (y/\nu)\sqrt{|\tau_w|_h/\rho} & \text{for } y \leq y_c \\ (h-y)/\nu\sqrt{|\tau_w|_0/\rho} & \text{for } y > y_c \end{cases}$$

The most common approach to the treatment of turbulence in hydrodynamic films is that of eddy viscosity postulated above. Consequently, the Reynolds equation for turbulent operation is of the same form as for the laminar case except that all viscosity terms contain appropriate modifiers, commonly denoted by G. Thus

$$\frac{\partial}{\partial x} \left[\frac{h^3}{\mu} G_x \left(\frac{\partial p}{\partial x} \right) \right] + \frac{\partial}{\partial z} \left[\frac{h^3}{\mu} G_z \left(\frac{\partial p}{\partial z} \right) \right] = 6U \left(\frac{\partial h}{\partial x} \right) \qquad (6.1)$$

The dissipation equation contains an additional function G_τ to account for the turbulent shear stress.

For cases in which Couette flow is the dominant velocity component, a linearized analysis developed by Ng and Pan, 1965, provides the values of the G coefficient as functions of the Reynolds number, as given in Fig. 1.4. In general, however, tribological devices are likely to depart from the Couette-dominated flow as, for example, when:

- The circumferential pressure gradients are very large.
- In hydrostatic bearings with high axial pressure gradients
- In seals and hybrid bearings where the velocities of the circumferential and axial cross flows may be of the same order.

Starting with the same eddy viscosity model but modified by the inclusion of the "law of the wall" behavior, Elrod and Ng, 1967, included in their analysis the effects of the axial and circumferential pressure gradients on the flow field. For Couette-dominated flows the analysis verified the correctness of the linear approach, with the exception of very high eccentricity cases ($\epsilon > 0.9$). Thus the linearized coefficients can be used in ordinary hydrodynamic bearings. However, in seals and hydrostatic bearings non-linear effects must be included. Some values of the non-linear G coefficients as functions of the Reynolds number and of the levels of the circumferential and axial pressure gradients are given in Fig. 6.2. In this figure the pressure gradients are given in the following normalized form:

$$\Delta_x = \frac{\rho h^3}{\mu^2} \left(\frac{dp}{dx} \right) \tag{6.2a}$$

$$\Delta_z = \frac{\rho h^3}{\mu^2} \left(\frac{dp}{dz} \right) \tag{6.2b}$$

6.2 TURBULENCE AND THD ANALYSIS

The reason that turbulence is of particular importance in THD analysis is that, as Fig. 6.3 illustrates, turbulent velocities induce extreme shear gradients at the walls. In addition to the increased viscous shear the apparent boost in viscosity reduces flow thus raising the temperature levels in the fluid film. In parallel to the postulate of mean and fluctuating components of velocity and pressure, the turbulent temperatures can likewise be represented by mean and fluctuating components, viz.

$$T = T_m + T'$$

Following Szeri, 1987, the (x, y) energy equation for turbulent flow can then be written as

$$u_m \frac{\partial T_m}{\partial x} + v_m \frac{\partial T_m}{\partial y} = \frac{k}{\rho c_p} \left[\frac{\partial^2 T_m}{\partial x^2} + \frac{\partial^2 T_m}{\partial y^2} \right] - \frac{\partial (v'T')_m}{\partial y} + \frac{\nu}{c_p} \Phi_m \tag{6.3}$$

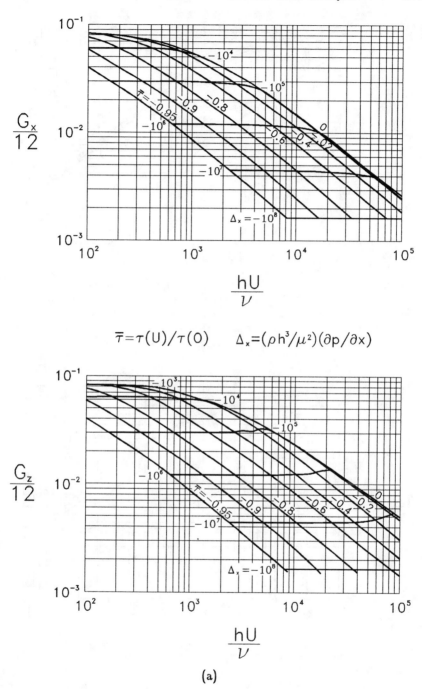

$$\overline{\tau}=\tau(U)/\tau(0) \qquad \Delta_x=(\rho h^3/\mu^2)(\partial p/\partial x)$$

(a)

Fig. 6.2 Non-linear turbulence coefficients, Elrod and Ng, 1967,
(a.) $(\partial p/\partial x) \gg (\partial p/\partial z)$; $\overline{\tau} = \tau(U)/\tau(0)$; $\Delta_x = (\rho h^3/\mu^2)(\partial p/\partial x)$,
(b.) $(\partial p/\partial x) \ll (\partial p/\partial z)$; $A = h(\partial p/\partial z)/|\tau|$; $\Delta_z = (\rho h^3/\mu^2)(\partial p/\partial z)$.

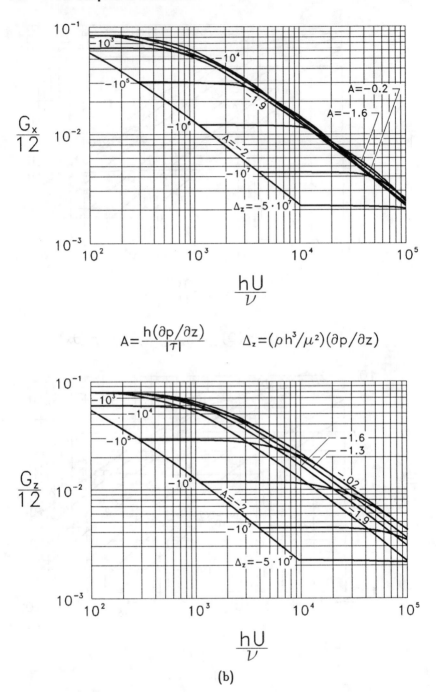

$$A = \frac{h(\partial p/\partial z)}{|\tau|} \qquad \Delta_z = (\rho h^3/\mu^2)(\partial p/\partial z)$$

(b)

Fig. 6.2 Non-linear turbulence coefficients, Elrod and Ng, 1967,
(a.) $(\partial p/\partial x) \gg (\partial p/\partial z)$; $\bar{\tau} = \tau(U)/\tau(0)$; $\Delta_x = (\rho h^3/\mu^2)(\partial p/\partial x)$,
(b.) $(\partial p/\partial x) \ll (\partial p/\partial z)$; $A = h(\partial p/\partial z)/|\tau|$; $\Delta_z = (\rho h^3/\mu^2)(\partial p/\partial z)$.

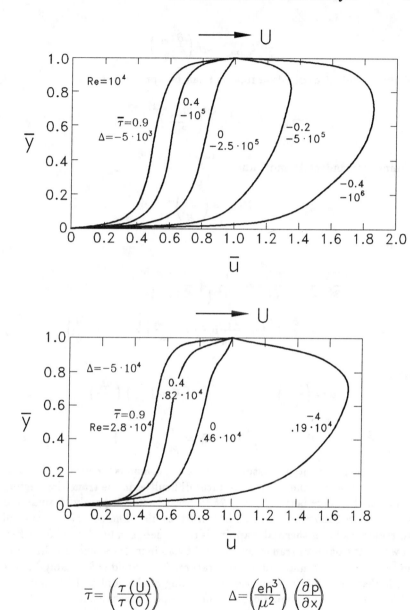

Fig. 6.3 Turbulent velocity profiles in fluid films, $\overline{\tau} = \tau(U)/\tau(0)$; $\Delta = (\rho h^3/\mu^2)(\partial p/\partial x)$, Elrod and Ng, 1967

If in analogy to momentum transport the velocity-temperature correlation is approximated by a thermal eddy diffusivity coefficient ϵ_H, we have

$$(-v'T')_m = \epsilon_H \left(\frac{\partial T_m}{\partial y} \right)$$

and the turbulent dissipation function can be approximated by

$$\Phi_m \simeq \left(1 + \frac{\epsilon_m}{\nu} \right) \left(\frac{\partial u_m}{\partial y} \right)^2 \Bigg|_{y=0}$$

Defining a turbulent Prandtl number

$$Pr_m = \left(\frac{\epsilon_m}{\epsilon_H} \right)$$

we have

$$\bar{u}_m \frac{\partial \overline{T}_m}{\partial \theta} + \bar{v}_m \frac{\partial \overline{T}_m}{\partial \bar{y}} = \frac{1}{Pe^*} \left\{ \left(\frac{C}{R} \right)^2 \frac{\partial^2 \overline{T}_m}{\partial \theta^2} \right.$$

$$\left. + \frac{\partial}{\partial \bar{y}} \left[\left(1 + \frac{P_r}{P_{rm}} \bar{\mu} \frac{\epsilon_m}{\nu} \right) \frac{\partial \overline{T}_m}{\partial \bar{y}} \right] \right\} + E_0 \left(\frac{\partial \bar{u}}{\partial \bar{y}} \right)^2 \qquad (6.4)$$

where

$$\bar{u} = \left(\frac{u_m}{U} \right), \quad \overline{T}_m = \frac{T_m}{T_0}, \quad \bar{v} = \left(\frac{R}{C} \right) \left(\frac{v_m}{U} \right)$$

$$Pe^* = \frac{\rho c_p \omega C^2}{k}, \quad E_0 = \frac{\mu_0 \omega}{\rho c_p T_0} \left(\frac{R}{C} \right)^2$$

The eddy viscosity is postulated to be a continuous function of the local Reynolds number alone. This introduces difficulties in the transition region, that is, in the range between $Re < 750$ where the flow is purely laminar and that of $Re > 1500$ where full turbulence prevails. The complex nature of this region, as far as thermal mapping is concerned, can be gleaned from Fig. 6.4 where the onset of transition induced by an increase in bearing diameter shows a decrease in maximum temperature. This could conceivably be due to a higher rate of heat transfer; to a mixing of the cold and hot portions of the lubricant; or to some other, more complex mechanism.

6.3 THE TRANSITION REGIME

It was pointed out above that the transition region represents a particular problem. The expressions used for this region consist mostly of sort of bridging of the laminar and turbulent regimes. The consequences of this ad hoc treatment of the transition region are aggravated when thermal effects are to be included. One aspect of these difficulties was shown in Fig. 6.4,

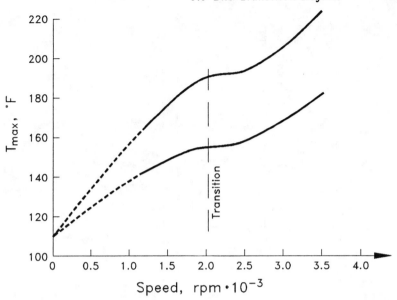

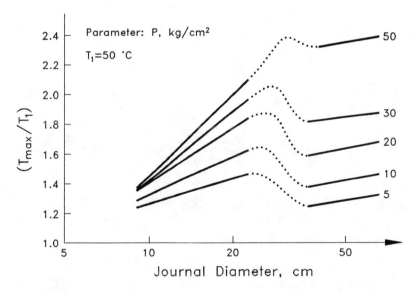

Fig. 6.4 Bearing temperatures in transition region, Szeri, TW 1988

where temperatures in the transition region were seen to fall below those of laminar flow. This inexplicable behavior seems to indicate that some major facet of incipient turbulent flow has been neglected as far as its thermal characteristics are concerned.

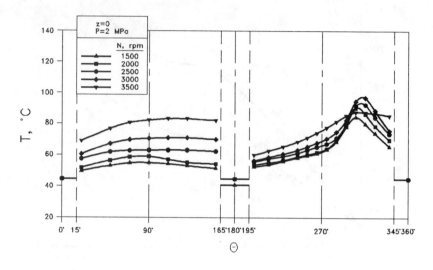

a) Elliptical Bearing

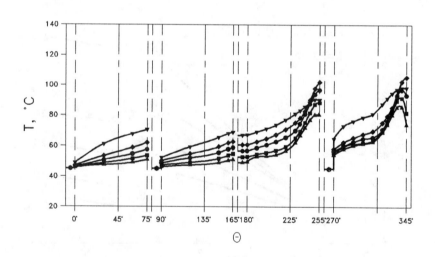

b) 4–Shoe Tilting Pad Bearing

Fig. 6.5 Bearing surface temperatures under various speeds, (a) elliptical bearing, (b) 4-shoe tilting pad bearing, Hopf and Schüler, TW 1988

This subject was discussed at some length by Hopf at the 1988 Thermal Workshop. He first corroborated this anomaly and expanded on it; he then went on to propose a rationale for this behavior supported by experimental data from tests on a large $(D = 50cm)$ turbine bearing. The occurrence of jumps in temperature upon crossing from one flow regime to the other was noted not only during a switch of the bearing's operating mode but even within a given bearing when parts of the film were laminar and others turbulent. The specific features of note, shown in Figs. 6.5, 6.6 and 6.7, are as follows:

- When maintaining constant load, the speed is progressively increased; the bearing surface temperatures at first rise, as expected, but then begin to drop. The latter occurs only in the vicinity of h_{min}.
- When maintaining constant speed, the load is progressively increased; there is at first very little change followed by a sudden rise in bearing surface temperature, again only in the vicinity of h_{min}.
- The above quasi discontinuities in temperature gradients occur only in the vicinity of h_{min} while the shaft temperature and the rest of the bearing surface continue to respond normally, that is the temperatures rise continuously with both speed and load.

This behavior was attributed to the fact that whereas laminar flow maintains an orderly stratification of the transverse temperatures, transition to turbulence causes both intense mixing and an increase in heat transfer rates, thereby lowering the high temperatures at the bearing surface. An increase in load, on the other hand, returns the flow to laminar conditions (by reducing h and thereby the Reynolds number), and stratifies the temperature layers between shaft and bearing, enabling the bearing surface to maintain its high temperatures. The reason that this occurs primarily near h_{min} is that it is in this region that large transverse temperature gradients are formed which would be susceptible to mixing. That such large thermal gradients do, in fact, exist near h_{min} while only mild variations prevail upstream, is shown in Fig. 6.8. These results were obtained from a 3-dimensional analysis of the experimental bearing which included the effects of the local Reynolds number on the momentum and energy equations; accompanying this figure are the local Reynolds numbers prevailing at each angular position. It can be seen that near h_{min}, where the flow is laminar, extremely steep transverse gradients are generated while upstream, where the flow is turbulent, there is only a mild variation in temperature across the film.

6.4 SLIDERS

The previously mentioned results by Launder and Leschziner were obtained from an isothermal analysis. Di Pasquantonio and Sala, 1984, used the same analytical approach but tried to account for the thermal effects accompanying turbulent flow. It is symptomatic of the state of affairs in the field of turbulence that despite the striking impact of inertia shown by the isothermal solution, the latter authors treat the THD case by explicitly

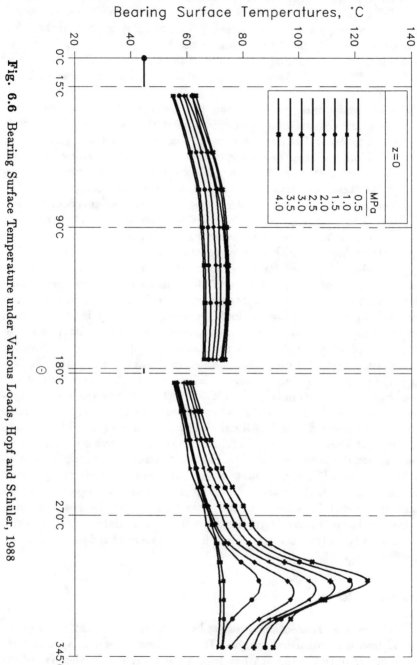

Fig. 6.6 Bearing Surface Temperature under Various Loads, Hopf and Schüler, 1988

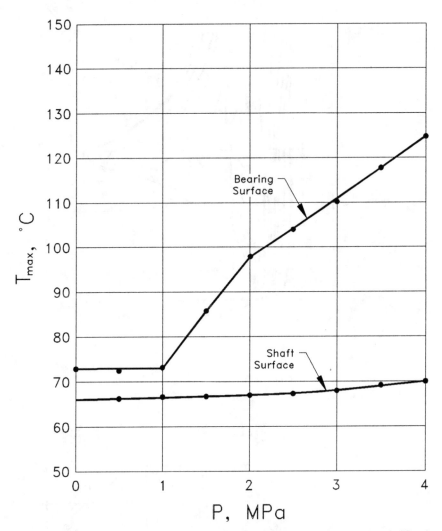

Fig. 6.7 Response of shaft and bearing surfaces to increase in load, Hopf and Schüler, TW 1988

excluding all inertia effects. The case they chose is that of an infinitely long adiabatic slider with temperature variations in both the x and y directions. The continuity, momentum, and energy equations written in terms of the mean values were (dropping the subscript m) as follows:

$$\frac{\partial u}{\partial x} + \frac{\partial v}{\partial y} = 0 \tag{6.5}$$

$$\frac{\partial(\rho u^2)}{\partial x} + \frac{\partial(\rho uv)}{\partial y} = -\frac{\partial p}{\partial x} + \frac{\partial}{\partial y}\left[(\mu + \mu_\tau)\frac{\partial u}{\partial y}\right] \tag{6.6}$$

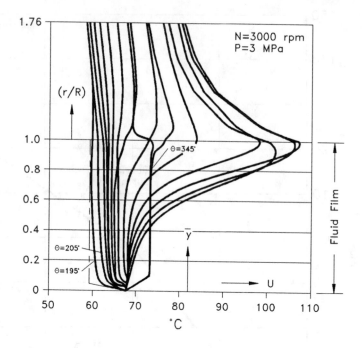

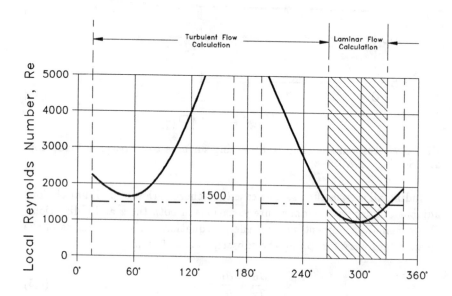

Fig. 6.8 Calculated transverse temperatures for mixed region, Hopf and Schüler, TW 1988

$$\frac{\partial(\rho u T)}{\partial x} + \frac{\partial(\rho u T)}{\partial y} = \frac{\partial}{\partial y}\left[\left(\frac{\mu}{Pr} + \frac{\mu_r}{Pr_r}\right)\frac{\partial T}{\partial y}\right] + \frac{\mu + \mu_r}{c_p}\left(\frac{\partial u}{\partial y}\right)^2 \quad (6.7)$$

The turbulent Prandtl number was assumed constant and equal to 0.9. The turbulent viscosity is expressed by:

$$\mu_r = F_1 \cdot \rho\frac{k^2}{\epsilon} \qquad (6.8)$$

where k and ϵ are the turbulent kinetic energy and the turbulent dissipative rates, respectively, and F_1 is a function of the local Reynolds number of turbulence

$$F_1 = \exp\left[-\frac{3H}{(1 + Re_T/50)^2}\right]$$

$$Re_T = \frac{\rho k^2}{\mu\epsilon}$$

Omitting the convection terms the equations for k and ϵ were taken to be

$$\frac{\partial}{\partial y}\left[\left(\mu + \frac{\mu_r}{\sigma_k}\right)\frac{\partial k}{\partial y}\right] + \mu_r\left(\frac{\partial u}{\partial y}\right)^2$$

$$- \rho\epsilon - 2\mu\left(\frac{\partial\sqrt{k}}{\partial y}\right)^2 = 0 \qquad (6.9a)$$

$$\frac{\partial}{\partial y}\left[\left(\mu + \frac{\mu_r}{\sigma_\epsilon}\right)\frac{\partial\epsilon}{\partial y}\right] + C_1\frac{\epsilon}{k}\mu_r\left(\frac{\partial u}{\partial y}\right)^2$$

$$- C_2\rho\frac{\epsilon^2}{k} + 2\frac{\mu\mu_r}{\rho}\left(\frac{\partial^2 u}{\partial y^2}\right)^2 = 0 \qquad (6.9b)$$

where the coefficients σ_k, σ_ϵ, C_1, C_2 can, according to Launder and Leschziner, be taken to be:

$$\sigma_k = 1 \quad , \quad \sigma_\epsilon = 1.3 \quad , \quad C_1 = 1.44$$

$$C_2 = 1.92[1 - 0.3\exp(-Re_r^2)]$$

The system of equations (6.5)–(6.9), together with the condition of overall mass conservation, provides a closed set of coupled nonlinear partial differential equations in the dependent variables, u, v, p, T, k, ϵ. The boundary conditions are:

$$y = 0: \quad u = U \quad , \quad v = k = \epsilon = \frac{\partial T}{\partial y} = 0$$

$$y = h: \quad u = v = k = \epsilon = \frac{\partial T}{\partial y} = 0$$

$$x = 0: \quad p = 0 \quad , \quad T = T_0$$

$$x = B: \quad p = 0$$

TABLE 6.1

Data for $L/B = \infty$ slider

Parameter	Value
a	2
(h_2 / B)	0.002
μ	$\dfrac{30.45}{T^{1.9}}$ for 26°C < T < 100°C
	4.7×10^{-3} for T > 100°C
	in Pascal-sec
k	$0.145 \dfrac{J}{m\text{-}s\text{-}°C}$
T_0	50°C
ρ	900 kg/cm^3
c_p	$1.67 \times 10^3 \dfrac{J}{kg\text{-}°C}$
F_1	Approximated by
	$F_1 = 0.09 - \dfrac{4.13}{Re_T + 47.46}$
c_2	1 for $Re_T > 1.5$
	$0.61 + 0.26\, Re_T$ for $0.34 < Re_T < 1.5$
	0.7 for $Re_T < 0.34$

The above equations were solved by finite difference methods for the specific geometry given in Table 6.1. A comparison of the relevant velocities given in Fig. 6.9 shows that inclusion of THD effects elevates still further the shear rates at the walls. The pressure profile obtained from the THD solution is compared to anisothermal and laminar profiles in Fig. 6.10. The temperature map, shown in Fig. 6.11, indicates that under turbulent conditions the transverse temperatures stays constant and considerably below the wall temperatures. The entire adjustment to the adiabatic requirement of $dT/dy = 0$ at the walls is accomplished across a very thin thermal layer, no more than 1%–2% distance from the wall.

6.5 JOURNAL BEARINGS

The next in complexity after the slider is that of an infinitely long

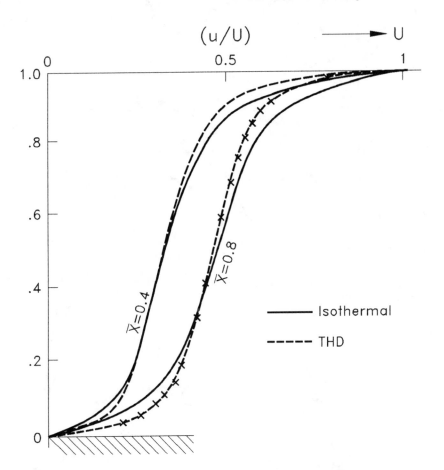

Fig. 6.9 Velocity profiles in infinitely wide slider under turbulent conditions, Di Pasquantonio and Sala, 1984

journal bearing, an analysis due to Safar and Szeri, 1974. As was done earlier our main attention will be given not to the the general issue of turbulent flows, but to the THD problematics inherent to it. One of them is that in parallel to the postulate of mean and fluctuating components of velocity and pressure, similar formulations are adopted for temperature and most notably for viscosity. Thus for the two-dimensional domain θ, y considered here, the viscosity is written in the form of

$$\mu = \mu(\theta, \bar{y}) \left[1 + \frac{\epsilon_m}{\nu}\right] = \mu_1 F_T(\theta, \bar{y})$$

where

$$\epsilon_m = \begin{cases} 0.4[y_w - 10.7\tanh\left(y_w/10.7\right)], & \text{at } \bar{y} = 0, 1 \\ 0.07, & \text{in the core} \end{cases}$$

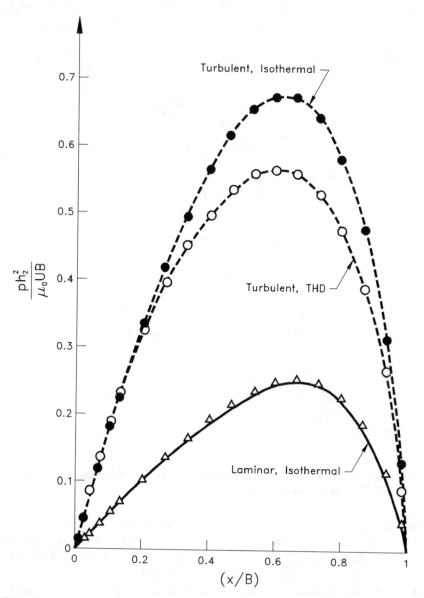

Fig. 6.10 Turbulent THD solution for infinitely wide slider, Di Pasquantonio and Sala, 1984

$$y_w = \frac{h}{\nu}\sqrt{\frac{|\tau|}{\rho}}$$

As a consequence, when the viscosity expression is inserted in the Reynolds equations, its form is no different from that of the laminar case with the variable viscosity a function of θ and y. The only difference here

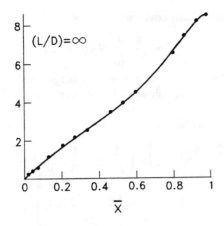

a) Temperatures along x

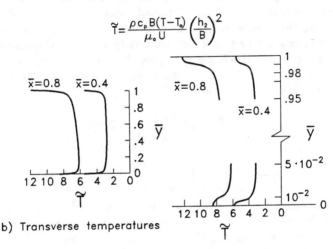

b) Transverse temperatures

c) Temperatures near the walls

Fig. 6.11 Temperature map in adiabatic turbulent slider, Di Pasquantonio and Sala, 1984

(a.) Temperatures along x, $\widetilde{T} = [\rho c_p B(T - T_0)/\mu_0 U](h_2/B)^2$,

(b.) transverse temperatures,

(c.) temperatures near walls.

is that the function F_T for turbulent flow will be more complex.

With regard to the energy equation a number of assumptions familiar from previous methods are introduced, viz.

- The shaft is isothermal at a temperature determined by the average of the bearing surface temperatures, namely

$$T_R = \frac{1}{R\beta} \int_0^\beta T_S(\theta, 0) R d\theta$$

- Fluid heat conduction in the θ direction is ignored.
- In the bearing pad heat flow in the circumferential direction is neglected.

The energy equation then reads

$$\overline{h}\overline{u}_m \left(\frac{\partial \widehat{T}_m}{\partial \theta} \right) = \frac{1}{Re^*} \left[\overline{\mu} \left(\frac{1}{Pr} + \frac{1}{0.769} \frac{\epsilon_m}{\nu} \right) \left(\frac{\partial \widehat{T}_m}{\partial \overline{y}} \right) \right.$$

$$\left. + \overline{\mu} \left(1 + \frac{\epsilon_m}{\nu} \right) \left(\frac{\partial \overline{u}_m}{\partial \overline{y}} \right)^2 \right]$$

where the subscript m refers to mean values. The constant 0.769 in the equation represents the turbulent Prandtl number. The analysis aims at incorporating the effects of heat transfer to the bearing pad by satisfying both temperature and heat flux continuity at the fluid-solid interface. However, a most radical scheme is employed to achieve this. The required differential equation and its concomitant boundary conditions are dispensed with and replaced by the following relationship:

$$\widehat{T} + \gamma(\theta) \frac{\partial \widehat{T}}{\partial \overline{y}} \bigg|_{\overline{y}=0} = 0$$

where

$$\gamma(\theta) = -\frac{1}{\overline{h}} \left(\frac{R}{C} \right) \left[\frac{1}{Nu} + \left(\frac{k}{k_s} \right) \ln \left(1 + \frac{t}{R} \right) \right]$$

As seen the entire heat transfer process is contained within the Nusselt number; a $Nu = 0$ denotes no heat flow to the bearing; with rising Nu the pad is made more conductive, all in the radial direction only.

The above expressions were applied to a full bearing and to a 100° partial journal bearing. A comparison is made between turbulent operation at $Re = 5,000$ with laminar conditions; and the adiabatic case, $Nu = 0$, is compared with that of $Nu = 1,000$. Figs. 6.12, 6.13, and 6.14 show the main results. The reduction of the temperature levels with a rise in Nusselt number is, of course, expected. However, the decrease in the dimensionless temperature $\widehat{T}$ with turbulence should not be construed as a drop in temperature. From the definition of the dimensionless $\widehat{T}$ we have for the actual temperature rise

$$(T - T_1) = \frac{\nu^2}{c_p C^2} \left(\frac{R}{C} \right) \cdot Re \cdot \widehat{T} \qquad (6.10)$$

Thus, the Reynolds number is a multiplier of $\widehat{T}$ in calculating the temperature rise. The actual values of $\widehat{T}$ can be obtained only for specific cases, as a rise in Re will significantly affect the right hand side of Equ. (6.10).

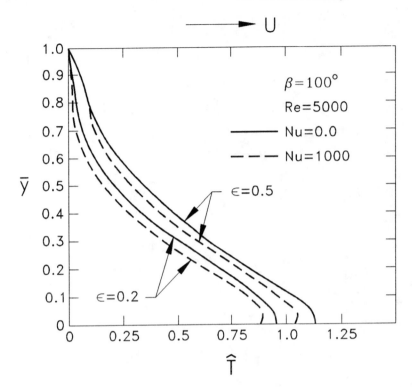

Fig. 6.12 Transverse temperatures in turbulent journal bearing film, Safar and Szeri, 1974

6.6 THRUST BEARINGS

The bearing geometry analysed by Huebner, 1974, is that of a sectorial pad with a taper in the circumferential direction only. The runner is assumed to be isothermal at a temperature equal to that of the inlet temperature. The R_1 and R_2 edges of the bearing are presumed to have a zero temperature gradient so that radially there is only convective heat flow. The energy equation used has the form of

$$\rho c_p \left[w \frac{\partial T}{\partial r} + \frac{u}{r} \frac{\partial T}{\partial \theta} + v \frac{\partial T}{\partial y} \right] = \frac{\partial}{\partial z} \left[k_0 \left(1 + \delta_0 Pr \frac{\epsilon}{\nu} \right) \frac{\partial T}{\partial y} \right]$$

$$+ \frac{\partial}{\partial r} \left[k_0 \left(1 + \delta_0 Pr \frac{\epsilon}{\nu} \right) \frac{\partial T}{\partial r} \right]$$

$$+ \mu \left(1 + \frac{\epsilon}{\nu} \right) \left[\left(\frac{\partial w}{\partial y} \right)^2 + \left(\frac{\partial u}{\partial y} \right)^2 \right]$$

and the Laplace equation for the bearing pad is

$$\frac{1}{r} \frac{\partial}{\partial r} \left[r \frac{\partial T_s}{\partial r} \right] + \frac{1}{r^2} \frac{\partial^2 T_s}{\partial r^2} + \frac{\partial^2 T_s}{\partial y_s^2} = 0$$

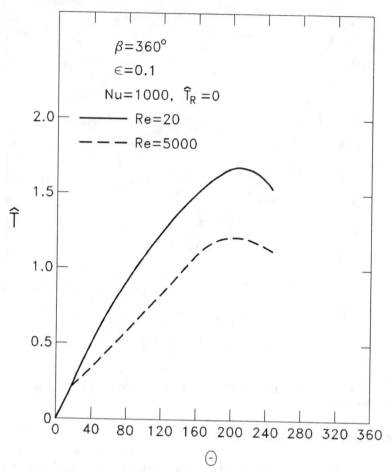

Fig. 6.13 Temperature distribution under laminar and turbulent conditions in full journal bearing, Safar and Szeri, 1974

The boundary conditions for the fluid film are taken to be

$$T_R = T_1$$

$$\left.\frac{\partial T}{\partial r}\right|_{R_1, R_2} = 0$$

$$k\left.\frac{\partial T}{\partial y}\right|_{y=h} = k_s\left.\frac{\partial T_s}{\partial y_s}\right|_{y_s=0}$$

On the solid

$$T_{\theta=0} = T_{\theta=\beta}$$

$$-k_s\left.\frac{\partial T_s}{\partial r}\right|_{R_2} = h_{sa}[T_s(R_2) - T_a]$$

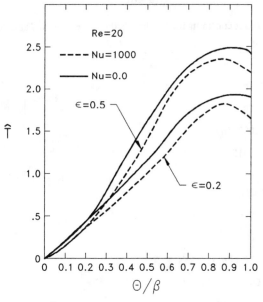

a) Laminar Conditions

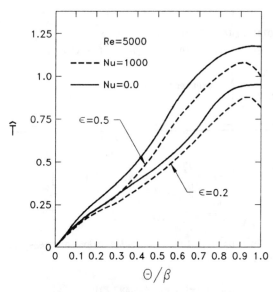

b) Turbulent Conditions

Fig. 6.14 Temperature distribution under laminar and turbulent conditions in 100° journal bearing, (a.) laminar conditions, (b.) turbulent conditions, Safar and Szeri, 1974

TABLE 6.2

Input constants for Huebner's thrust bearing of Table 6.3

$\beta = 30°$	
$R_1 = 1$ in.	
$R_2 = 2$ in.	$\dfrac{\rho R_1 \omega h_2}{\mu_1}\left(\dfrac{h_2}{R_1}\right) = 9.223$
$h_1 = 0.003$ in.	
$h_2 = 0.0015$ in.	
$t = 1$ in.	$Re_1 = \dfrac{\rho R_1 \omega h_2}{\mu_1} = 6153$
$T_1 = 100°F$	
$T_a = 100°F$	
$R_1\omega = 700$ ft/sec	$Pr = 13.91$
$\nu_1 = 1.32$ centistokes @ $100°F$	
$C = 5.602,\ D = 14.80$	$Ec = 0.4075$
$\rho = 42$ lbm/ft^3	
$c_p = 0.48$ Btu/lbm-$°F$	$Nu = 0.02780$
$h_{sa} = 10$ Btu/hr-ft^2-$°F$	
$k = 0.0742$ Btu/hr-ft-$°F$	
$k_s = 30$ Btu/hr-ft-$°F$ (0.5 percent carbon steel)	
$\left.\begin{array}{l} k_0 = 0.3 \\[4pt] \delta_0 = 10.7 \end{array}\right\}$ turbulence constants	

$$-k_s\left.\frac{\partial T_s}{\partial y_s}\right|_{y_s=t} = h_{sa}[T_s(t) - T_a]$$

$$T_s(R_1) = T_1$$

For the $\mu - T$ relationship a practical but somewhat messy relationship has been used, namely

$$\nu(T) = -0.6 + 10^{[10(D-C\log_{10}(459.69+T))]}$$
$$C = \log_{10}[\log_{10}(\nu_1 + 0.6)/\log_{10}(\nu_2 + 0.6)]/\log_{10}(669.7/559.7)$$
$$D = \log_{10}[\log_{10}(\nu_2 + 0.6) + c\log_{10}(669.7)$$
$$\nu_1 = \text{viscosity at} 100°F \text{ (centistokes)}$$
$$\nu_2 = \text{viscosity at} 210°F \text{ (centistokes)}$$

TABLE 6.3
Comparison with laminar and isothermal solutions for thrust bearing

		Laminar flow		Turbulent flow	
		ISO	THD	ISO	THD
$\dfrac{W}{\mu_1 \omega R_1^2}\left(\dfrac{h_2}{R_1}\right)^2 \longrightarrow$		0.1030	0.09864	0.4765	0.4926
$\dfrac{F_T}{\mu_1 \omega R_1^2}\left(\dfrac{h_2}{R_1}\right) \longrightarrow$		1.154	1.127	12.246	11.49
	(In)	1.224	1.229	1.220	1.268
$\dfrac{Q}{\omega R_1^2 h_2}$	(Out)	0.9046	0.9161	0.8690	0.9213
	(Side)	0.3191	0.3125	0.3511	0.3463

Huebner, 1974

These equations were solved for the specific bearing given in Table 6.2. The range of Reynolds numbers spanned by the bearing was from a $Re_{min} = 6150$ to a $Re_{max} = 24,610$, reaching thus well into the turbulent regime. The laminar cases were obtained by simply setting $\epsilon_m = 0$. Whereas in laminar flow an isothermal solution overestimated load capacity and power loss vis-a-vis the THD solution, the reverse is true in turbulent flow, as shown in Table 6.3. This is due to the fact that, unlike laminar flow where a drop in viscosity leads directly to a loss of load capacity, in turbulent flow this is overtaken by the simultaneous increase in the Reynolds number and the level of turbulence. In general, since a drop in viscosity has two opposite effects on performance, the THD analysis yielded results not far from the isothermal case.

Fig. 6.15 shows the expected increase in the rate of viscous shear at the walls for turbulent conditions. Quite unexpected is the pronounced increase in the transverse velocity profiles plotted in Fig. 6.16. All of this, however, pales in comparison with the powerful effects of turbulence on the pressures and temperatures given in Figs. 6.17, 6.18, and 6.19. This strong impact holds for both the isothermal and THD solutions. According to these results, the load capacity of the turbulent bearing at the Reynolds numbers used in the present example would be nearly an order of magnitude higher than laminar operation. Likewise the ΔT's would be nearly 5–10 times higher—and this despite the substantial drop in viscosity. To what extent this is real and how much of it is due to the limitations of turbulence theory, remains to be determined.

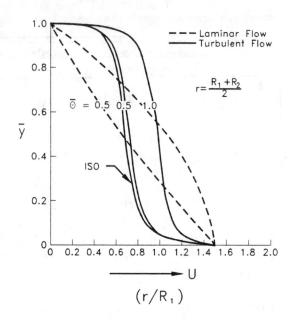

Fig. 6.15 Laminar and turbulent THD velocity profiles, Huebner, 1974

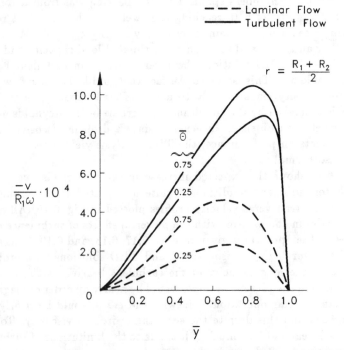

Fig. 6.16 Transverse velocities under THD conditions, Huebner, 1974

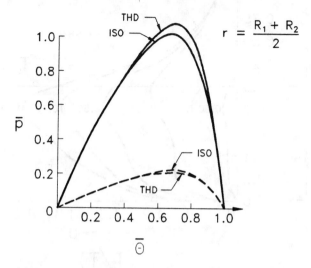

a) Circumferential Pressure Distribution

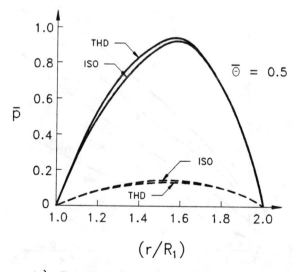

b) Radial Pressure Distribution

Fig. 6.17 Laminar and turbulent pressure profiles under THD conditions, Huebner, 1974

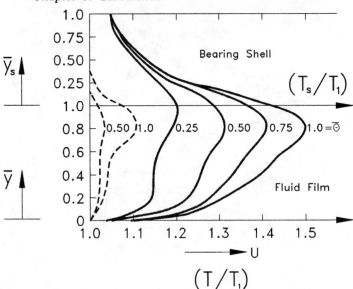

Fig. 6.18 Transverse temperature profiles in thrust bearing under THD conditions, Huebner, 1974

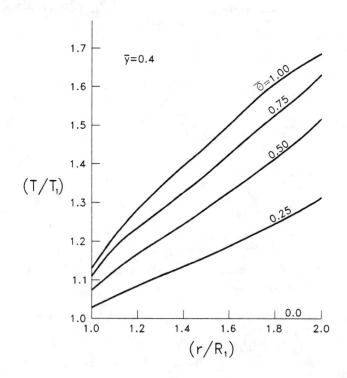

Fig. 6.19 Radial temperature distribution in thrust bearing, Huebner, 1974

Chapter 7

ELASTIC AND THERMAL DISTORTIONS

The elastohydrodynamic element in tribology assumes enhanced significance when the thermal aspect is added to the analysis. This is in the first place due to the presence of thermal distortion induced by the prevailing temperature gradients. These thermal stresses sometimes overshadow the deflections due to mechanical forces. EHD considerations are usually required in processes with a high level of complexity, such as in traction studies, piston rings, rolling element bearings, gears, and most prominently, in the technology of metalworking. The processes of rolling, shaping, drawing, extruding, etc. involve extreme tribological conditions, including often changes of state of the materials. Without its thermal aspects it would be difficult to quantify or even qualitatively represent the ongoing process and its likely outcome.

Another element, new in the present context, is that EHD processes imply very high pressures. Consequently viscosity has to be taken as a function not only of temperature, as has been the practice thus far, but likewise of pressure, as shown in Fig. 7.1. In certain situations as in gears, or Hertzian contacts, the effects of pressure may be more telling than that of temperature, acting, of course, in a direction opposite to the thermal effect. This prevents the lubricant from either leveling off into a nearly iso-viscous fluid at the high temperatures or from becoming a quasi-solid due to the high pressures, thus tending to maintain the strongly variable nature of viscosity in the fluid film. The high pressures would also impact the density of the lubricant, which may become variable along with the viscosity. Furthermore, when the high rates of shear prevailing in these extremely thin films are added to the effects of high pressure and temperature these are likely to convert the lubricant into a complex rheological fluid for which the postulate of the Reynolds equation $\tau = \mu(du/dy)$ may no longer hold.

In general, in an EHD contact, three kinds of motion contribute to dissipation, and thus also to thermal effects. One is the rolling velocity which is a squeeze film type of heating; a relative translational velocity, due to the different radii of the mating surfaces, referred to as sliding or slip; and there could also be spin.

261

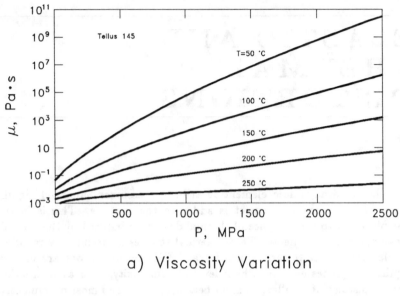

a) Viscosity Variation

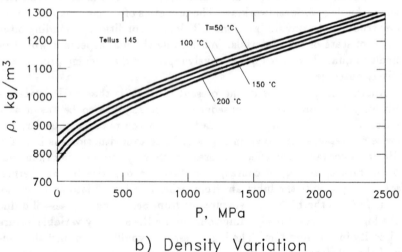

b) Density Variation

Fig. 7.1 Example of viscosity and density variations with temperature and pressure.

Here, as in the case of turbulence, there are inherent constraints on the ability of the tribologist to resolve many of these complex problems. Other branches of science are intimately involved such as eutectics, changes of state, and possibly two-phase flow. In all of these fields our knowledge is fragmentary. The same holds true of the non-Newtonian or viscoelastic models required to properly characterize the lubricant under the high stresses encountered in EHD problems. From a purely tribological standpoint the chief handicap here is that one is dealing with fluid films which are one or two orders of magnitude smaller than those encountered in conventional cases. The film thickness in traction drives, the contacts in rolling element bearings, gears, and so on, are extremely small, and their measurement is important enough and complex enough to be considered in a separate chapter, later on. Oftentimes, it is even difficult to ascertain whether we are dealing with extremely thin but still legitimate fluid films, or with boundary lubrication—a process which would take us outside the hydrodynamic field altogether.

From a mathematical standpoint, even were the conceptual hurdles mastered, there remains the inherent complexity of having to deal with one more differential equation, and, with it, a syndrome of additional boundary conditions. In general, the Reynolds, energy, and elasticity equations must be solved simultaneously, with the geometry of the bearing, due to elastic and thermal deflections, no longer known a priori. In parallel to thermal analysis where in addition to the energy equation we need expressions for heat transfer and equations of state—so, too, with the elasticity equation we need supplementary equations in order to fully define the problem.

7.1 ELLIPTICAL CONTACTS

In many investigations of contacts formed by rolling element bearings, gears, traction drives, etc., the interface is represented as a line contact. This is equivalent to analyzing an infinitely long configuration. However, an elliptical contact represents the more realistic geometry formed by such interfaces. The elasticity equation underlying EHD analysis must then be a function of both planar coordinates. Thus the deflection due to elastic stresses is

$$\delta(x,z) = \frac{2}{\pi E'} \int \int_r \frac{p(\xi,\lambda)}{\sqrt{(x-\xi)^2 + (z-\lambda)^2}} d\xi d\lambda \qquad (7.1)$$

where

$$E' = \frac{2}{[(1-\nu_1^2)/E_1 + (1-\nu_2^2)/E_2]}$$

This equation is utilized in all of the approaches discussed below and, as with the Reynolds equation, it will not be elaborated upon here, leaving all discussion aimed specifically at the thermo-hydrodynamic implications of the problem. The several analytical approaches taken up below will progress from the simplest case to the relatively more complete treatment of the subject.

7.1.1 Quasi-Hertzian Pressure Field

This initial treatment is characterized by the postulate that the contact area is dominated essentially by a Hertzian pressure distribution, modified at the inlet and outlet by hydrodynamic and thermal considerations. As such this approach is valid for heavily loaded contacts where the elastic stresses dominate the field. This work, due to Brüggemann and Kollmann, 1981, introduces several additional constraints. Referring to Fig. 7.2 these are

- The two mating surfaces are isotropic and of the same material.
- The velocities u and w are constant and are known. Also regardless of the prevailing deflections the cylinder radii remain constant.
- Even though the fluid viscosity is a function of pressure and temperature and exhibits viscoelastic effects, the lubricant is taken to be a Newtonian fluid.

With thermal effects included there are five independent variables to solve for, namely, u, w, p, T and δ, the latter being the change in fluid film geometry due to the exerted pressures. To solve for these variables we have the momentum equations in x and z; the continuity equation; the elasticity equation given above, Equ. (7.1); and finally the energy equation.

It can be shown by an order of magnitude analysis that in the thin films prevailing under EHD conditions the convective and conductive heat fluxes in the x and z directions are small compared to the transverse heat flow. Consequently the energy equation assumes the form

$$k\frac{\partial^2 T}{\partial y^2} = -\mu \left[\left(\frac{\partial u}{\partial y}\right)^2 + \left(\frac{\partial w}{\partial y}\right)^2 \right] \tag{7.2}$$

The lubricant used is that given in Fig. 7.1 and its behavior can be represented by

$$\mu = 10^{-3} \cdot \exp[0.0167 \cdot p \cdot \ln(10^3 \mu_T) + \sqrt{\ln(10^3 \mu_T)^3}\,]^{2/3}$$

where μ_T is a function of temperature only. If the rate of pressure application is very high, as it is likely to be in EHD contacts, the lubricant reaction is retarded. If t_1 is the time in which the pressure builds up, the viscosity can be calculated from the following equations:

$$E_i(-s_1) - E_i(-s) + 0.05 \cdot [e^{-s} - e^{-s_1}] = \frac{t_1}{\tau_2} \tag{7.3}$$

where $E_i(s)$ is the exponential integral

$$s_1 = \ln(\mu_2/\mu_1), \quad s = \ln(\mu_2/\mu)$$

and

$$\tau_2 = \frac{50 \cdot \mu_2}{(2400 + 9 \cdot p) \cdot 10^6}$$

The boundary conditions at the two surfaces are as follows. At the interface $y = 0$, moving with velocity U_1 the temperature is

$$T_{(x,0)} = \frac{1}{\sqrt{k_1 \cdot \rho_1 \cdot c_1 \cdot U_1 \cdot \pi}} \cdot \int_{x_1}^{x} k \cdot \frac{\partial T}{\partial y}\bigg|_{x',0} \cdot \frac{dx'}{\sqrt{x - x'}} + T_1 \tag{7.4a}$$

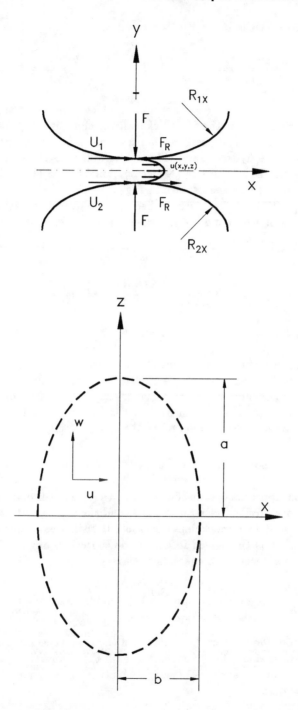

Fig. 7.2 The fluid film in direction of motion for a ball on a roller

At $y = h$ moving with a velocity U_2

$$T_{(x,h)} = \frac{1}{\sqrt{k_2 \cdot \rho_2 \cdot c_2 \cdot U_2 \cdot \pi}} \cdot \int_{x_1}^{x} k \cdot \frac{\partial T}{\partial y}\Big|_{x',h} \cdot \frac{dx'}{\sqrt{x - x'}} + T_1 \qquad (7.4b)$$

As said, the solution proceeded from the postulate of an essentially Hertzian pressure distribution and the film thickness, therefore, is constant over most of the contact area. The pressures depart from a Hertzian profile at the inlet x_1 where the hydrodynamic film starts and near the exit where the characteristic EHD pressure spikes occur, along with a sharp decrease in film thickness, and these are shown in Fig. 7.3.

Equations (7.1) – (7.4) were applied to an elliptical contact formed by the interaction of a ball and a roller having the following characteristics:

$$D_1 = 40\,\text{mm} \quad ; \quad p_{max} = 2,000\,\text{MPa}$$

$$D_2 = 96\,\text{mm} \quad ; \quad \frac{U_1 + U_2}{2} = 4\,\text{m/s}$$

$$s = \frac{U_1 - U_2}{2} = 0.2 \quad ; \quad T_1 = 50°C \quad ; \quad \mu_1 = 58 \cdot 10^{-3}\,\text{Pa} \cdot \text{s}$$

In general the results indicate that the temperatures in the contact zone follow more or less the pressure distribution. These can be seen from Fig. 7.4, where T_{max} occurs at $z = 0$ and near the spike. The effect of the various parameters on T_{max} is shown in Fig. 7.5. It rises with both load and slip but for a given slip T_{max} is lower when the mean velocity $(U_1 + U_2)/2$ rises.

7.1.2 Thin Non-Hertzian Contacts

The next approach is that of discarding the quasi-Hertzian assumption for the pressure profile, but retaining the condition of a very thin film. The latter permits Blahey and Schneider, 1986, to utilize essentially the same energy equation as that given in the previous section, augmented only by the compression term. The equation then reads

$$k\frac{\partial^2 T}{\partial y^2} = \mu\left[\left(\frac{\partial u}{\partial y}\right)^2 + \left(\frac{\partial w}{\partial y}\right)^2\right] + p\left[\left(\frac{\partial u}{\partial y}\right) + \left(\frac{\partial w}{\partial y}\right)\right] \qquad (7.5)$$

Thus here, too, the primary heat flow is in the transverse direction. In the moving solid, however, heat flow exists in all three directions and, due to the movement of the surfaces, Jaeger's equation for a moving heat source has been used, viz.

$$\frac{\partial^2 T_s}{\partial x_s^2} + \frac{\partial^2 T_s}{\partial z_s^2} + \frac{\partial^2 T_s}{\partial y_s^2} = \frac{1}{\sigma_s}\left(\frac{\partial T_s}{\partial t}\right) \qquad (7.6)$$

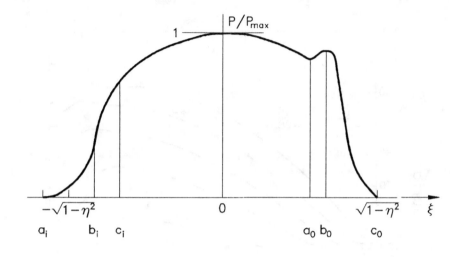

$$\xi = x/b \qquad \eta = z/a$$

$$a_i = -1.07 \cdot \sqrt{1-\eta^2}$$

$$b_i = -0.87 \cdot \sqrt{1-\eta^2}$$

$$c_i = -0.75 \cdot \sqrt{1-\eta^2}$$

$$a_0 = 0.59 \cdot \sqrt{1-\eta^2}$$

$$b_0 = 0.68 \cdot \sqrt{1-\eta^2}$$

$$c_0 = \sqrt{1-\eta^2}$$

Fig. 7.3 Modifications of the Hertzian pressure field, Brüggemann and Kollmann, 1981

The viscosity and density functions to be used in the above equations are

$$\mu = \mu_1[1 + C_1 p]^{C_2} \cdot e^{-\alpha(T-T_1)}$$

$$\rho = \rho_1 \left[1 + \frac{C_3 p}{1 + C_4 p} \right]$$

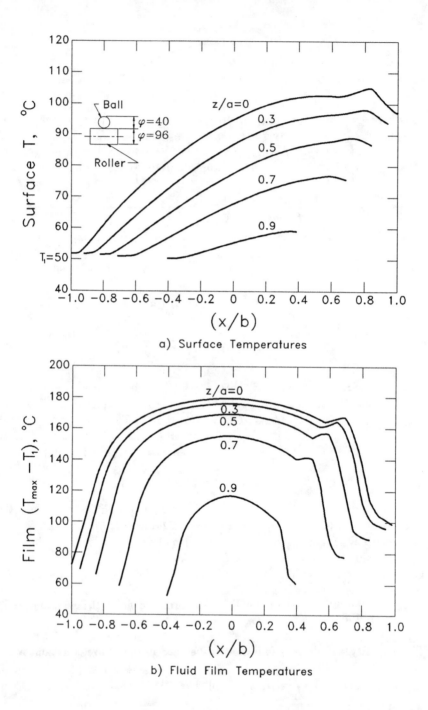

a) Surface Temperatures

b) Fluid Film Temperatures

Fig. 7.4 Temperatures as function of position, Brüggemann and Kollmann, 1981

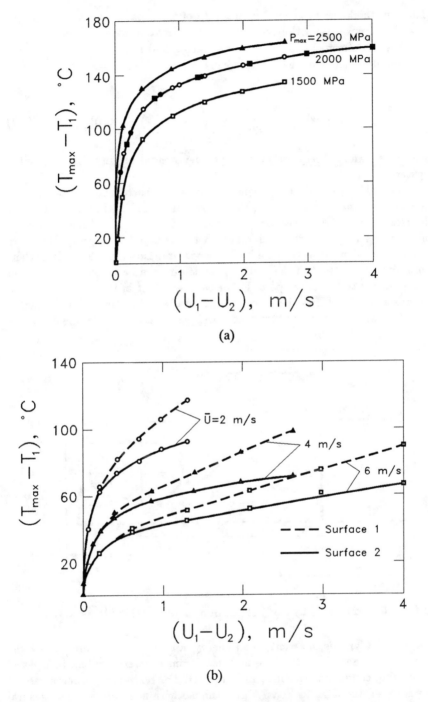

(a)

(b)

Fig. 7.5 Effect of surface velocities on temperatures, (a) lubricant T_{max}, (b) surface T_{max}, Brüggemann and Kollmann, 1981

The film thickness in general is made up of three parts:
- the separation at the origin of the coordinate system
- the undeformed geometry of the mating bodies
- the deflections of the surfaces caused by the elastic stresses

The film then is written as

$$h(x,z) = h_c + \frac{x^2}{2R_x} + \frac{z^2}{2R_z} + \delta(x,z) - \delta(0,0) \qquad (7.7)$$

where here and subsequently subscript c denotes the origin of the coordinate system.

In EHD contacts a problem always arises as to the points of commencement and termination of the hydrodynamic film, that is, of the fully lubricated region. In section 7.1.1 these questions have been settled by using essentially a Hertzian domain. Here the problem is open. Based on some experimental work done by Hamrock and Dowson, 1977, on isothermal films, the extent of the hydrodynamic film can be taken to correspond more or less to the envelope of Fig. 7.6. The extent in the direction of motion is some 5–8 times the elliptical axis in the x direction with the higher numbers applicable to higher speeds; and the axial extent is some 3–4 times the z axis of the ellipse.

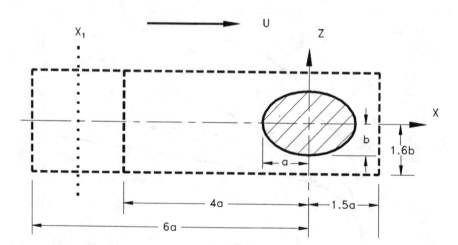

Fig. 7.6 Domain of hydrodynamic film in elliptical contacts

The Reynolds, elasticity, and energy equations, supplemented by the appropriate equations of state and heat transfer, were applied to a set of operating conditions specified in Table 7.1. For reference purposes cases 1 and 2 in Table 7.2 and Fig. 7.7 give the isothermal solutions. The general statement that can be made about the thermal effects is that the temperatures follow closely the behavior of the pressures. Thus the largest ΔT's

TABLE 7.1

Input parameters for solution Section 7.1.2

Parameter	Value		Parameter	Value	
$\bar{W}$	$W/2E'R_x^2$	0.74×10^{-6}	b/a	6	
G	$\alpha E'$	4.52×10^3	c_p	$1.88 \cdot 10^3 \frac{N\text{-}m}{kg\text{-}°C}$	
$(\frac{X\ in}{a})$		-4.06	k	$0.145 \frac{W}{m\text{-}°C}$	
S	$\frac{2(U_2 - U_1)}{U_2 + U_1}$	0.25	α	$0.036 \frac{1}{°C}$	
μ_1		4.11×10^{-2} Pa-s	k_s	$46 \frac{W}{m\text{-}°C}$	
ρ_1		888 kg/m^3	c_{ps}	$460 \frac{W}{kg\text{-}°C}$	
ρ_s		7865 kg/m^3	α_s^*	$1.27 \times 10^{-5} \frac{m^2}{s}$	
τ_1, τ_2		0.7 mm	T_1	$30°C$	
$\frac{1}{\rho}\left(\frac{\partial\rho}{\rho T}\right)$		$-0.7 \times 10^{-3} \frac{1}{°C}$	ν	0.3	
			T_a	$20°C$	
			E	$200 \cdot$ GPa	
			$\bar{U} = \frac{\mu_1(U_1 + U_2)}{2E'R_x}$	0.168	

occur at the central portion, $z = 0$, with the maximum temperature occurring just ahead of the outlet zone, i.e. near the pressure spike. These results are shown in Figs. 7.8 and 7.9. Figs. 7.10 and 7.11 show that for pure rolling the values of THD pressures and h_{min} are very close to the isothermal cases. Even for appreciable levels of sliding the pressures of

TABLE 7.2

Parameters used in sample solution of Section 7.1.2

Case	D i m e n s i o n l e s s　　p a r a m e t e r s					
	$\bar{U} = \dfrac{\mu_1 U_a}{E'\alpha}$	$G = \dfrac{E'}{\bar{\alpha}**}$	$\bar{W} = \dfrac{W}{E'a^2}$	$\bar{k} = \dfrac{b}{a}$	$g = \dfrac{W^{8/3}}{e \cdot \dfrac{-2}{U_a}}$	$g = \dfrac{GW^3}{\nu \cdot \dfrac{-2}{U_a}}$
1*,3	5.049E-11	4.522E+03	7.369E-07	6.0	1.738E+04	7.100E+05
4	1.683E-11	4.522E+03	7.369E-07	6.0	1.564E+05	6.390E+05
2*,5	8.414E-12	4.522E+03	7.369E-07	6.0	6.258E+05	2.556E+07

* Isothermal solution
**$\bar{\alpha} = C_1 [C_2 - 1] = 9.31 \times 10^{-10}$ [21.7]

the isothermal and THD solutions are not far apart, except near the spike where thermal effects reduce the height of the spike. h_{min} does decrease as a result of THD effects, dropping some 25% at very high levels of sliding.

7.1.3 EHD Films with Full Energy Equation

A more elaborate analysis on elliptical contacts was conducted by Dong and Shi-zhu, 1984, for the geometry shown in Fig. 7.12; the physical data for the system are those given in Table 7.1. The only term neglected in the fluid film is that of conduction in the x direction, yielding the expression

$$c_p\rho\left(u\frac{\partial T}{\partial x} + v\frac{\partial T}{\partial y} + w\frac{\partial T}{\partial z}\right) = k\left(\frac{\partial^2 T}{\partial y^2} + \frac{\partial^2 T}{\partial z^2}\right)$$
$$+ \mu\left[\left(\frac{\partial u}{\partial y}\right)^2 + \left(\frac{\partial w}{\partial y}\right)^2\right] - \frac{T}{\rho}\cdot\frac{\partial\rho}{\partial T}\left(u\frac{\partial p}{\partial x} + w\frac{\partial p}{\partial z}\right) \qquad (7.8)$$

For the moving surfaces the Jaeger formulation is retained for the solids, except for conduction in the x direction, or

$$\frac{\partial^2 T_i}{\partial y^2} + \frac{\partial^2 T_i}{\partial z^2} = \frac{\rho_i c_{pi} U_i}{k_i}\cdot\frac{\partial T_i}{\partial x} \qquad (i = 1, 2)$$

For viscosity and density, expressions close to the ones used in the previous analysis were employed, viz.

$$\mu = \mu_1 e^{-[\alpha(T-T_1)-C_1 p]}$$

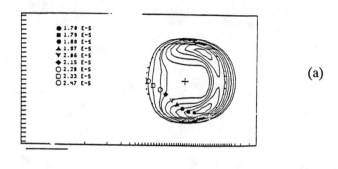

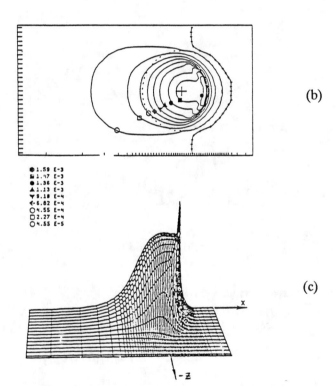

Fig. 7.7 Isothermal solution for elliptical contact: Case 2 of Table 7.2, (a) film thickness, (h/R_x), (b) pressure contours, (p/E'), (c.) pressure profiles, Blahey and Schneider, 1987

$$\rho = \rho_1 \left[1 + \frac{C_1 p}{1 + C_2 p} + C_3(T - T_1) \right]$$

The boundary conditions used were as follows:

• At x_1, $T = T_1$ for the fluid film and both solids

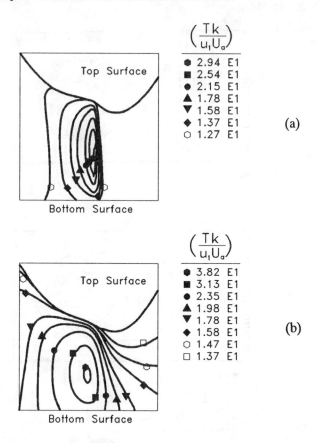

Fig. 7.8 Temperature profiles in fluid film: Case 3 of Table 7.2, (a) $[(U_1 - U_2)/U_1] = 0.5$, (b) $[(U_1 - U_2)/U_1] = 2$, Blahey and Schneider, 1987

- $\partial T/\partial z = 0$ at $z = \pm L/2$
- $k(\partial T/\partial y) = k_s(\partial T_s/\partial y_s)$ for both $y = 0$ and $y = h$
- On surfaces $AA'B'B$ and $AA'B'B$

$$h_{sa}(T_s - T_a) = k_s(\partial T_s/\partial y_s)$$

- All other surfaces on the two mating bodies are assumed to be at $T = T_1 = constant$.

The elliptical contact was run for a series of variable mean velocities and slip, as specified in Table 7.3. The results can be summarized as follows:

1. Film thickness and pressures are only sightly reduced by THD effects— see Fig. 7.13 as well as Table 7.4.

2. The temperatures follow closely the behavior of the EHD pressures as shown in Figs 7.14, 7.15, and 7.16 as well as in Table 7.5. Highest temperatures occur at the center of film. This result conforms with the conclusions of the analyses given in the previous sections.

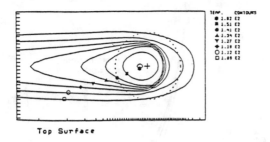

Top Surface

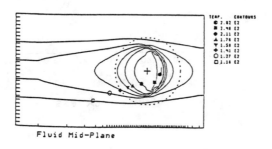

Fluid Mid-Plane

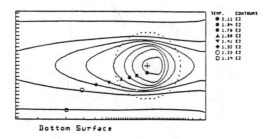

Bottom Surface

Fig. 7.9 Temperature contours in elliptical contacts: Case 4 of Table 7.2, $s = 2$. Numbers refer to $(Tk/\mu_1 U_a^2)$, Blahey and Schneider, 1987

3. Large sliding velocities produce lower ΔT's on the surfaces but a steep rise in temperatures at the mid-plane of the fluid film.

4. For constant slip there is a strong effect of the mean surface velocity $\overline{U}$, a result opposite to that of obtained in section 7.1.1. This is probably due to the fact that in the Brüggemann-Kollmann solution the pressures are primarily Hertzian and thus almost independent of sliding velocity. The temperature behavior merely follows that linkage.

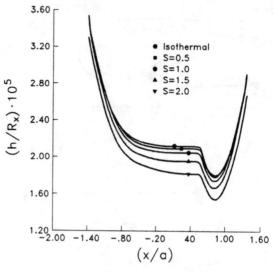

a) Film Thickness

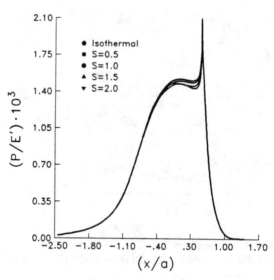

b) Pressure Profiles

Fig. 7.10 Values of p and h_{min} for various sliding ratios: Case 5 of Table 7.2, Blahey and Schneider, 1987

5. Despite the considerable difference in format of the energy equations the present solutions are not much different from those obtained in the

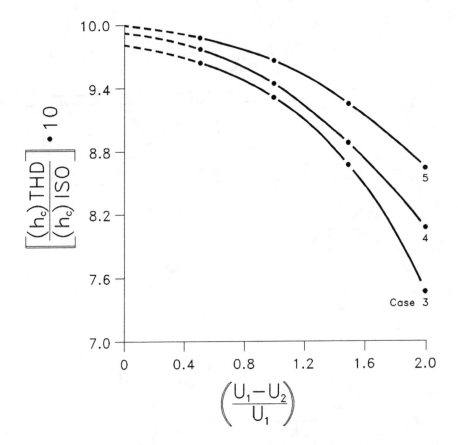

Fig. 7.11 Effect of sliding on film thickness, Blahey and Schneider, 1987

preceding section. Thus even when using different values for slip ($s = 0.5$ in the present versus $s = 0.25$ in the Blahey-Schneider solution) the two results are very close; a 1.3% reduction in h_{min} in going from isothermal to an THD solution in the analysis of section 7.1.2 whereas here the reduction amounted to 2.1%.

Some overall performance characteristics of a THD elliptical contact are given in Fig. 7.17.

7.1.4 General Elliptical Contacts

Additional data of the temperature distribution in an elliptical contact presented by Conry, TW 1988, is shown in Fig. 7.18. It confirms the fact that the temperatures follow the pressure distribution and that the maximum temperature occurs at the mid-plane of the fluid film.

TABLE 7.3

Runs for Section 7.1.3
(for input data see Table 7.1)

Parameters	C_1	C_2	C_3	C_4	C_5	C_6	C_7	C_8
					$\longleftarrow$ Constant $\longrightarrow$			
$\bar{U} \times 10^{10}$	0.084	0.168	0.337	0.540	0.168	0.168	0.168	0.168
	$\longleftarrow$ Constant $\longrightarrow$							
S	0.25	0.25	0.25	0.25	0.05	0.15	0.4	0.6

$$\bar{U} = \frac{U_1 + U_2}{2} \qquad S = \frac{U_1 - U_2}{U_1}$$

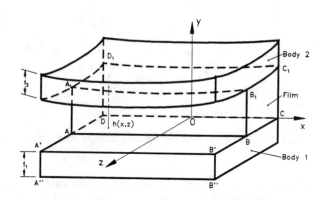

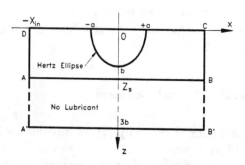

Fig. 7.12 Configuration for analysis of section 7.1.3

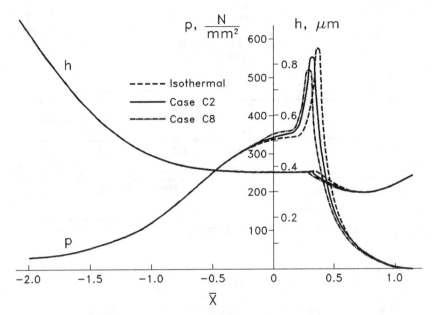

Fig. 7.13 Comparision between isothermal and THD cases in an EHD slider, Dong and Shi-zhu, 1984

7.2 INFINITELY LONG ROLLERS

In a contact between two infinitely long rollers with radii R_1 and R_2, there is a geometry curvature in the direction of motion only. The film thickness in such a domain, including the elastic deflection, can be written as

$$h = h_2 + \frac{x^2 - x_2^2}{2R'} - \frac{2}{E'} \int_{-\infty}^{x_2} \ln \left| \frac{x' - x}{x' - x_2} \right| p(x') dx' \qquad (7.9)$$

where

$$\frac{1}{R'} = \frac{1}{R_1} + \frac{1}{R_2}$$

The analysis, due to Cheng, 1965, locates the coordinate delineating the entry region at $-0.875 < x_A < -0.75$. The corresponding energy equation in the x, y coordinates is portrayed by

$$\rho c_p \left[u \frac{\partial T}{\partial x} + v \frac{\partial T}{\partial y} \right] - k \frac{\partial^2 T}{\partial y^2} = \mu \left(\frac{\partial u}{\partial y} \right)^2 - u T \left(\frac{\partial \rho / \partial T}{\rho} \right) \left(\frac{dp}{dx} \right) \qquad (7.10)$$

The surface boundary conditions are those of a moving heat source given by the Jaeger integral, viz.

$$\pi (\rho_s c_{ps} U k_s)^{\frac{1}{2}}_{1,2} T_{0,h} = \pm \int_{-\infty}^{x} k \frac{\partial T}{\partial y} \bigg|_{0,h} \frac{dx'}{(x - x')^{1/2}} \qquad (7.11)$$

TABLE 7.4

Isothermal and THD solutions
in elliptical contacts

		Case			
		C1	C2	C3	C4
Isothermal	h_c	0.2365	0.3714	0.5727	0.7243
solution	h_{min}	0.1965	0.3090	0.4764	0.6092
Thermal	h_c	0.2334	0.3665	0.5661	0.7226
solution	h_{min}	0.1919	0.2921	0.4719	0.6038

Dimensions in μm

TABLE 7.5

Temperature rises in elliptical contact

Parameters		Temperatures			
		Film		Solids	
$\bar{U} \times 10^{10}$	$\dfrac{U_1 - U_2}{U_1}$	ΔT_c	ΔT_{max}	$(\Delta T_1)_{max}$	$(\Delta T_2)_{max}$
0.3366	0.25	22.01	89.76	32.49	32.12
0.1683	0.50	25.72	48.31	26.34	25.93

The viscosity and density functions used by Cheng are

$$\frac{\mu}{\mu_1} = \exp\left[-\left(\alpha\left(\frac{1}{T} - \frac{1}{T_1}\right) + \beta_p + \gamma\frac{p}{T}\right)\right]$$

$$\frac{\rho}{\rho_1} = 1 + \frac{C_1 p}{1 + C_2 p} + C_3(T - T_1)$$

with the geometry of the sample solutions given in Table 7.6. The data under group A represent fairly heavy loading and Figs. 7.19 and 7.20 give the pertinent THD solutions for this case; the data in group B are for a case with unequal roller materials, one of steel, the other of glass, and for these Figs. 7.21 – 7.24 give the results. The following generalizations about the thermal effects can be extracted from these various plots:

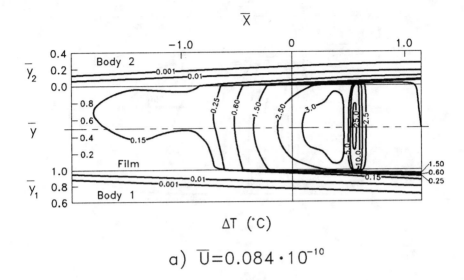

a) $\overline{U} = 0.084 \cdot 10^{-10}$

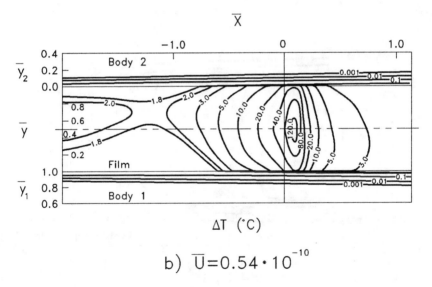

b) $\overline{U} = 0.54 \cdot 10^{-10}$

Fig. 7.14 Temperature rise in elliptical contact for $s = 0$ at $z = 0$, Dong and Shi-zhu, 1984

- The ΔT's for $-2 < (x/x_a) < -1/2$ are due mostly to heating by rolling i.e. squeeze film action.
- For pure rolling isothermal and THD solutions yield practically the same h and p. This and the previous conclusion agree with the results obtained for elliptical contacts.
- With slip present THD effects are quite significant and in the following

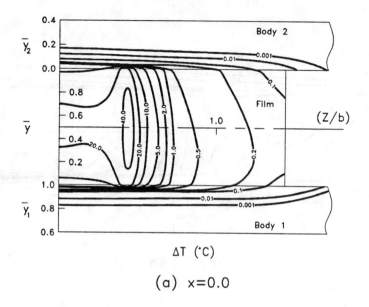

(a) x=0.0

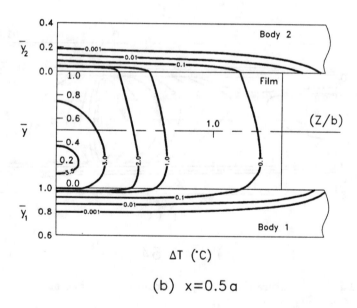

(b) x=0.5a

Fig. 7.15 Transverse temperature rise in elliptical contact for case no. 3, Dong and Shi-zhu, 1984

way:

- At high speeds they reduce the peak pressure
- At low speeds they increase the peak pressure

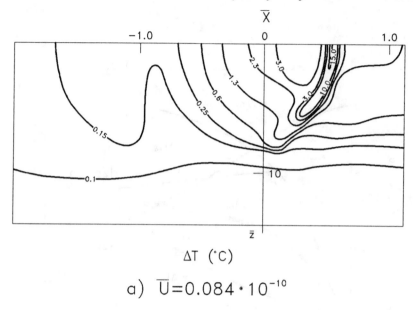

ΔT (°C)

a) $\overline{U} = 0.084 \cdot 10^{-10}$

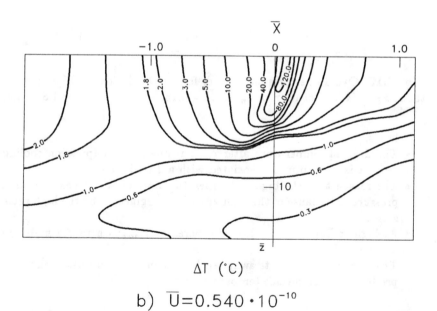

ΔT (°C)

b) $\overline{U} = 0.540 \cdot 10^{-10}$

Fig. 7.16 Temperature rises at midplane of fluid film in elliptical contact, Dong and Shi-zhu, 1984

- Film thickness is little affected by slip.
- Viscous heating plays the dominant role in raising the temperature

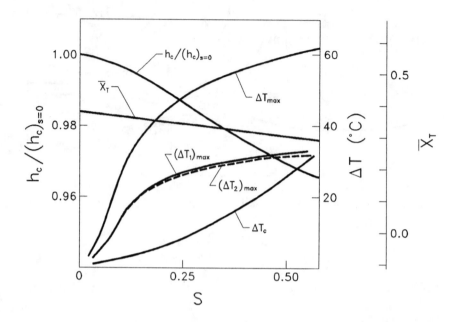

Fig. 7.17 Overall performance curves for THD solution of elliptical contacts, $x_T = (x/a)$, location of T_{max}, $\overline{U} = 0.168 \cdot 10^{-10}$, Dong and Shi-zhu, 1984

levels.

- The drop in temperature after p_{max} is due to decompression cooling — that is, a drop in T associated with a drop in p.
- The reason that the highest temperatures are obtained near the high pressures is because of the high viscosities engineered by the high pressures.
- At high rolling speeds a low slip increases T_{max}; whereas a high slip decreases T_{max}.
- The glass roller conducts away less heat than the steel surface and thus produces higher surface temperatures.

7.3 SLIDERS

The slider bearing analyzed by Rohde and Oh, 1975, included in addition to elastic deflections also distortions due to thermal gradients. While the deflections were kept two-dimensional depending on both x and z, the temperature gradients $(\partial T/\partial z)$ were ignored. In previous analyses this

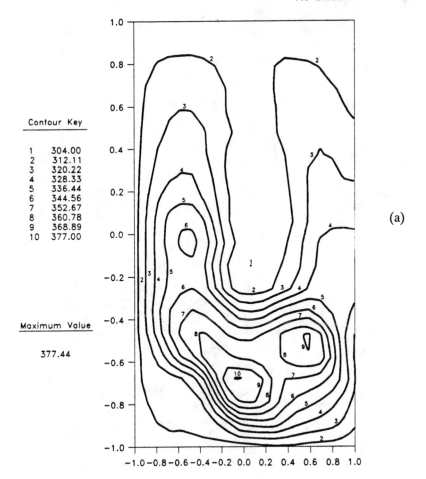

Fig. 7.18 Temperatures in an elliptical contact, (a) top surface, (b) midplane of fluid film, Conry, TW 1988

dependence of T on x only was affirmed in most cases. However, in the case where there are elastic and thermal deflections this independence of z is a much more severe approximation. Conduction in the x direction as well as any density effects either on the viscosity or compression work was neglected. Thus, the energy equation used had the relatively simple form.

$$\rho c_p \left[u \frac{\partial T}{\partial x} + w \frac{\partial T}{\partial y} \right] = k \left(\frac{\partial^2 T}{\partial y^2} \right) + \mu \left(\frac{\partial u}{\partial y} \right)^2$$

For the heat flow in the solids the Laplace equation $\nabla^2 T = 0$ was used. The models considered are shown in Fig. 7.25. One is that of a massive bearing support represented by a semi-infinite solid; the other a relatively thin plate simply supported at both ends. These two models called for radically different elasticity equations as follows:

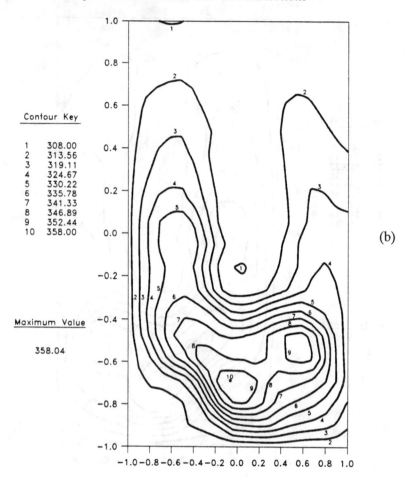

Contour Key

1	308.00
2	313.56
3	319.11
4	324.67
5	330.22
6	335.78
7	341.33
8	346.89
9	352.44
10	358.00

Maximum Value

358.04

(b)

Fig. 7.18 Temperatures in an elliptical contact, (a) top surface, (b) mid-plane of fluid film, Conry, TW 1988

A) Semi-Infinite Solid

Its representation was handled by assigning to a small discrete area of dimensions $(2a \times 2b)$ a constant load and writing for the displacement at its center, due to any point in the domain, the equation.

$$\delta = \frac{1 - \nu^2}{\pi E} \int_{-b}^{b} \int_{-a}^{a} \frac{dx'dy'}{\sqrt{(x - x')^2 + (y - y')^2}}$$

which yields

$$\delta = \frac{1 - \nu^2}{\pi E} p(a - x) \left[\frac{1}{2} \sinh^{-1} \eta + \log \left| \frac{1 + \sqrt{1 + \eta^2}}{\eta} \right| \right] \Bigg|_{\frac{-a-x}{+b-x}}^{\frac{a-x}{b-x}}$$

TABLE 7.6

Input data for L/D = ∞ rollers

C_p - specific heat of the oil	0.4 Btu/lb-°F
ω - specific weight of oil	0.0325 lb/in.^{3}
E_1 - modulus of elasticity of roller no. 1	12 X 10^6 lb/in.^{2}
E_2 - modulus of elasticity of roller no. 2	30 X 10^6 lb/in.^{2}
ν_1 - Poisson's ratio of roller no. 1	0.33
ν_2 - Poisson's ratio of roller no. 2	0.33
k_{s1} - thermal conductivity of roller no. 1	0.605 Btu/ft-°F-hr
k_{s2} - thermal conductivity of roller no. 2	26.6 Btu/ft-°F-hr
w_{s1} - specific weight of roller no. 1	.0938 lb/in.^{3}
w_{s2} - specific weight of roller no. 2	0.283 lb/in.^{3}
C_{s1} - specific heat of roller no. 1	0.199 Btu/lb-°F
C_{s2} - specific heat of roller no. 2	0.11 Btu/lb-°F
C_1 - pressure-density exponent	0.0000045 in.^{2}/lb
C_2 - pressure-density exponent	0.000013 in.^{2}/lb

	Series "B"	Series "A"
Diameter of rollers	6 in.	3 in.
Material of roller no. 1	glass	steel
Material of roller no. 2	steel	steel
Width of rollers	1/4 in.	1/4 in.
Lubricant	2190T	AEI OM-100
Load	520 lb/in.	715 lb/in.
β - pressure viscosity exponent	-.000273 in.^{2}/lb	-.0001069 in.^{2}/lb
α - temperature viscosity exponent	8,067°F	8,497°F
γ - pressure-temperature viscosity exponent	0.254 in.^{2}/°F-lb	0.1366 in.^{2}/°F-lb
μ_1- inlet viscosity	1.8x10^{-5} lb-sec/in.^{2}	0.58x10^{-5}lb-sec/in.^{2}
T_1- inlet temperature	85°F	110°F
k - thermal conductivity of oil	0.08 Btu/ft-°F-hr	0.1 Btu/ft-°F-hr

$$+ \frac{1-\nu^2}{\pi E}p(a+x)\left[\frac{1}{2}\sinh^{-1}\eta + \log\left|\frac{1+\sqrt{1+\eta^2}}{\eta}\right|\right]\Bigg|_{\frac{a+z}{b+z}}^{\frac{a+z}{-b+z}} \quad (7.12a)$$

provided the point $(x, z, 0)$ lies outside the rectangle. If the point $(x, z, 0)$ is at the center of the rectangle, the displacement becomes

$$\delta = \frac{4(1-\nu^2)}{\pi E}p[a\sinh^{-1}\frac{b}{a} + b\sinh^{-1}\frac{a}{b}] \quad (7.12b)$$

B) Simply Supported Plate

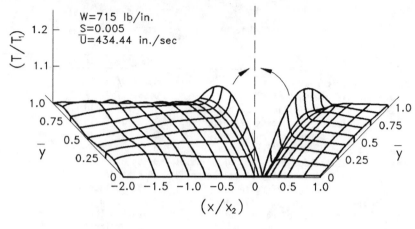

a) Low S/U_a Ratio

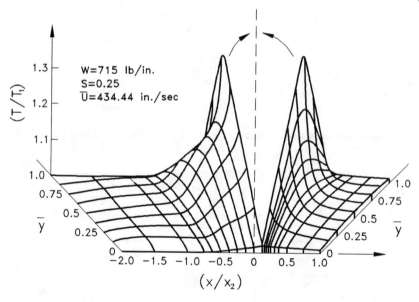

b) High S/U_a Ratio

Fig. 7.19 Temperature map in fluid film. Note: Film split at T_{max} for purposes of presentation, Cheng, 1965

The differential equation which governs the deflection δ of a plate of thickness t is the biharmonic equation

$$\nabla^4 \delta = \frac{p(x,z)}{D} - \frac{M_t}{(1-\nu)D} \qquad (7.13)$$

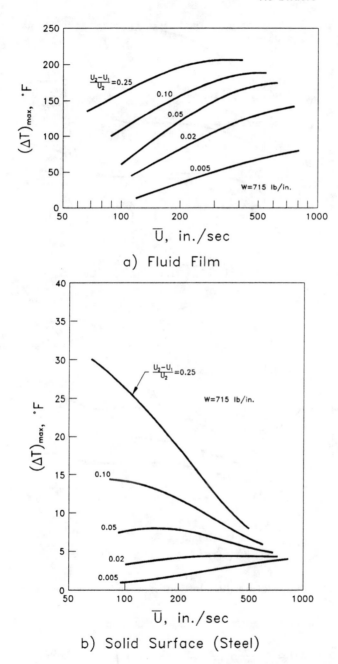

Fig. 7.20 Maximum temperatures in infinitely long rollers, Cheng, 1965

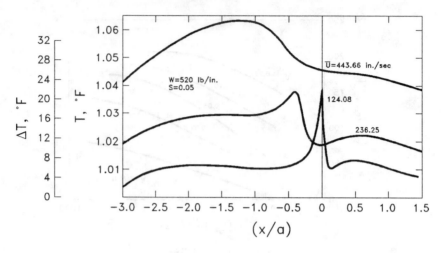

a) Fluid Film at Low s

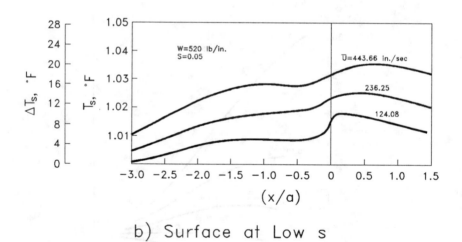

b) Surface at Low s

Fig. 7.21 Temperature distribution in steel-glass rollers at low values of
s, Cheng, 1965

Above

$$D = \frac{Et^3}{12(1-\nu^2)} \quad , \quad M_t = E\gamma \int_{-t/2}^{t/2} \nabla^2 T y_s' dy_s'$$

$y_s' = 0$ coincides with the middle surface of the plate and γ is the coefficient
of expansion of the solid. As seen, Equ. (7.13) includes the effects of thermal
distortion. For model B the conditions of zero deflection and zero moment
were imposed at both ends.

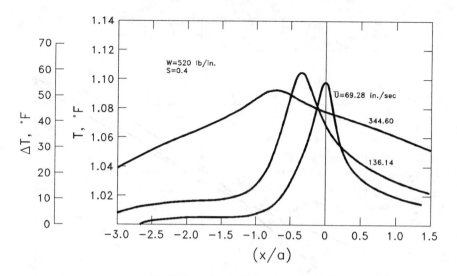

a) Fluid Film

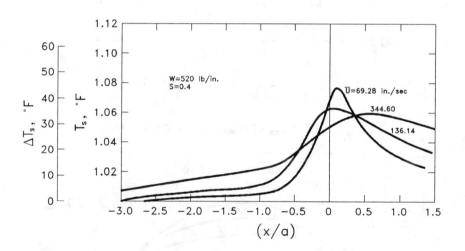

b) Surface

Fig. 7.22 Temperature distribution in steel-glass rollers at high values of s, Cheng, 1965

The boundary conditions on the energy equation were as follows:
1. At $x = 0, T = T_1$
2. On the runner $T = T_1 = const$
3. On the bearing surface two different conditions were tried
   a. $T = T_1 = $ constant

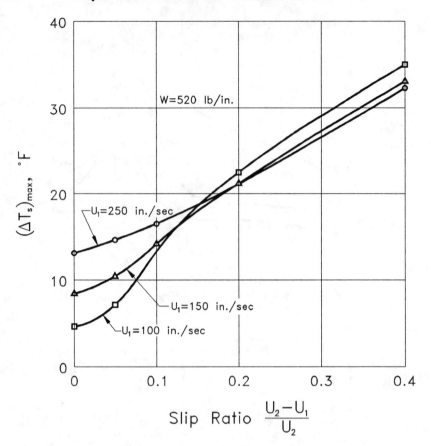

Fig. 7.23 T_{max} of glass surface, Cheng, 1965

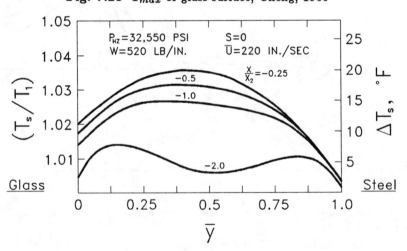

Fig. 7.24 Transverse temperature profile in infinitely long rollers, Cheng, 1965

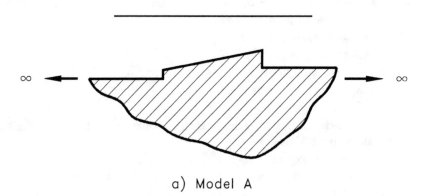

a) Model A

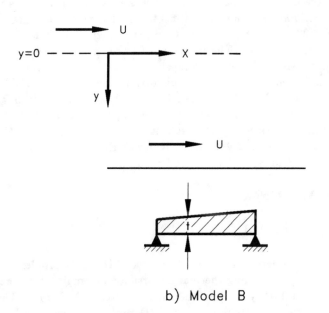

b) Model B

Fig. 7.25 Slider support geometries

b.

$$k\frac{\partial T}{\partial y}\bigg|_{h} = -k_s\frac{\partial T_s}{\partial y_s}\bigg|_{y_s=0}$$

4. At the solid surfaces not exposed to the fluid film

$$-k_s\frac{\partial T_s}{\partial y}\bigg|_{y_s\neq0} = -h_{sa}(T_s - T_a)$$

The simple exponential viscosity-temperature relation given by Equ. (2.11) was used. The other operating data are given in Table 7.7. These include the following permutations:

1. Isothermal bearing surface
   - Rigid surface
   - Elastic effects only
   - Thermal distortions only
   - Combined effects
2. Heat transfer at bearing surface with parametric combinations similar to those given above.

One of the unexpected results obtained was that rigid surfaces, case 3, yielded lower load capacities than a deflected plate, case 9; this was true both when isothermal and THD solutions were run. The authors attribute this to the fact that, as seen in Fig. 7.26 in case 4, h_{min} was allowed to increase whereas in cases 7 and 9 it was kept constant. It is well known that the ratio h_1/h_2 is a most critical factor in determining bearing behavior. Figs. 7.27 and 7.28 give results for case 11 which includes both elastic and thermal distortions. The thermal distortion, as can be seen from Fig. 7.28, is quite insignificant when compared to the elastic deformation. Furthermore, the full elastic and thermal distortions solution did not differ very much from that of the THD analysis for a rigid surface. Nearly in all cases the effect of a variable viscosity was more important than the EHD effect. For example, ignoring THD effects increases W by 135%, 115%, and 116% for cases 1, 3, and 4 whereas ignoring EHD effects—but not $\mu(T)$—yielded only small variations in load capacity.

7.4 THRUST BEARINGS

7.4.1 Line Supported Pad

Robinson and Cameron, 1975, analyzed an EHD-THD problem for sector thrust bearings, including thermal distortion. The major deficiency of the analysis is that they ignore the transverse term $\partial^2 T/\partial y^2$. This is a particularly serious shortcut because EHD problems arise mostly in very thin films where transverse temperature gradients are all-important. For this reason the present discussion will not go into much detail as far as the energy equation is concerned. However, the authors' work is still of considerable relevance because the deformations are given special attention.

There are four components in the total deflection of a THD-EHD surface: bending and compression due to the pressure forces; and bending and expansion due to the temperature gradients. In addition, the authors treat the problem of a flat plate, providing data for its hydrodynamic action as a consequence of deformation. The subject of thermal deflections will thus be treated here in some detail.

TABLE 7.7

Slider with elastic and thermal deflections

Operating conditions

B, L = 3.6 in. (0.0915 m)
k   = 0.075 Btu/hr-°F-ft (0.039 W/(m-K)
k_s = 30.0 Btu/hr-°F (15.6 W/(m-K))
c_p = 0.48 Btu/lb-°F (580 J/(kg-K))
E   = 30.0 x 10^6 psi (20.6 k-Pa)
h_{min} = 2.56 x 10^{-3} in. (6.50 x 10^{-5}m)
h_{max} = 5.12 x 10^{-3} in. (13.0 x 10^{-5}m)
t   = 0.75 in. (1.91 x 10^{-2} m)
h_{sa} = 100 Btu/(hr-ft^2-°F)(175 W/(m^2-K)
T_1 = 100°F (37°C)
T_a = 80°F(26.8°C)
U   = 100 ft/sec(30.48 m/8)
γ = 6.5 x 10^{-6} 1/°F (1.17 x 10^{-5}K)
ν = 0.3 (Poisson Ratio)
ω = 54.6 lb/ft^3 (8560 N/m^3)
α = 0.025 1/°F (0.045 1/K)

Solutions

Case	T_s	Fluid Film	+ δ_{EL}	+ δ_{TH}	Model	- W	(x_c/B)
1*	T_1	const	no	no	rigid	0.068	0.58
2	T_1	var	no	no	rigid	0.041	0.56
3	var	var	no	no	rigid	0.029	0.57
4*	T_1	const	yes	no	inf. solid	0.056	0.59
5	T_1	var	yes	no	inf. solid	0.037	0.56
6	var	var	yes	no	inf. solid	0.027	0.58
7*	T_1	const	yes	no	plate	0.065	0.62
8	T_1	var	yes	no	plate	0.040	0.58
9	var	var	yes	no	plate	0.030	0.60
10	var	var	no	yes	plate	0.029	0.57
11	var	var	yes	yes	plate	0.030	0.60

---

*Isothermal cases
+ δ_{EL} - elastic deflection; δ_{TH} = thermal deflection

    In the present formulation it is assumed that there is a linear temperature distribution across the pad thickness, commencing with the variable surface temperature at the interface with the fluid film and extending to the back of the pad, assumed to be at the constant temperature of the entering lubricant. As far as thermal bending of a pad is concerned, the tempera-

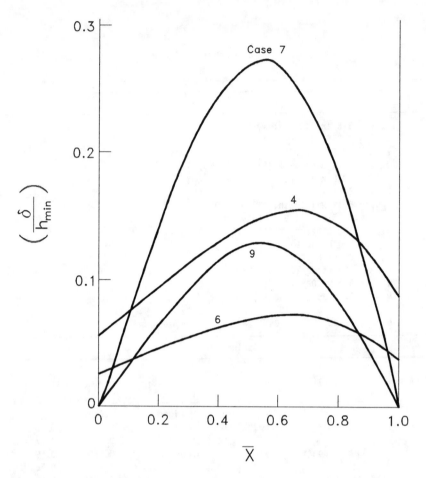

Fig. 7.26 Centerline elastic deformation of pad, Rohde and Oh, 1975

ture distribution, it is shown, can be replaced by an equivalent mechanical loading. This simulated pressure load can be written as

$$q = \frac{D(1+\nu)\gamma}{t}\nabla^2(\Delta T) \tag{7.14}$$

with $D = Et^3/12(1-\nu^2)$ and γ the coefficient of linear expansion of the pad material. There is also an edge loading on the boundary, represented by a moment and an edge reaction. In polar coordinates the bending moments are

$$M_r = M_\theta = \frac{D(1+\nu)\gamma}{t}\Delta T \tag{7.15}$$

and the edge reactions per unit distance are

$$F_r = \frac{D(1+\nu)\gamma}{t}\frac{\partial}{\partial r}(\Delta T) \tag{7.16a}$$

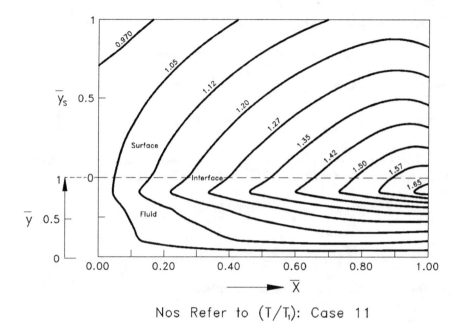

Nos Refer to (T/T_1): Case 11

Fig. 7.27 Temperature distribution in slider fluid film and bearing, Rohde and Oh, 1975

$$F_\theta = \frac{D(1+\nu)\gamma}{t}\frac{1}{r}\frac{\partial}{\partial\theta}(\Delta T) \qquad (7.16b)$$

Three more conditions are needed to fix the distorted surface in space. Two such conditions are to assure that the deflections at each edge of the pad support are zero; and the third to make the centers of the two radial boundaries have the same deflections. Having substituted a pressure load for the temperatures and having provided the required edge moments and reactions, the thermal deflections can be calculated in the same manner as deflections due to bending caused by a pressure profile. With regard to the thermal bending of the runner first its temperature profile has to be postulated. It is assumed that due to rotation there is uniformity of temperature along any given circumference and that, therefore, the only temperature gradient is in the radial direction. If G is the shear force at radius r (per unit of circumferential length), the deflection is given by the differential equation.

$$\frac{d}{dr}\left[\frac{1}{r}\frac{d}{dr}\left(r\frac{d\delta}{dr}\right)\right] = \frac{G}{D} \qquad (7.17)$$

The shear force corresponding to thermal loading is

$$G = \frac{D(1+\nu)\gamma}{t}\frac{d}{dr}(\Delta T)$$

If ΔT is expressed as a cubic in r, i.e.

$$\Delta T = B_0 + B_1 r + B_2 r^2 + B_3 r^3$$

the differential equation can be integrated to give

$$\delta = \frac{(1+\nu)\gamma}{t} \left(\frac{B_1}{9}\gamma^3 + \frac{B_2}{16}\gamma^4 + \frac{B_3}{25}\gamma^5 + D_1 + D_2 \ln r + D_3 r^2 \right) \quad (7.18)$$

The three constants of integration D_1, D_2, and D_3 may be found from the three boundary conditions which, with reference to the previous section, are

$$\delta = 0 \quad \text{and} \quad M_r = \frac{D(1+\nu)\gamma}{t}(\Delta T)_{r=R_1} \quad \text{at} \quad r = R_1$$

and

$$M_r = \frac{D(1+\nu)\gamma}{t}(\Delta T)_{r=R_2} \quad \text{at} \quad r = R_2$$

The bending moment can then be found from

$$M_r = D\left(\frac{d^2\delta}{dr^2} + \frac{\nu}{r}\frac{d\delta}{dr} \right) \quad (7.19)$$

Thermal expansion can be treated in a manner similar to compression. For linear temperature gradients through the pad and runner the thermal growth of a thickness element dy_s a distant y_s from the unloaded side will be

$$d\delta = \gamma \left[T_1 + (T - T_1)\frac{y_s}{t} \right] dy_s$$

The total growth of the thrust pad is, therefore,

$$\delta = \gamma \int_0^t \left[T_1 + (T - T_1)\frac{y_s}{t} \right] dy_s = \frac{1}{2}\gamma t(T + T_1) \quad (7.20)$$

Only the variation in thermal growth will affect the film shape and since T_1 is constant for both pad and runner, it can be omitted. The full term for direct thermal expansion then becomes

$$\delta = -\frac{1}{2}\left[(\gamma T t)_{s1} + (\gamma T t)_{s2}\right] \quad (7.21)$$

The negative sign indicates that direct thermal expansion decreases film thickness whereas other modes of deflection increase it.

The bearing analyzed was a pad supported by a pivot whose shape is given in Fig. 7.29. The figure also gives the relevant properties of the lubricant and materials used. In line with the previous remarks, our discussion will be confined to the pressures and film shape induced by the thermal and elastic deflections. Some of these are shown in Table 7.8 and Fig. 7.30.

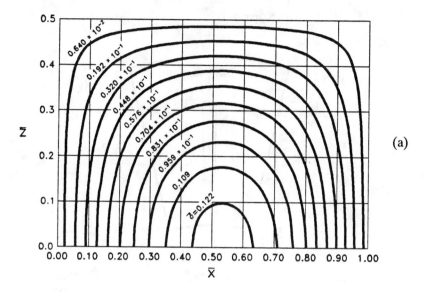

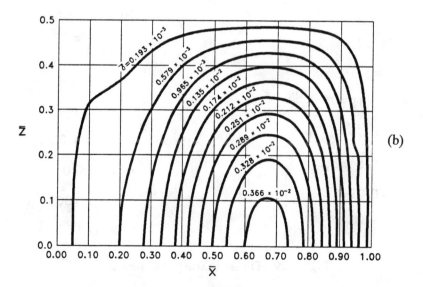

Fig. 7.28 Comparison of slider elastic and thermal deflections: Case 11, (a) total deflections, (b) thermal deflections, Rohde and Oh, 1975

First it can be seen that the axial distortion is larger than the circumferential. With rising speed there is an increase in deflection due to the higher

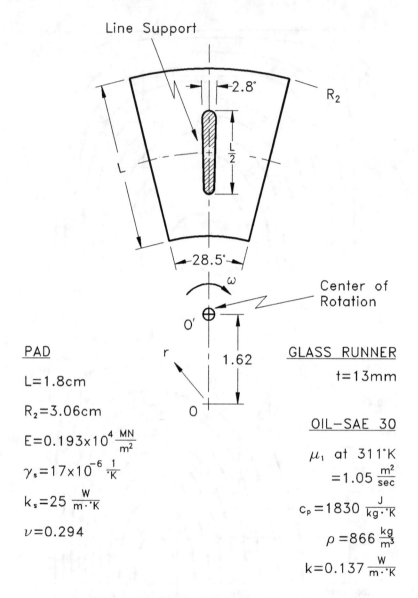

Fig. 7.29 Thrust bearing configuration

temperatures induced by the higher dissipation. With load, however, there are two opposing trends. Load increases bending but at the same time also increases the compressive deflection. Consequently, there is little change in δ with a change in loading. However, the most significant outcome of this work, perhaps, is that, contrary to what was concluded in the previous section, thermal distortion was found here to contribute markedly to load capacity, its magnitude being about twice that of the elastic deflections.

TABLE 7.8

Effects of elastic and thermal distortions in thrust bearings

ω & P	Parameter*	With δ	δ = 0
211 rad/s 1.37 MN/m²	minimum film thickness maximum film temperature average outlet temperature	12.59 332.60 327.52	7.16 340.39 332.99
644 rad/s 1.37 MN/m²	minimum film thickness maximum film temperature average outlet temperature	15.60 353.46 349.29	12.07 370.84 358.99
214 rad/s 6.89 MN/m²	minimum film thickness maximum film temperature average outlet temperature	3.56 347.64 342.70	2.85 361.96 346.29
610 rad/s 6.89 MN/m²	minimum film thickness maximum film temperature average outlet temperature	5.54 382.61 374.10	3.90 414.55 389.78

*h in μm; T's in K.   Robinson and Cameron, 1975

A pad identical to that shown in Fig. 7.29 was used to study flat plate behavior. In comparison with the pivoted pad the flat land geometry has its h_{min} closer to the center; the higher the speed and the lower the load the nearer does h_{min} move to the center. This can be seen from the plots of Fig. 7.31. Again as with tilting pad, distortion rises with speed but is insensitive to load, emphasising the strong thermal effect. Ultimately when distortion is high enough cavitation sets is in he diverging portions, as shown in Fig. 7.32 making the flat land equivalent to a crowned pad. At low loads there is an optimum with regard to speed, induced by a reduction in viscosity and a loss of effective area due to cavitation. The levels of distortions as a function of load and speed and the resulting values of h_{min} are shown in Fig. 7.33. The distortions here are almost double those found in the tilting pad bearing but the functional behavior is similar. The overall importance of the thermal bending can be gleaned from Table 7.8 where the h_{min} is seen to rise anywhere from 20% to 90% as a result of the thermal bending. The individual contributions of the various forms of bending are detailed in Table 7.9.

7.4.2 Effect of Support Geometry

It should be clear from the foregoing results for the line-supported pad that the shape and size of this support is likely to have a telling effect on the elastic and thermal distortions of the pad, and consequently on its performance. Such a scrutiny was undertaken by Ettles, 1986, for both conventional circular stems of various diameters, and for a novel type of

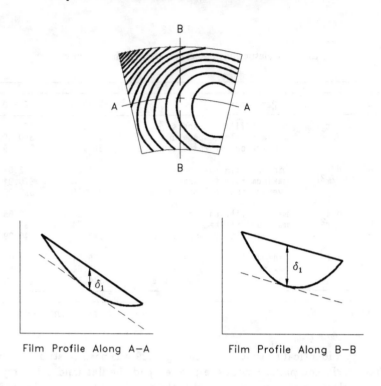

Film Profile Along A—A   Film Profile Along B—B

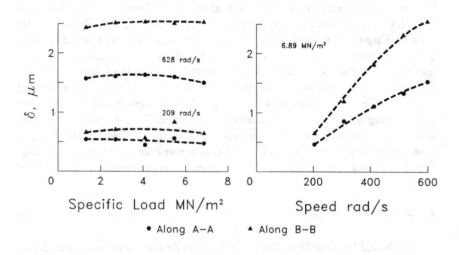

Fig. 7.30 Deflections as function of load and speed in thrust bearings, Robinson and Cameron, 1975

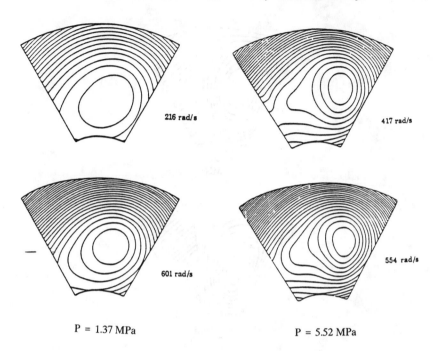

Fig. 7.31 Film shapes in flat plate due to thermal distortion $h_{min} = 0.51\,\mu$m; each contour spaced $\Delta h = 0.51\mu m$ apart, Robinson and Cameron, 1975

support consisting of a set of two partial arcs, as shown in Fig. 7.34. These supports were applied to a large-size thrust bearing of a hydro-electric plant, where deflections play a particularly important role. The operational characteristics of this bearing were:

$$R_2 = 1524\,\text{mm} \qquad \text{Pivot offset} - 55\%$$
$$R_1 = 762\,\text{mm} \qquad N = 150\,\text{rpm}$$
$$\text{No. of pads} - 12 \qquad \mu \text{ at } 40°\text{C} = 150\,\text{MPa·s}$$
$$t = 170\,\text{mm}$$

Figs. 7.35 and 7.36 show the effect of various support geometries on minimum film thickness and maximum pad temperature. Below 4 MPa the size of the stem has a negligible effect on performance, but above it rapid deterioration sets in with the small sizes. This is brought about by the severe crowning when the inefficient film shape aggravates the thermal distortion, leading rapidly to excessive heating and possible failure. The thermal and elastic deformations can be made to cancel each other by the use of large size supports as shown by curves c and d in the figures.

Fig. 7.37 shows isobars, maximum pad temperatures, and film thickness for a very small stem; a large size stem; and a design with two partial arcs of large diameter. For the small stem, the crowning is clearly excessive as evidenced by the shape of the isobars. With the larger stem, the

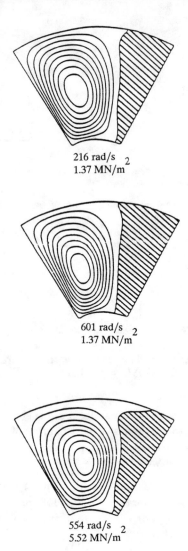

216 rad/s
1.37 MN/m^2

601 rad/s
1.37 MN/m^2

554 rad/s
5.52 MN/m^2

Fig. 7.32 Cavitation in flat land pads due to thermal distortion, Robinson and Cameron, 1975

crowning is moderate and the pressure distribution quite reasonable. A set of arcs relieved between 1 and 5 o'clock and between 7 and 11 o'clock, however, shows a much better performance and can be made to approximate the effectiveness of a two stem support, known to yield good results.

7.5 JOURNAL BEARINGS

The problem of elastic and thermal deflections is much more serious in

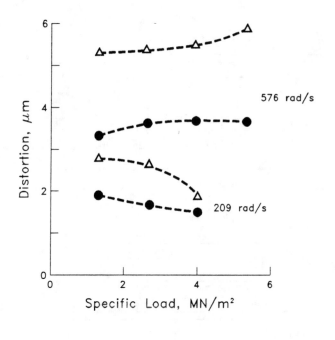

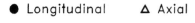

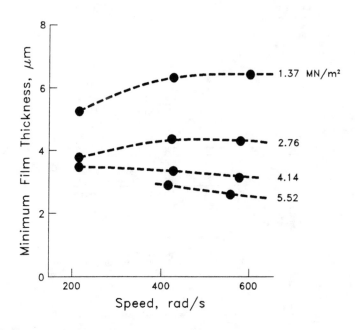

Fig. 7.33 Effects of load and speed in flat land pads, Robinson and Cameron, 1975

TABLE 7.9

Modes of deflection in flat land pad

		δ in θ direction, μm					
	Specific	Thermal bending		Pressure bending			Compression
Speed	load					Thermal	due to
rad/s	MN/m²	Pad	Collar	Pad	Collar	expansion	pressure
281	1.37	1.412	1.050	neglected	-0.069	-0.104	-0.514
600	1.37	1.787	1.952	"	-0.066	-0.107	-0.590
213	2.76	1.767	1.182	"	-0.147	-0.108	-0.890
582	2.76	2.187	2.066	"	-0.147	-0.124	-0.120
416	5.52	2.062	1.878	"	-0.310	-0.081	-2.014
554	5.52	2.680	2.237	"	-0.315	-0.141	-2.232

TABLE 7.10

Input data for Figs. 7.39-7.42

	β			Pivot Position, %			θ_p		
Geometry	Pad 1	Pad 2	Pad 3	Pad 1	Pad 2	Pad 3	Pad 1	Pad 2	Pad 3
A	104	104	57	0.55	0.55	0.56	125	230	0
B	104	57	57	0.55	0.55	0.56	160	265	30

Length = 0.56 m
Diameter = 0.750 m
Pad thickness = 0.16 m
Radial clearance = 0.53 x 10⁻³ m
Load 5 x 10⁵ N < W < 2 x 10⁶ N
Rotational speed = 1500 rpm
Oil viscosity: μ = 32 cst (T = 40°C). μ = 6.5 cst (T = 100°C)
Pad material: A72323-3 ASTM standard
Shaft surface temperature: 60°C
Outer pad surface temperature: 50°C
Oil inlet temperature: 40°C

TABLE 7.11

Performance of tilting pad bearings including
elastic and thermal restrictions
W = 10⁶ N

Geometry	Deformation	Load per pad 10⁴ (N)			Minimum film thickness (μm)			Maximum temperature (°C)			Average distortion (μm)
		Pad 1	Pad 2	Pad 3	Pad 1	Pad 2	Pad 3	Pad 1	Pad 2	Pad 3	Pad 1
A	Pressure and temperature	105	105	24	81	90	145	94	94	79	73
	None	119	119	36	91	91	138	94	94	78	0
B	Pressure and temperature	119	52	23	71	98	145	96	91	79	94
	None	123	55	25	79	96	144	98	93	80	0
	Pressure only	117	50	22	84	98	143	75	70	73	87

Brugier and Pascal, 1989

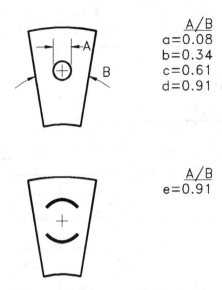

$$\frac{A/B}{}$$
a=0.08
b=0.34
c=0.61
d=0.91

$$\frac{A/B}{}$$
e=0.91

Fig. 7.34 Varieties of pad support

journal bearings than in thrust bearings discussed previously. This stems from the fact that unlike with thrust bearings, the journal has no freedom to move away from the bearing it faces when it expands. There is thus, in addition to changes in the geometry of the two mating surfaces, also the danger of overall loss of clearance. In extreme cases, this can lead to shaft seizure, a form of failure which results in damage not only to the bearing but also to the shaft and possibly to the whole installation.

7.5.1 Tilting Pad Bearings Including Deformations

A fairly comprehensive analysis of two varieties of tilting pad journal bearings, shown in Table 7.10, was performed by Brugier and Pascal, 1989, which included not only the distortion of the pad but also of the pivot. The analysis considers the effect of mixing inlet temperatures in which the difference between the journal and pad termperatures was joined by a parabolic distribution. In the energy equation the transverse convection term $v(\partial T/\partial y)$ was retained but, interestingly, it was not possible to satisfy the $v = 0$ boundary condition on the surfaces, which the authors attribute to the uncertainties in the numerical procedures. In the region of reverse flow the solution was handled in the manner described in Section 5.1.2. Otherwise, the usual conditions of continuity of heat flux at the pad-film interface, and of an isothermal shaft, were retained.

The results of this analysis are of particular interest in that they present separately the contributions of the elastic and thermal deformations. Thus, Fig. 7.38 shows the distortion of the pad and pivot due to one or both of

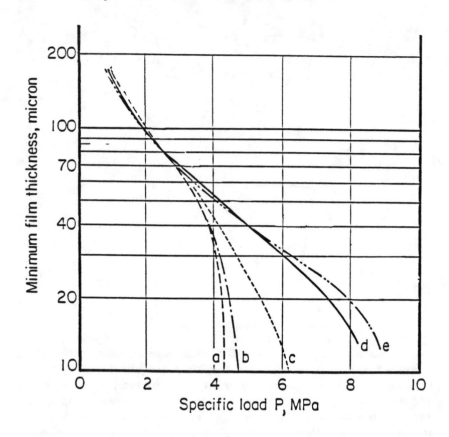

Fig. 7.35 Minimum film thickness as function of support geometry, Ettles, 1986

these effects. It can be seen that the thermal component tends to restore the pad and pivot to their original configurations, although this does not necessarily mean that the total displacement or departure from the original position is less.

Another item of particular interest are the maps of Fig. 7.39 showing separately the effects of the elastic and thermal terms on the pad temperatures. As seen, the temperatures are considerably higher when the thermal effects are accounted for, yielding a T_{max} of 20°C higher than for a solution with elastic distortion only. Moreover, a negative slope, i.e., a zone of declining temperatures, is generated at the pad outlet. This provides some corroboration for the observed phenomenon of declining temperatures in the cavitating or trailing portions of journal bearings, as well as to the results of Hopf and Schuller in Section 6.3. Thus such temperature drops may be due to a combination of heat transfer, turbulence, as well as the now noted effects of elastic and thermal distortions.

The general performance of tilting pad bearings including deformations

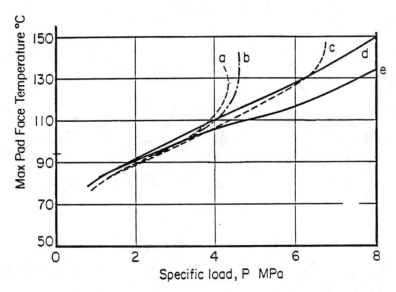

Fig. 7.36 T_{max} of pad as function of support geometry, Ettles, 1986

are given in Table 7.11 and Figs. 7.40 and 7.41. These data are all for the same loading of 10^6 N which corresponds to a unit loading of 24 atm, or some 350 psi, a moderate load. While the impact of thermal distortion depends on the particular bearing geometry and mode of loading—see the difference between A and B in Table 7.11—it can be said that in general the thermal effects further reduce the value of h_{min} and raise that of T_{max}. There is seen to be no effect on the spring and damping constants, whether in the direction of (xx) or normal to it (yy). However, it should be kept in mind that these are normalized quantities and that the actual K's and B's are to be multiplied by the viscosity which would be affected by thermal distortion. This analysis concludes with the observation that both elastic and thermal deformations play a part, a conclusion which falls in between the assertions of Section 7.3 where the elastic effects were said to be dominant, and Section 7.4 where the thermal components seemed predominant.

7.5.2 Thermal Growth of Bearing and Journal

The thermal growth of a journal and of a corresponding full journal bearing was conducted by Fillon et al, 1987, for the bearing discussed in Section 4.3.4 whose operational characteristics are those given in Table 4.8. If the outside of the bearing shell is at a constant temperature, the changed bearing internal surface can be represented by a single circle, and the main effects of such a growth are a change in the radial clearance and the eccentricity ratio. Fig. 7.42a shows the changed diameters of both bearing and journal due to the prevailing rise in temperatures. These results are for

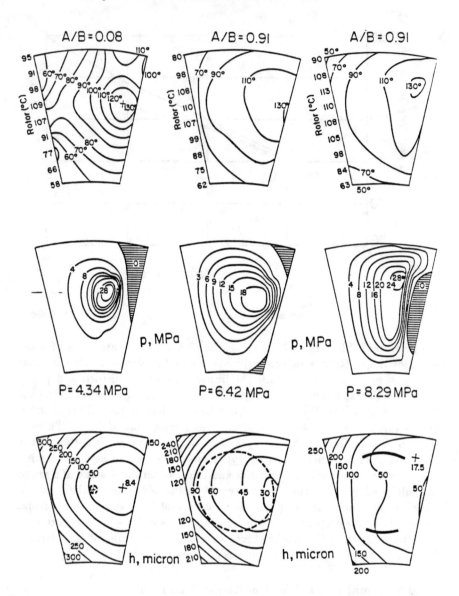

Fig.   7.37 Mapping of isobars, isotherms, and thickness for various support geometries, Ettles, 1986

an initial $\epsilon = 0.8$ and $N = 2,000$ rpm. Here the calculations show actually an increase in general clearance, including an increase in h_{min} as a result of the thermal expansions, but this is not always the case. Fig. 7.42b shows the variation of h_{min} for a range of ϵ's and as seen, up to an $\epsilon = 0.4$ there is a decrease in h_{min} because of the expansion of the two members. These results are for steady state conditions in which the growth of both journal

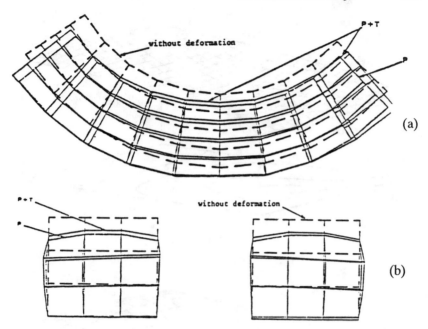

Fig. 7.38 Pad and pivot deformations due to elastic and thermal distortions, (a) pad, (b) pivot, Brugier and Pascal, 1989

and bearing have reached equilibrium. However, if for whatever reason the rates of growth of the two surfaces are not the same and the journal grows faster than the bearing, serious difficulties may ensue, as discussed in the next section.

7.5.3 Shaft Seizure

Shaft seizure can happen at the start-up of a cold machine wherein the journal is free to expand in response to the sudden generation of thermal energy while the bearing, being constrained usually by a massive support system consisting of housing, pedestals, etc., cannot expand outwardly as fast as the journal expands into the fluid film. As the initial clearance decreases, the process is accelerated because the rate of heat generation rises and with it the differential expansion. In tilting pad bearings this is additionally aggravated by the fact that the pivots are the only solid heat path between the pads and housing. When the original clearance is sufficiently reduced, hydrodynamic action breaks down, and seizure occurs.

Martin, TW, 1988, presented detailed experimental data on shaft seizure as a function of shaft acceleration, the latter a fair indicator of the rate of viscous dissipation. This is shown in Table 7.12 and Fig. 7.43. As seen, at 4,000 rpm a bearing with an initial clearance of 3.2 mils lost within 4 minutes of start-up some 90% of its machined clearance, but still managed

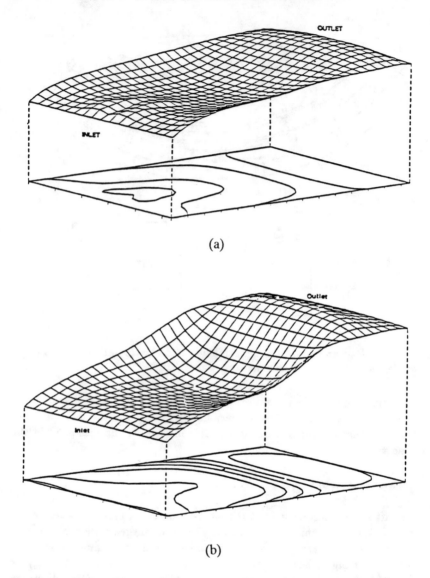

(a)

(b)

Fig. 7.39 Effect of pad distortion on temperature field, (a) elastic deformation only: $T_{max} = 75°C$, (b) elastic and thermal deformations: $T_{max} = 95°C$, Brugier and Pascal, 1989

to pull through; subsequently, as the bearing started expanding, most of the lost clearance was retrieved. However, when an attempt was made to accelerate the shaft over the same period of time to 5,000 rpm, seizure occurred. The appearance of the seized pads is shown in the photograph of Fig. 7.44.

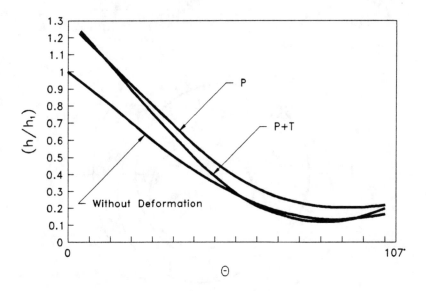

Fig. 7.40 Shape of film thickness, Brugier and Pascal, 1989

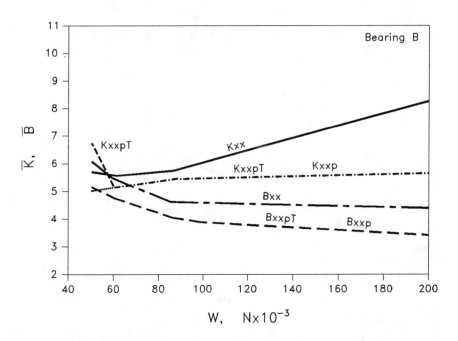

Fig. 7.41 Effect of deformation on the vertical dynamic coefficients, $P =$ pressure deformation only, PT =pressure and temperature deformations, Brugier and Pascal, 1989

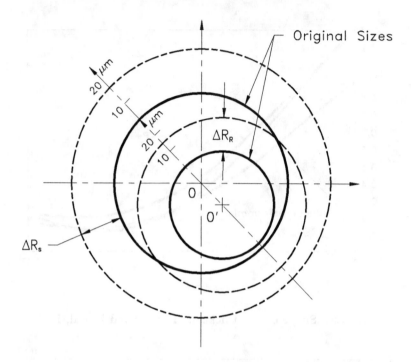

a) Bearing and Journal Thermal Expansion

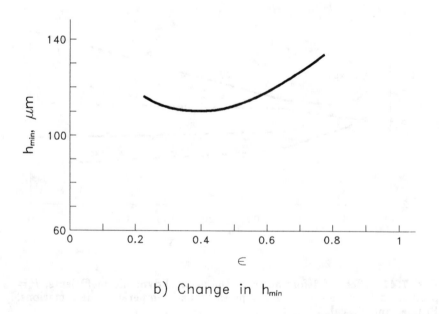

b) Change in h_{min}

Fig. 7.42 Impact of thermal growth in journal bearings, Fillon et al, 1987

TABLE 7.12

Transient bearing surface temperatures in
tilting pad journal bearing starting from cold

			Temperature °C					
Time from			1	2	3	4	5	6
Start Min	Sec	Operating condition	Oil inlet	Oil outlet	Pad before bottom pad	Bottom pad	Pad after bottom pad	Housing
4,000 rpm								
0	0	Before starting	30	Not connected	34	34	34	35
-	55	Full speed	30	"	46	70	68	35
2	0	"	30	"	52	80	80	38
3	0	"	30	"	54	84	84	40
4	0	"	30	"	56	86	88	41
5	0	"	31	"	58	90	90	43
7	0	"	35	55	62	94	94	47
8	0	"	35	56	64	95	95	48
10 minutes later								
18	0	"	42	58	63	92	92	52
5000 rpm								
0	0	Before starting	22	22	21	21	21	21
2	5	Full speed	23	46	65	102	103	29
3	0	"	23	48	77	113	115	33
4	0	"	24	49	87	118	120	37
5	0	"	24	51	100	125	127	40
6	0	"	26	52	120	135	137	43
6	35	Just before wiping	27	52	135	145	145	44
6	43	Max temp at wiping	27	53	200	195	170	45
6	54	Temp at trip	28	40	170	160	145	45

Conway-Jones and Leopard, 1975

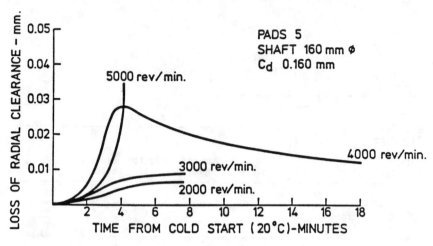

Fig. 7.43 Shaft acceleration and loss of clearance, Conway- Jones and Leopard, 1975

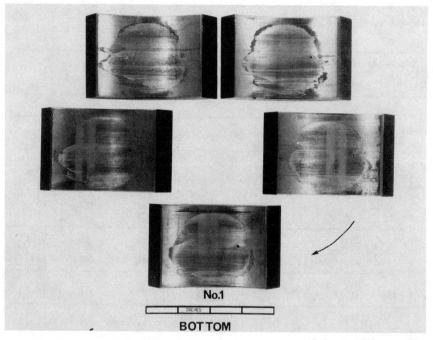

Fig. 7.44 Photograph of Seized tilting pad bearing, $D = 6.3$ in., $C_{pivots} = 0.00315$ in., Glacier Metal Co., Ltd.

Chapter 8

TWO-PHASE REGIMES

Two-phase lubrication analysis represents probably the clearest example of the impediments a tribologist faces due to the inadequate state of knowledge of certain branches of fluid dynamics in general. These particular difficulties stem from two causes. One is that two-phase fluid flow, vague as it is, becomes even more so when thermal effects are introduced. Heat transfer is an empirical discipline, but with isotropic media at least rational deductions or extrapolations can be made on the basis of existing axioms; no such clear extensions or generalizations can be made for liquid-gas or liquid-solid mixtures. The second difficulty is that given the small clearances of tribological elements a two-phase lubricant poses special problems. Among these one can mention the uncertainty of how to accommodate the large volume changes attending vaporization, or how to treat solid particles larger than the bearing clearance. Consequently the appearance of a two-phase lubricant poses not only conventional problems such as what is an effective viscosity of such mixtures or what is the effect on flow, dissipation, load capacity, etc.—but more fundamental questions as to the mode of response and viability of the affected devices. Clearly many of these answers will have to lie in basic experiments aimed specifically at ascertaining the mechanism of two-phase THD fluid films.

The difficulties arise at the most mundane level, such as in the formulation of an effective viscosity for gas-liquid mixtures, for example. In a review of this issue Feng and Hahn, 1986, collected the following assembly of expressions as used by various investigators:*

$$\mu = \mu_f \quad \text{(Owens)} \tag{8.1a}$$

$$\mu = \frac{1}{(\lambda/\mu_g) + (1-\lambda)/\mu_f} \quad \text{(Isbin)} \tag{8.1b}$$

---

* The references cited in Equs. (8.1) are all given in the paper by Feng and Hahn, 1986.

$$\mu = \mu_g + (1 - \lambda)\mu_f \quad \text{(Cichitti)} \tag{8.1c}$$

$$\mu = \lambda\mu_g(\rho/\rho_g) + (1 - \lambda)\mu_f(\rho/\rho_f) \quad \text{(Dukler)} \tag{8.1d}$$

Equ. (8.1a) obviously is bound to fail as the quality increases, while (8.1b) is merely a reciprocal of Equ. (8.1c). Even though more elaborate, Equ. (8.1d) is in fact similar to 8.1c, except that it enlarges the effect of the lower viscosity term and reduces the effect of the higher viscosity by applying a multiplying factor greater than unity to the former and less than unity to the latter. Owens' correlation overpredicts, whereas Isbin's underpredicts the viscosity drop in two phase flow, i.e. the viscosity of the mixture is in fact less than that evaluated by the former and greater than that evaluated by the latter.

The above differences, however, do not exhaust the range of uncertainties. Whatever their shortcomings Equs. (8.1a) – (8.1d) all stipulate a decrease in gas-liquid viscosity with an increase in gas content. Another proposed expression, that of Hayward, reverses this seemingly obvious trend. Based on experimental data obtained for oil-air mixtures, Hayward offers the formula

$$\mu = (1 + 0.015\,X_a)\mu_f \quad \text{(Hayward)} \tag{8.1e}$$

where X_a is the percentage volume occupied by air bubbles and μ_f the viscosity of the oil. Equ. (8.1e) thus differs qualitatively from the other expressions by asserting that the viscosity of a mixture actually increases with an increase in gas content.

From the tribological viewpoint there are at least three generic families of two-phase lubrication, each likely to have a different impact on the behavior of the system. The first is what may be called the steady state family of homogenous two-phase flows. This would pertain to lubricants which, though consisting of two or more phases, are homogenous throughout the fluid film. An example of this would be a foaming oil, or a fluid carrying a suspension of fine solid particles. The other group would consist of processes involving a change of state such as vaporization, liquefaction, or solidification of the lubricant. These changes brought about by the lubricant reaching a saturated thermodynamic state would occur inside the narrow confines of the fluid film. The radical changes in specific volume and the release or absorption of the high enthalpies inherent in such transitions call into question the very ability of a bearing or a seal to cope with such radical transformations.

Finally, there is the case of cavitation which represents essentially the case of liquids mixed with a certain quantity of gases, released from solution or sucked in from the environment. Also, should cavitation pressures be low enough to fall below the vapor pressure of the lubricant, vaporization of the lubricant will take place.

8.1 LIQUID-SOLID PHASES

In considering a liquid lubricant carrying a suspension of solid particles one obvious distinction between possible domains would be whether the particles are larger or smaller than h. Admittedly, even when the particles are smaller than h their presence will affect the dynamics of the fluid film, but with particles larger than h a completely new mode of operation may be expected. This division has been adapted by Khonsari and Esfahanian, 1988, who essentially assign two different stress-strain relationships to the two domains, as follows:

$$For \quad \delta < h \qquad \tau_1 = \mu(c, T, \rho) \left(\frac{\partial u}{\partial y} \right) \qquad (8.2a)$$

$$For \quad \delta > h \qquad \tau_2 = \mu(c, T, \rho) \left(\frac{\partial u}{\partial y} \right) \pm c\tau_p \qquad (8.2b)$$

where c is the solid fraction and τ_p the extra stress caused by the presence of the solids. As seen, the first case retains essentially the Newtonian relation, except that it makes the viscosity dependent also on the solid fraction. The second model, however, represents a non-Newtonian fluid, accounting for the stress due to the shearing of the solids at the mating surfaces. The non-Newtonian model is restricted by the requirement that c be low, of the order of a few percent.

Since

$$\frac{\partial \tau_2}{\partial y} = \frac{\partial}{\partial y} \left[\mu \left(\frac{\partial u}{\partial y} \right) \right] = \left(\frac{\partial p}{\partial x} \right)$$

a form identical to that of a conventional fluid, the Reynolds equation for the liquid-solid mixture, whether of the first or second kind, will be no different from that of ordinary analysis. In the energy equation, however, an additional term will appear accounting for the added dissipation engendered by the solid phase. This term, used in an appropriately normalized energy equation, can be written as

$$\frac{c\tau_p R^2}{kT_1} \left| \frac{\partial u}{\partial y} \right|$$

The problem is then solved in the usual way using the viscosity formula

$$\mu = C_1 \exp \left[-(C_2 T_1 + C_3) \left(\frac{T}{T_1} \right) \right] + C_4$$

where the constants are given in Table 8.1, along with the other pertinent data to which the energy and Reynolds equations have been applied. An

TABLE 8.1

Constants used in liquid-solid analysis

Parameter	Symbol	English Units	SI Units
Journal radius	R	1.084 in.	2.753 cm
External bearing radius	R_s	1.5084 in.	3.831 cm
Radial clearance	C	1.0×10^{-3} in.	2.54×10^{-3} cm
Bearing length	L	1.0 in.	2.54 cm
Speed	N	3500 rpm	-------
Bearing thermal conductivity	k_s	28.8 Btu/hr-ft-°F	49.84 W/m-°C
Inlet temperature	T_1	140.33°F	60.18°C
Oil supply pressure	P_s	18 psi	124×10^3 N/m²
Ambient temperature	T_a	77.4°F	25.22°C
Viscosity at inlet temperature	μ_1	3.967×10^{-6} Reyns	0.273 poise
Lubricant specific heat	C_p	0.46 Btu/lbm-°F	1730 J/kg-°C
Lubricant thermal conductivity	k	0.079 Btu/hr-ft-°F	0.137 W/m-°C

Constants used in μ-T expression

$$C_1 = 0.26 \text{ Reyns} = 1.79 \times 10^4 \text{ poise}$$

$$C_4 = 0.26 \times 10^{-6} \text{ Reyns} = 1.79 \times 10^{-2} \text{ poise}$$

$$C_2 = 0.0335 \ (1/°C)$$

$$C_3 = 9.1512$$

additional refinement used here is that the inlet temperature T_1 is the inlet mixing temperature obtained on the basis of formula (2.9).

The particular two-phase lubricant chosen for study is an SAE 30 oil carrying one of two kinds of particles, viz.

$$\text{MoS}_2: \quad c = 1\%; \ \delta = 18 \,\mu\text{m}; \ \tau_p = 8 \times 10^5 \text{N/m}^2 \ (118\,\text{psi})$$

$$\text{PTFE(Polytetrafluoroethylene)}: \quad c = 1\%; \ \delta = 24 \,\mu\text{m};$$

$$\tau_p = 6.9 \times 10^5 \text{N/m}^2 \ (100\,\text{psi})$$

The values of τ_p listed above were taken from experiments by Rylander, 1966, and Figs. 8.1 and 8.2 show the friction coefficients as obtained

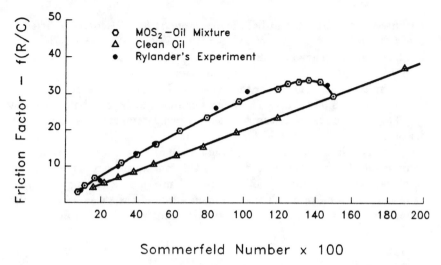

Fig. 8.1 Friction due to MoS₂ particles in lubricant, Khonsari and Esfahanian, 1988

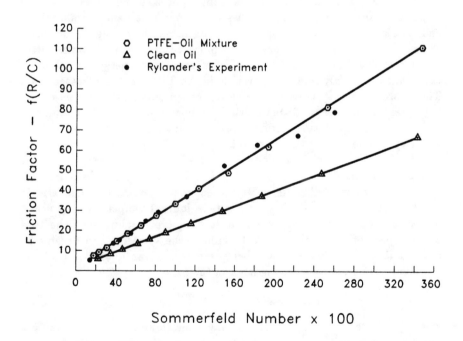

Fig. 8.2 Friction due to PTFE particles in lubricant, Khonsari and Esfahanian, 1988

from his tests and from the present THD analysis. With the smaller MoS₂ particles the friction curve is higher than that for a clean oil over low S numbers, but approaches that of a clean oil at an $S = 1.5$. This can be attributed to the fact that at light loads (large S) h_{min} becomes sufficiently large to permit easy passage of the particles. With the larger PTFE particles, however, the curve for the two-phase lubricant stays above that of a clean oil throughout the entire range of loads.

A similar trend is noted for the behavior of shaft temperature and T_{max} plotted in Figs. 8.3 and 8.4. The figures show an unexpected minimum at $\epsilon = 0.3$. Also unusual is the fact that journal temperatures are nearly identical with the mean fluid film temperature. Most likely both are due to the artificiality of some of the boundary conditions, as, for example, the requirement that the net heat flux into the journal be zero. Finally, Fig. 8.5 shows the load capacity for liquid-solid mixtures as being somewhat lower than for a clean oil. Thus the main effect of the presence of a solid suspension seems to be a rise in maximum temperature which in terms of ΔT_{max} is of the order of $25 - 30\%$.

8.2 LIQUID-GAS MIXTURES

The case of a bubbly oil differs from liquid-solid mixtures in several respects. Clearly there is no subdivision here into small and large bubbles, and no discrete increase in stress upon contact with the mating surfaces. From a mathematical standpoint, the presence of gaseous bubbles introduces compressibility into both the momentum and energy equations, a facet thus far of only minor importance in the THD analyses treated here. The bubbles, as they traverse the clearance space, will shrink due to the high pressure and expand due to the rise in temperature. Consequently an analysis of liquid-gas mixtures must account for the variation of both viscosity and density of the lubricant.

One such attempt was made by Abdel-Latif et al., 1985, applicable to a circular thrust pad shown in Fig. 8.6. The energy and heat transfer equations in their familiar form were used, minus the conduction term in the θ direction, i.e. $(\partial^2 T / \partial \theta^2) = 0$; and the journal temperature, as usual, was taken to be constant. The basic technique of dealing with the two-phase aspect of the problem was to modify the viscosity and density by a dependence on a term c, which represents the gas/liquid ratio of the lubricant. Thus viscosity was written in the form

$$\mu = \mu_f \left[1 - c(p, T)\right]$$

where μ_f is the viscosity of the homogenous liquid. Next, perfect gas equations were utilized to arrive both at a formulation of c and of the variable density. The effect of surface tension forces on the bubbles was included. Thus for a bubble of radius r_0 surrounded by a liquid at a pressure p_l, the gas pressure is given by

$$p_g = p_l + \left(\frac{2\sigma}{r_0}\right) \tag{8.3}$$

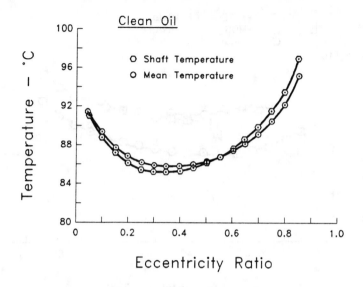

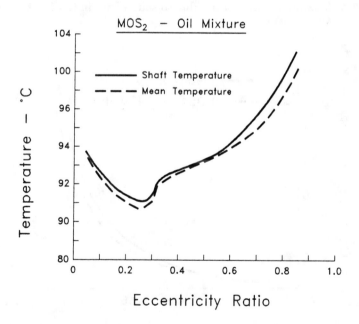

Fig. 8.3 Mean and shaft temperatures due to solid phase in lubricant, Khonsari and Esfahanian, 1988

where σ is the surface tension.

With the help of the polytropic relation $pv^n = const$ the function c

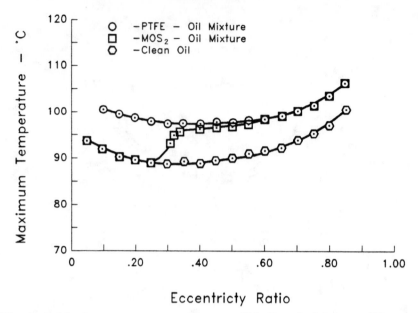

Fig. 8.4 Maximum temperature due to solid phase in lubricant, Khonsari and Esfahanian, 1988

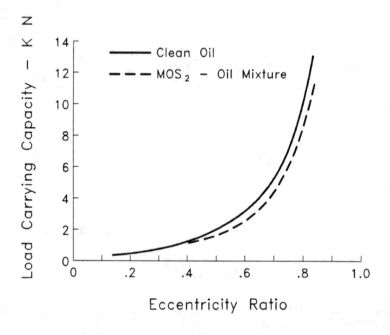

Fig. 8.5 Effect of solid suspension on load capacity, Khonsari and Esfahanian, 1988

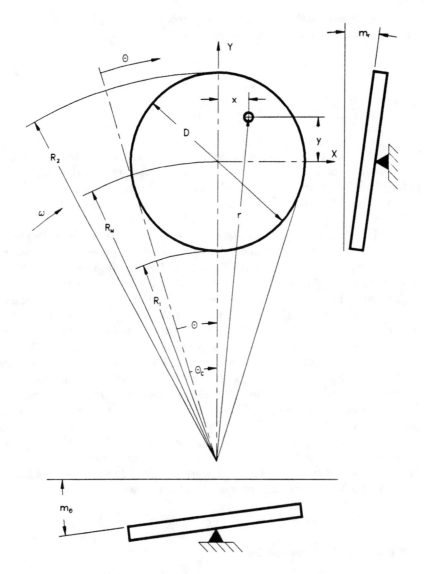

Fig. 8.6 Configuration of circular thrust pad

was found to assume the form

$$c = \left(\frac{v_g}{v_l}\right)_a \left[\frac{\bar{p}_a + 2}{A\bar{p}_l + \bar{p}_a + (2D/r)}\right]^{1/n} \tag{8.4}$$

where subscript a refers to ambient conditions, and

$$\bar{p} = \frac{p}{(\sigma/r_0)}; \quad n = 1.2; \quad A = \frac{\mu_1 R_1 D r_0 \omega}{\sigma h_2^2}$$

The density of the mixture, as affected by changes in bubble volume with pressure and temperature, is given as

$$\rho = \rho_a \left[1 + \frac{\sigma_a}{(1 + \gamma \bar{p}_l)^{\frac{1}{n}} (T/T_1)} \right] \tag{8.5}$$

where

$$\gamma = \frac{\mu_1 R_1 D \omega}{h_2^2}$$

In the use of the above relationships such effects as bubble collection and possible solubility were ignored. Thus solved, the results showed no perceptible departure either in load capacity, friction, or temperature levels from results obtained for a clean liquid. The only effect seemed to be a shift of the center of pressure toward the trailing edge of the pad.

This lack of any significant departure from isotropic lubricant behavior tallies with a series of experiments conducted by the present author where in an attempt to simulate severe oil foaming, large quantities of air bubbles were forced into the clearance of a journal bearing, the injection being as high as 90% of air by volume. This aeration made no perceptible impression on the performance of the bearing either in the value of eccentricity, power loss, or stability of the system. One observation made during the test, which goes far toward explaining the indifference of the bearing to the injection of air, is that much of the air was expelled very early, in the upstream portions of the fluid film. The remainder that did enter the clearance space was radically reduced in volume by the high pressures prevailing near h_{min}. The temperature should have only a minor effect since the volume is proportional to the absolute T and would thus have only a marginal effect on the percentage change. From this viewpoint a lightly loaded bearing may thus be more susceptible to the effects, positive or negative, of a bubbly lubricant than would a bearing at high eccentricity. The analysis conducted above seems to corroborate some of the above intuitive interpretations.

8.3 CHANGE OF PHASE

8.3.1 General Characteristics

There are a number of tribological devices whose operation involves, potentially at least, a change of phase of the fluid. Among these are seals in petroleum refineries, in hot water boiler feed pumps, and in reactor coolant pumps, as well as seals, bearings and dampers in cryogenic turbo-pumps. Difficulties attendant the phenomenon of change of phase include a negative stiffness over certain ranges of operation, leading to unstable equilibrium and collapse of film. Even when operation is possible, self-sustained oscillations may lead to chatter, ultimately causing destruction of the seal. All of these phenomena, as well as the routine performance characteristics, are critically dependent on the thermal state of the fluid.

Fig. 8.7 shows the fluid path through a seal on a $T - s$ diagram. The distance from saturated liquid to saturated vapor in the seal can vary from zero to the entire length of the seal. For low leakage seals the points will be close together and vaporization will take place at a discrete distance; for high leakage seals the boiling may be spread over the entire seal. The closer to isothermal operation, the shorter the region over which boiling occurs.

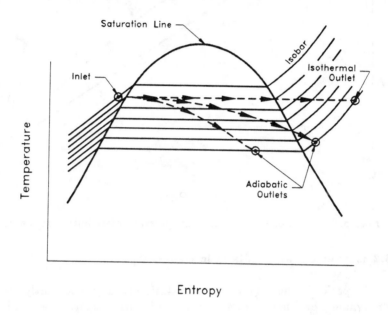

Fig. 8.7 Isothermal and adiabatic vaporization of a liquid

Typical pressure drops through a seal are given in Fig. 8.8. If the fluid in the seal is all liquid, the pressure is nearly linear; if all gas it is nearly quadratic. However, if vaporization occurs, then for a given film thickness the leakage is reduced and the pressure is higher than either for an all-liquid or all-gas seal. The opening force engendered by this pressure as a function of h produces the curve of Fig. 8.9 where the positive slope represents negative axial stiffness. The instability picture thus generated is that given in Fig. 8.10. Operation on the right hand side of the curve in Fig. 8.10 is stable but the left side is unstable. The behavior depends critically on the state of the sealed liquid, particularly its degree of subcooling. As the sealed fluid nears saturation conditions, the stiffness curve becomes entirely positive and the seal is unstable. Hence, a seal operating satisfactorily at one set of conditions may become unstable if the temperature is changed, as shown in Fig. 8.11 for a fluid whose saturation temperature is 453°K. Thus the temperature distribution along the seal is a crucial determinant in the performance and stability of two-phase seals.

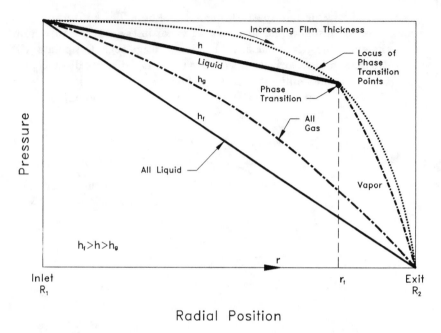

Radial Position

Fig. 8.8 Effect of phase change on pressure distribution in a seal

8.3.2 Isothermal and Adiabatic Cases

The path of a fluid in seals, hydrostatic bearings, and partly also in hydrodynamic bearings is toward a state of higher temperatures and lower pressures. Therefore the most likely path is that of a liquid flashing into vapor rather than the condensation of a gas. The two thermal extremes in such a process are isothermal and adiabatic trajectories, portrayed in Fig. 8.7. The isothermal model corresponds to a system of low flow rates while the adiabatic one corresponds to high flows in which convection is the dominant mode of heat transfer. In the introductory remarks the change of phase was visualized as possibly occurring instantaneously, raising serious questions about the ability of a seal or bearing to accommodate such radical transitions. This mode has not been yet tackled by experiments, and analysts, too, prefer to avoid such a scenario. The more common approach deals with vaporization as a gradual process. In this mode the liquid-vapor mixture while remaining a homogenous substance continuously rises in quality* with the properties, such as density and viscosity, varying accordingly. Such a gradual process would presumably allow the flow to accommodate itself to the changing conditions. An additional assumption inherent in this scenario is that the heat of vaporization be low relative to

---

* The term quality is here used in its thermodynamic sense, denoting the mass ratio of vapor to total mass of fluid

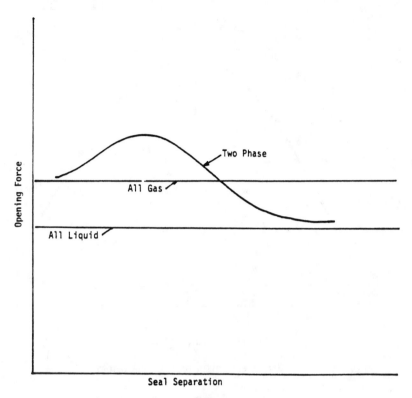

Fig. 8.9 Force produced in two phase flow in a seal

the dissipation energy. Otherwise, with large amounts of heat being taken from the fluid film to vaporize the liquid the properties of the mixture may become a function of the heat transfer process; moreover, temperature discontinuities may occur at the interfaces, if discrete boiling is assumed. Still, as will be seen later on, even with slow vaporization instabilities are triggered which can lead to the destruction of the device.

In the work of Hughes and Chao, 1979, the lubricant considered was steam, with its thermodynamic and transport properties taken from appropriate "real" fluid tables. This was applied to a face seal which had no variations in the circumferential direction but otherwise could be flat, convergent, or divergent in the radial direction. Depending on its initial and operating conditions the liquid could start vaporizing at any radial position.

Since tabulated thermodynamic properties are to be used, the convective term in the energy equation is here written in terms of its enthalpy i, namely

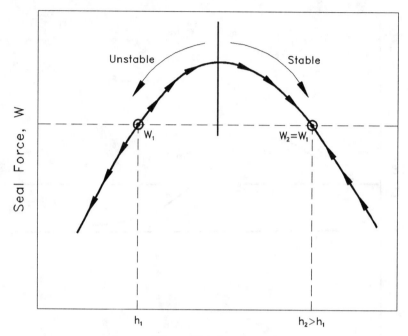

Fig. 8.10 Mechanism of seal instability

$$\rho w \left(\frac{\partial i}{\partial r} \right) = w \left(\frac{dp}{dr} \right) + k \left(\frac{\partial^2 T}{\partial y^2} \right) + \mu \left[\left(\frac{\partial u}{\partial y} \right)^2 + \left(\frac{\partial w}{\partial y} \right)^2 \right] \qquad (8.6)$$

Aside from ignoring all θ variations, conduction in the r direction was ignored. Since the mass flow, a constant, is given by

$$m = 2\pi r \int_0^h \rho w \, dy$$

and

$$\int_0^h w \left(\frac{\partial p}{\partial r} \right) dy = \int_0^h \mu w \left(\frac{\partial^2 u}{\partial y^2} \right) dy$$

$$= \int_0^h \mu \frac{\partial}{\partial y} \left(u \frac{\partial w}{\partial y} \right) dy - \int_0^h \mu \left(\frac{\partial w}{\partial y} \right)^2 dy$$

we have for Equ. 8.6

$$\frac{m}{2\pi r} \left(\frac{di}{dr} \right) = \int_0^h \mu \frac{\partial}{\partial y} \left[w \frac{\partial w}{\partial y} \right] dy + \int_0^h \mu \left(\frac{\partial u}{\partial y} \right)^2 dy + \int_0^h \frac{\partial^2 T}{\partial y^2} dy \quad (8.7)$$

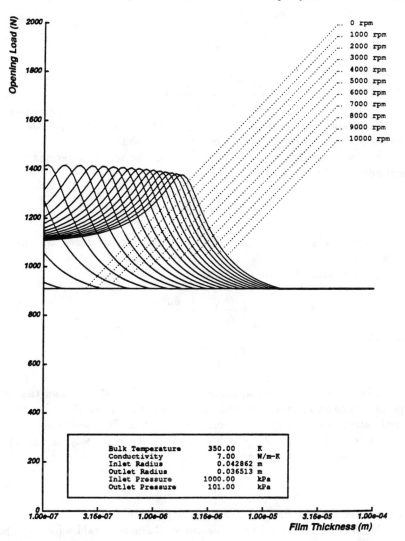

Fig. 8.11 Load versus film thickness for different fluid temperatures, Hughes and Basu, TW 1988

From $w = 0$ at $y = 0$ and $y = h$, the first term on the right hand side above is zero. Also

$$k \int_0^h \frac{\partial^2 T}{\partial y^2} dy = k \frac{\partial T}{\partial y}\bigg|_h - k \frac{\partial T}{\partial y}\bigg|_0 = -Q$$

and since for $\left(\frac{\partial p}{\partial \theta}\right) = 0$ we have

$$\mu \left(\frac{\partial^2 u}{\partial y^2}\right) = 0$$

or

$$u = \frac{r\omega}{h} y$$

we get

$$\frac{m}{2\pi r}\left(\frac{di}{dr}\right) = \frac{\mu(r\omega)^2}{h} - Q \qquad (8.8)$$

Equation (8.8) holds for either single or two phase fluids provided values for μ, k, etc. appropriate for mixtures are used.

For isothermal cases for which $di/dr = 0$, the heat transfer quantity Q is simply equal to that of dissipation or

$$Q = \mu\frac{(r\omega)^2}{h}$$

This heat is absorbed by the surfaces whose temperatures are thereby raised.

For the adiabatic case $Q = 0$, and

$$\left(\frac{dp}{dr}\right) = -\frac{6\mu m}{\pi r \rho h^3}$$

and we thus have

$$\left(\frac{di}{dr}\right) = \frac{2\pi r^3 \mu \omega^2}{mh}$$

Above, the quantities μ, ρ and i are functions of p, T and the local quality. The equations must thus be solved simultaneously using appropriate state functions for the involved parameters. It is worth noting that μ as a function of quality λ is here treated the same way as any other thermodynamic property, that is

$$\mu = \lambda\mu_g + (1 - \lambda)\mu_f \qquad (8.9)$$

the subscript g denoting vapor and f the liquid.

With these functional relationships available from tables (or appropriate equations of state) the two initial conditions p_1 and T_1 and the exit pressure p_2 suffice to determine the state of the fluid, including its two-phase composition, at each station r. In the present case T_1 was set equal to T_2 and the seal model used was that of a flat geometry with the flow occurring inwards, from the larger to the smaller radius, a so-called outside seal.

The mechanism of incipient instability is illustrated for both the isothermal and adiabatic cases, in Fig. 8.12. As seen, there are two values of h which produce the same total load W, the adiabatic case exhibiting this double-valued behavior at higher values of h than the isothermal case. As indicated in Fig. 8.10 the larger value of h represents a stable equilibrium, whereas any excursion about the lower value of h is unstable. A perturbation causing an increase in h leads towards a stable value of h; however,

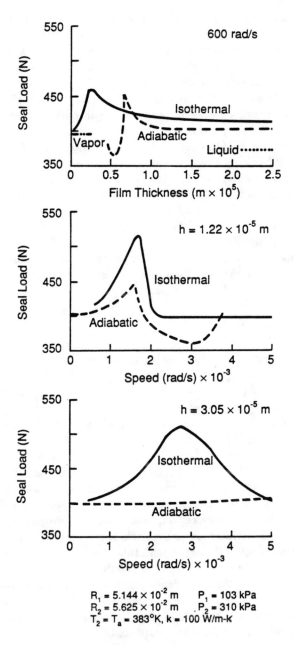

Fig. 8.12 The double valued function W, Hughes and Chao, 1979

a decrease in h leads to ever smaller values of h and collapse of the seal. The associated thermal behavior of the seal depends on the time constants of the transients involved. Generally, the excursion transients are at much shorter duration than any associated thermal transient. Thus an unstable collapse generates more heat than can be carried away. Consequently the W versus h behavior is more accurately represented by the adiabatic mode for all values of h, even though under steady state conditions for small h's the isothermal approach is more correct.

Fig. 8.13 shows plots of temperature and pressure throughout a typical seal for the adiabatic case. The temperature in part (a) increases in the liquid region, then drops through the two-phase region; the sharp knee in the temperature profile occurs at the saturation line. In part (b) the two-phase fluid becomes vapor before leaving the seal, hence the two sharp bends in the temperature profile. Here, in the vapor region, the temperature actually rises. Mass flow rate is shown in Fig. 8.14. It is important to note that the leakage rate is reduced significantly as vaporization occurs. The all liquid and isothermal gas flow limits are also shown. The adiabatic gas flow rate will generally be less than the isothermal limit because of the higher kinematic viscosity due to the higher temperature. As compared to an all liquid seal the mass leakage rate may decrease by one or two orders of magnitude because of vaporization.

8.3.3 Heat Transfer Effects

While low and high leakage seals can be represented by isothermal and adiabatic processes respectively, most seals actually fall in the middle region, thus involving both convection and conduction in the thermal path. A particularly untenable proposition in two-phase flow is to postulate equality of wall and fluid mean temperatures which, even if it held for the liquid zone, could not possibly prevail once vaporization commenced.

The three different regions—liquid, mixed, and vapor—all have different characteristics. In the liquid region heat generation is high, convection is low, and heat is conducted away from the liquid. Consequently liquid temperatures are higher than those of the wall, although the difference is very small. In the two-phase region, heat generation drops with an increase in quality and convection reaches the highest level in that region. Since the heat generation is not sufficient to maintain vaporization, heat is conducted into the fluid film from the walls. Here the fluid temperature is lower than the wall temperature, their difference reaching a maximum inside the two-phase domain. Over the vapor region heat generation is minimal and since the fluid temperature at the end of the two-phase region is low, conduction to the vapor takes place from the walls and the film temperature rises. A diagram of wall and fluid temperatures is shown in Fig. 8.15.

Since heat transfer is involved, a rigorous approach would require a solution for the tranverse temperature distribution, $T(y)$. Shying away from this complexity Hughes and Basu, TW 1988, approached the problem by what they called the "film-coefficient model," in which they aimed at a one-dimensional, radial solution only, accounting for heat conduction by

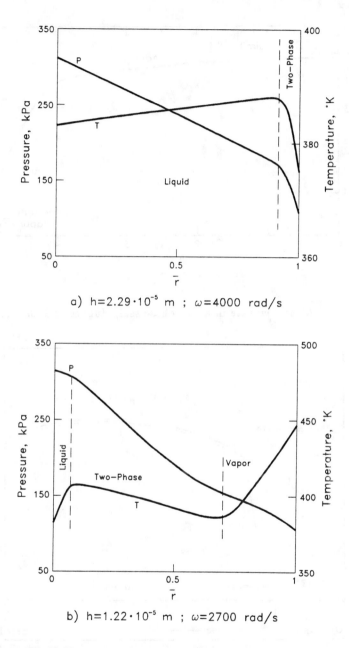

a) $h=2.29 \cdot 10^{-5}$ m ; $\omega=4000$ rad/s

b) $h=1.22 \cdot 10^{-5}$ m ; $\omega=2700$ rad/s

Fig. 8.13 Fluid film properties for adiabatic expansion, Hughes and Chao, 1979

an assumed Nusselt number distribution along the film. The seal geometry and the fluid are the same as in the previous section. Starting with a

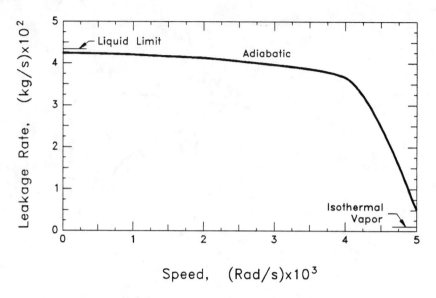

Fig. 8.14 Mass flow rate in a two-phase seal, Hughes and Chao, 1979

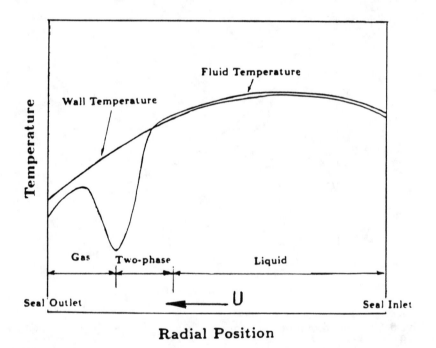

Fig. 8.15 Fluid and wall temperatures in seal, Hughes and Basu, TW 1988

momentum equation in which centrifugal forces are included we have

$$-\frac{dp}{dr} + \mu \frac{\partial^2 w}{\partial y^2} = -\frac{\rho u^2}{r} \tag{8.10a}$$

and integrating across the film twice, this becomes

$$\frac{dp}{dr} = -\frac{6m\mu}{\pi \rho r h^3} + \frac{3\rho r \omega^2}{10} \tag{8.10b}$$

For the energy equation a form integrated across y is used, namely

$$\int_0^h \rho w \frac{di}{dr} dy = \int_0^h w \frac{dp}{dr} dy + \mu \int_0^h \left[\left(\frac{\partial u}{\partial y} \right)^2 + \left(\frac{\partial w}{\partial y} \right)^2 \right] dy$$
$$+ \int_0^h k \left(\frac{\partial^2 T}{\partial r^2} + \frac{1}{r} \frac{\partial T}{\partial r} + \frac{\partial^2 T}{\partial y^2} \right) dy$$

Multiplying the momentum equation (8.10a) by w and integrating across the film, we have

$$\int_0^h w \frac{dp}{dr} dy = \mu \int_0^h w \frac{\partial^2 w}{\partial y^2} dy + \rho \int_0^h \frac{w u^2}{r} dy$$
$$= \mu \left[w \frac{\partial w}{\partial y} \Big|_0^h - \int_0^h \left(\frac{\partial w}{\partial y} \right)^2 dy \right] + \rho \int_0^h \frac{w u^2}{r} dy$$

Combining with the energy equation and neglecting conduction along the film the result is

$$\frac{m}{2\pi r} \left(\frac{di}{dr} \right) = \mu w \frac{\partial w}{\partial y} \Big|_0^h + \rho \int_0^h \frac{w u^2}{r} dy + \int_0^h \mu \left(\frac{\partial w}{\partial y} \right)^2 dy + \int_0^h k \frac{\partial^2 T}{\partial y^2} dy$$

Since $w = 0$ at $y = 0, h$, the first term on the right-hand side goes to zero. The term

$$\int_0^h k \frac{\partial^2 T}{\partial y^2} dy$$

integrates to

$$k \frac{\partial T}{\partial y} \Big|_{y=h} - k \frac{\partial T}{\partial y} \Big|_{y=0} = -q$$

where q is the heat conduction into the seal from the fluid.

Using the proper expressions for u and w the integrated energy equation is obtained

$$\frac{m}{2\pi r} \frac{di}{dr} = \mu \frac{r^2 \omega^2}{h} + \frac{3\omega^2 m}{20\pi} + \frac{\rho^2 r^2 h^3 \omega^4}{700\mu} - h_s [T - T_{S,R}] \tag{8.11}$$

with

$$q = h_s[T - T_{S,R}] \tag{8.12}$$

Above, h_s is the heat transfer coefficient between the fluid film and the seal walls and is related to the assumed Nusselt number distribution along the flow path via

$$Nu = \frac{h_s d}{k}$$

with $d = 2h$, the hydraulic diameter.

Thus far the equations have been stated without regard to the two-phase nature of the fluid. This is now introduced as follows: specific volume, enthalpy, and thermal conductivity are expressed by their partial properties

$$v = v_f + \lambda v_{fg}$$

$$i = i_f + \lambda i_{fg}$$

$$k = k_f + \lambda k_{fg}$$

The viscosity is modelled according to the volume fraction

$$\mu = (1 - \lambda)\frac{v_f}{v}\mu_f + \lambda\frac{v_g}{v}\mu_g$$

or

$$\nu = \nu_f + \lambda\nu_{fg}$$

Substituting the above relationships for the two-phase thermodynamic and transport properties in Equs. (8.10) and (8.11) we obtain

$$\frac{dp}{dr} = -\frac{6m(\nu_f + \lambda\nu_{fg})}{\pi rh^3} + \frac{3r\omega^2}{10(v_f + \lambda v_{fg})} \tag{8.13}$$

$$\frac{d\lambda}{dr} = \frac{1}{i_{fg}}\left[\frac{2\pi\omega^2 r^3}{mh}\left(\frac{\nu_f + \lambda\nu_{fg}}{v_f + \lambda v_{fg}}\right) + \frac{3\omega^2 r}{10}\right.$$
$$+ \frac{\pi h^3 \omega^4 r^3}{350m}\frac{1}{(v_f + v_{fg})(\nu_f + \nu_{fg})} - \frac{2\pi rh_s[T - T_{S,R}]}{m}$$
$$\left. - \frac{di}{dp}\left\{-\frac{6m(\nu_f + \nu_{fg})}{\pi rh^3} + \frac{3r\omega^2}{10(v_f + \lambda v_{fg})}\right\}\right] \tag{8.14}$$

For the energy equation the two seal faces are assumed to be backed up by semi-infinite solids, as shown in Fig. 8.16. The temperature at any r due to a heat source at r' may be expressed as

$$T(r) = \int_A \frac{q(\vec{r'})dA}{4\pi k|\vec{r} - \vec{r'}|}$$

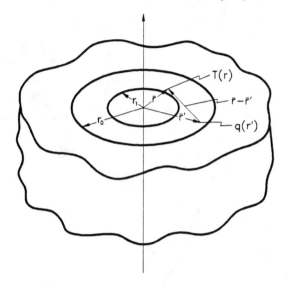

Fig. 8.16 Geometry of seal face

where A is the area over which heat is generated. The energy equation can be written in matrix form

$$\{T\} = [C]\{q\} + \{T_a\}$$

where q is given by Equ. (8.12).

In the sample solutions presented below convection in the liquid region has been neglected. Nusselt numbers for an infinite aspect ratio cross-section like seals, with either an incompressible liquid or perfect gas with no heat generation, are calculated to be 8.235 for a constant wall heat flux and 7.54 for a constant wall temperature. Thus a value of $Nu = 8$ was used.

$(F - h)$ and leakage curves are presented in Figs. 8.17 and 8.18. The opening force F is over-predicted by the isothermal model while stiffness and leakage rates are the same for both methods of analysis. The variation of the fluid properties along the seal interface are shown in Fig. 8.19 for a film thickness of $4\,\mu$m. Vaporization starts close to the outlet. Since heat generation is not sufficiently high, complete vaporization does not occur. The fluid leaves the seal in a two-phase state with a quality of about 0.05.

Curves for a very low film thickness, $0.5\,\mu$m, are presented in Figs. 8.20 and 8.21. Because of the small h, the leakage rate is lower, heat generation is higher, and complete vaporization occurs here. Because of the high fluid pressures, this vaporization occurs close to the outlet. As seen, the fluid and wall temperatures in the liquid region differ by a fraction of a degree only. With vaporization taking place, the fluid temperature drops in the two-phase region with T reaching a minimum at the end of the

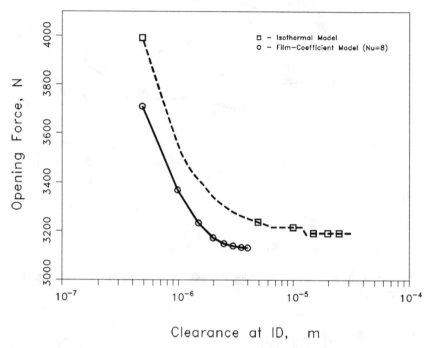

Clearance at ID,   m

Fig. 8.17 Opening force in seal as function of film thickness, Hughes and Basu, TW 1988

two-phase domain. As a result, at the beginning of the vapor region there is a high rate of heat conduction into the fluid from the walls, and since vaporization is already complete the gas temperature rises in a very short distance. Over the remainder of the gas region, the fluid temperature is lower than at the wall and some amount of heat is conducted into the fluid. This conductive heat, together with whatever heat dissipation occurs, goes mainly into expansion work by the gas. Vaporization takes place over a very small distance, 1.5% of the interface length, which is only 0.07 mm in this case. The energy required to vaporize the fluid film is some 5% of the heat generated. Therefore, vaporization at a discrete point may not be a bad assumption for this and similar cases.

8.4 MELT LUBRICATION

Melt lubrication consists of a process where the lubricant is provided by melting one of the interacting surfaces. Thus what we have here is a phase change from solid to liquid. The heat of fusion required for this

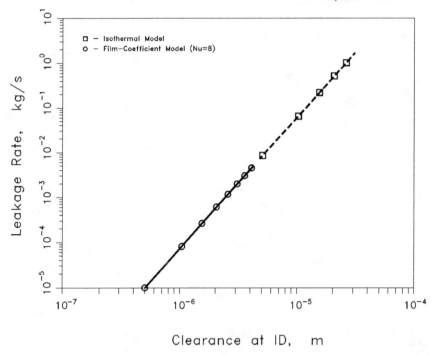

Fig. 8.18 Leakage rate in two-phase seals, Hughes and Basu, TW 1988

transformation can come by heat transfer from the non-melting surface; from the intense viscous dissipation induced by the sliding surfaces; or it can, of course, involve both of these processes.

Such a melt process has a much wider range of practical application than would seem at a first look. It ranges, in fact, as far as the ordinary sliding and skidding on ice or snow. The original interpretation of this mechanism given by Reynolds himself saw the formation of a liquid film between the skate and ice as a result of a lowered melting point due to the elevated pressure. However, more recent studies showed that the liquid film is, in fact, the product of sliding friction. Other common applications are in the field of internal ballistics where the surface of a projectile is made to melt on its passage through the gun barrel; test sleds on rails whose high velocities are made possible by melt lubrication at the interface; or more recently, the coating of metals with a low melting point overlay to facilitate the manufacturing process.

The following analysis is based on the presentation by Bejan at the 1988 Thermal Workshop as well as on his 1989 paper on the same subject. Bejan visualized two possible models, one a block of melt material pressed against

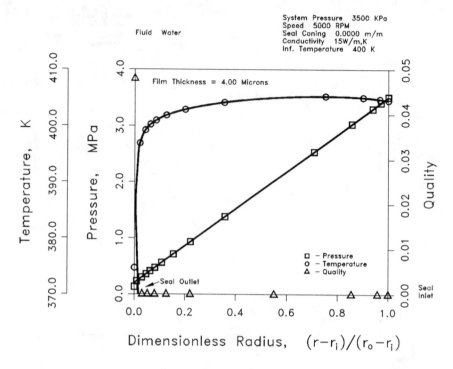

Fig. 8.19 Variation of fluid properties along seal interface, Hughes and Basu, TW 1988

a moving surface, Fig. 8.22a; or a solid block pressed against a moving melt material, Fig. 8.22b. The first model yields a constant-thickness liquid film with melt material translating downwards with a velocity V—its melting rate. In the second model the stationary solid block bites into the melt material. Here a slight tilt is produced which means that effectively the solid block does translate vertically relative to the melt material. This velocity V is one of the unknowns of the problem, as is the frictional force F_r causing fusion of the melt material.

In the analysis the melt material is taken to remain at the constant fusion temperature T_m, that is an isothermal surface, whereas the opposite surface is taken to be adiabatic.

In the energy equation, the effects of convection, conduction, and dissipation are represented by the groupings $(U\Delta T/B)$, $(\sigma\Delta T/h^2)$ and $(\mu/\rho c_p)(U/h)^2$. In the present case convection is neglected leaving the dissipation to be carried away by transverse conduction. Thus the temperature rise in the film is $\Delta T \sim (\mu/k)U^2$ and the temperature of the non-melting surface is $(T_m + \Delta T)$.

The next assumption made is that of an infinitely long slider. The

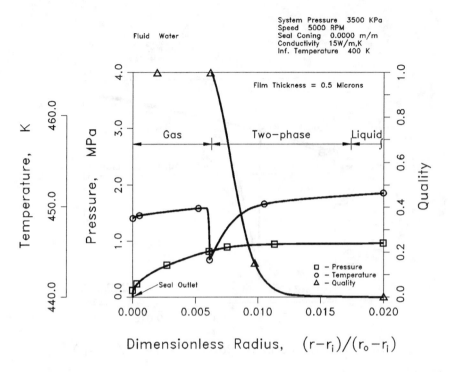

Fig. 8.20 Variation of fluid properties near seal outlet, Hughes and Basu, TW 1988

energy equation assumes then the extremely simple form

$$\frac{\partial^2 T}{\partial y^2} = -\frac{\mu}{k}\left(\frac{\partial u}{\partial y}\right)^2$$

Using the familiar expression for $\left(\frac{\partial u}{\partial y}\right)$, integrating twice, and applying the boundary conditions $\Delta T = 0$ at $y = h$ and $\left(\partial T/\partial y\right) = 0$ at $y = 0$, we obtain

$$T = \frac{\mu}{k}\left[\frac{G^2 h^4}{24\mu^2}(1-\overline{y})(2\overline{y}^3 - 2\overline{y}^2 + \overline{y} + 1)\right.$$
$$\left. - \frac{Gh^2 U}{6\mu}(1-\overline{y})^2(2\overline{y}+1) + \frac{U^2}{2}(1-\overline{y}^2)\right] \qquad (8.15)$$

where $G = -(dp/dx)$. We also have

$$Q_x = \frac{Gh^3}{12\mu} + \frac{Uh}{2} \qquad (8.16a)$$

$$\frac{dQ_x}{dx} = V \qquad (8.16b)$$

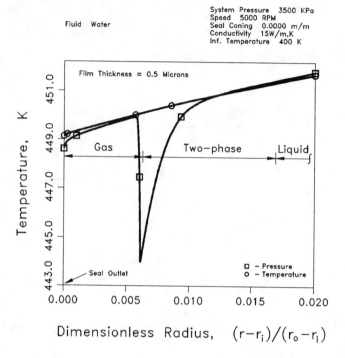

Fig. 8.21 Variation of temperatures near seal outlet, Hughes and Basu, TW 1988

$$Q_x = -Q_0 + V_x \qquad (8.16c)$$

where Q_0 is the back flow at $x = 0$. This yields $(VB - Q_0)$ as the amount of molten material leaving the slider at $x = B$. We then have

$$-\frac{Gh^2}{6\mu U} = 1 - \frac{2}{H}(\overline{x} - \tilde{Q}_0) \qquad (8.16d)$$

where

$$H = \frac{hU}{BV} \qquad (8.16e)$$

$$\tilde{Q}_0 = \frac{Q_0}{BV} \qquad (8.16f)$$

The melting rate that fixes the position of the upper surface of the film is controlled by the amount of heat conducted to that surface, or

$$-k\left(\frac{\partial T}{\partial y}\right)_{y=h} = \rho h_{sf} V$$

where h_{sf} is the heat of fusion. The above, combined with the results of equations (8.15) and (8.16), yields for the film thickness

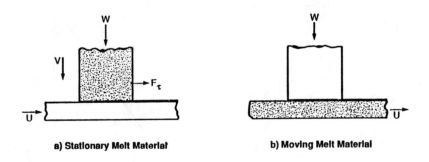

a) Stationary Melt Material   b) Moving Melt Material

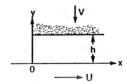

c) Coordinate System

Fig. 8.22 Modes of melt lubrication

$$H = 3M \left[1 - \frac{2}{H}(\bar{x} - \tilde{Q}_0) \right]^2 + M \qquad (8.17a)$$

where

$$M = \frac{\mu U^3}{\rho h_{sf} B V^2} \qquad (8.17b)$$

Above $\tilde{Q}_0$ and M are interrelated through the boundary conditions $p = 0$ at $x = 0$ and $x = B$. This can be expressed as

$$\int_0^B \left(\frac{dp}{dx} \right) dx = 0$$

which means

$$\int_0^1 \left[\frac{1}{H^2} - \frac{2}{h^3}(\bar{x} - \tilde{Q}_0) \right] d\bar{x} = 0$$

Figs. 8.23 and 8.24 show the relationship between $\tilde{Q}_0$ and M and the gap shape $H(x)$. M decreases monotonically with $\tilde{Q}_0$ which means that as V increases an increasing fraction of melt flows backwards. The fluid film has a peculiar converging-diverging shape whose minimum film thickness

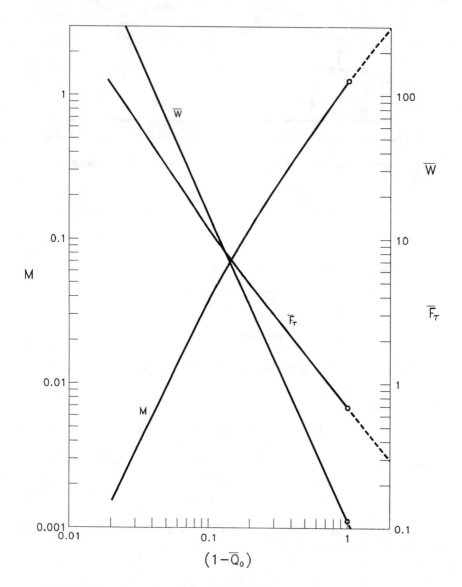

Fig. 8.23 Behavior of melt lubrication parameters, Bejan, 1989

occurs past the halfway x distance. Since the cubic for H has only one real root the transition between the two sections of H is smooth.

The pressure distribution illustrated in Fig. 8.25 plots the dimensionless quantity

$$\bar{p} \equiv \frac{pBV^2}{\mu U^3} \equiv I_p^{(}\bar{x}) = \int_0^{\bar{x}} 6\left[\frac{1}{H^2} - \frac{2}{H^3}(\bar{x}' - \tilde{Q}_0)\right] d\bar{x}' \qquad (8.18)$$

where we see that the peak pressure occurs just downstream of h_{min}. This

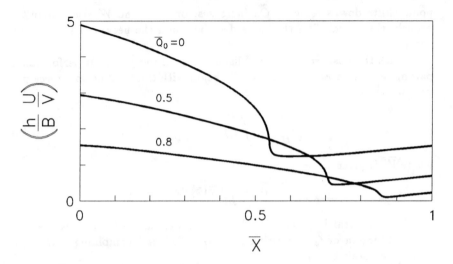

Fig. 8.24 Shape of fluid film in melt lubrication, Bejan, 1989

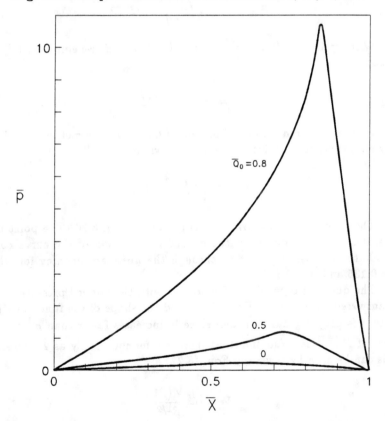

Fig. 8.25 Pressure distribution in melt lubrication, Bejan, 1989

point shifts downstream as $\tilde{Q}_0$ increases, or as V and W rise. Simultaneously the overall film thickness decreases and the necking becomes less conspicuous.

With the basic solution established some of the pertinent performance parameters can now be obtained. To begin with the total normal force per unit width is

$$\left(\frac{W}{L}\right) = \int_0^B p\, dx = \frac{\mu U^3}{V^2} \cdot \overline{W} \tag{8.19}$$

where $\overline{W}$ is given by

$$\overline{W} = \int_0^1 \overline{p}(\bar{x})\, d\bar{x}$$

This integral is plotted in Fig. 8.23 which shows that $\overline{W}$ is a single valued function of $\tilde{Q}_0$. Writing $P = W/(BL)$ and combining with Equ. (8.19) we obtain

$$\overline{P} \equiv \frac{PB^2}{\sigma\mu} = \left(\frac{U}{V}\right)^2 \cdot Pe \cdot \overline{W} \tag{8.20}$$

Writing $M = (U/V)^2 m$ where $m = (mU/\rho h_{sf}B)$ we eliminate $(U/V)^2$ to obtain

$$\beta = m\frac{\overline{P}}{Pe} \equiv \overline{W}M \tag{8.21}$$

where the right hand side is a function of $\tilde{Q}_0$ only. The melting speed can now be formulated using Equ. (8.20) and writing

$$V = U\left(\frac{Pe}{\overline{P}}\right)^{\frac{1}{2}} \overline{V}(\beta) \tag{8.22}$$

The function $\overline{V}(\beta) = [\overline{W}(\beta)]^{\frac{1}{2}}$ is plotted in Fig. 8.26. The point that separates the solid line from the dashed-line portion of the curve corresponds to the case of zero flow through the upstream opening for which $\beta = 0.135$ and $\beta = 0.333$.

The question of whether V increases with the applied pressure $\overline{P}$ can be answered by analyzing Equ. (8.21) and the shape of the function $\overline{V}(\beta)$. Since β is proportional to $\overline{P}$ and since V increases faster than $\beta^{\frac{1}{2}}$ the net effect of $\overline{P}^{-\frac{1}{2}}\overline{V}$ is a function that increases monotonically as $\overline{P}$ increases. This can be seen by rewriting Equ. (8.22) as

$$V = U_m^{\frac{1}{2}} \frac{\overline{V}(\beta)}{\beta^{1/2}}$$

and by examining the curve labelled $(\overline{V}/\beta^{1/2})$ in Fig. 8.26.

The total tangential force per unit width in the z direction is

$$\frac{F_\tau}{L} = \mu \frac{U^2}{V} \cdot \overline{F}_\tau \tag{8.23}$$

where $\overline{F}_\tau$ stands for the integral

$$\overline{F}_\tau = \int_0^1 \left[\frac{1}{H} + \frac{H}{2} \left(\frac{d\bar{p}}{d\bar{x}} \right) \right] d\bar{x}$$

As shown in Fig. 8.23 this integral is a function of $\tilde{Q}_0$ or, via Equ. (8.20), a function of β. The friction factor is obtained by dividing Equ. (8.23) by Equ. (8.19) and using the result obtained for V. This yields

$$f = \left(\frac{Pe}{\overline{P}} \right)^{\frac{1}{2}} \left(\frac{\overline{F}_\tau}{\sqrt{\overline{W}}} \right) \tag{8.24}$$

The friction coefficient f is plotted in Fig. 8.26. It increases slowly as β or as $\overline{P}$ increases.

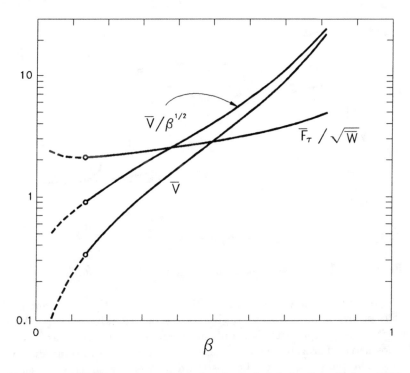

Fig. 8.26 Melting rate and friction in melt lubrication, Bejan, 1989

The remaining question is how much hotter than the melt material must the slider surface become in order to sustain the frictional melting.

The answer follows from setting $y = 0$ in Equ. (8.15) and using Equ. (8.16) to write

$$T = \frac{\mu}{k} U^2 \widetilde{T} \tag{8.25}$$

$$\widetilde{T} = \frac{3}{2} \left[1 - \frac{2}{H}(\bar{x} - \widetilde{Q}_0) \right]^2 + \frac{3}{2} - \frac{2}{H}(\bar{x} - \widetilde{Q}_0) \tag{8.26}$$

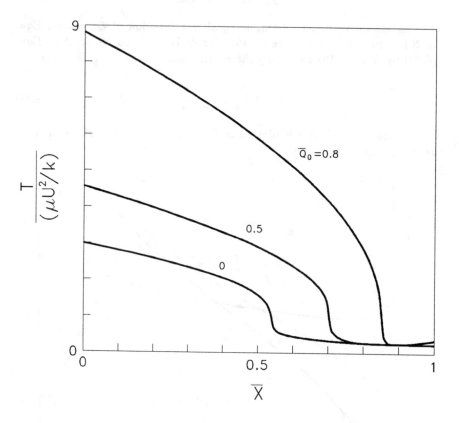

Fig. 8.27 Temperature distribution in melt lubrication, Bejan, 1989

This function is given in Fig. 8.27 which shows that the "hot spot" always occurs at $x = 0$. For a given value of the parameter $\widetilde{Q}_0$ the required excess temperature is significantly higher in the upstream than in the downstream section. As the pressure increases, $\widetilde{Q}_0$ and β increase, and so does the temperature of the upstream part of the slider. The surface temperature in the downstream section is insensitive to changes in the applied load. The surface temperature distribution resembles in shape the gap profile H. The effect of load on the excess temperature of the slider surface is summarized in Fig. 8.28. Both the maximum and the average temperatures increase as β increases. The average temperature is defined as

$$\tilde{T}_{av} = \int_0^1 \tilde{T}(\bar{x}, \beta)d\bar{x}$$

Finally, it is instructive to look at the transverse field. Fig. 8.29 shows the temperature profiles at three stations. The common characteristic of these profiles is the temperature gradient at the upper surface which must be the same at any $\bar{x}$, because the melting speed V is independent of longitudinal position. It is because of the constancy of this temperature gradient that the slider surface temperature must vary, more or less, as H. In general, however, the transverse temperature profiles are not "similar."

8.5 CAVITATION

The essence of cavitation as it appears in tribological processes is that upon reaching a sufficiently low pressure level gases and vapor are released from the lubricant to form cavities within the liquid. The pressure at which dissolved gases will emerge, say, p_g, and the differential $(p - p_g)$ that a lubricant can sustain before rupturing depends on mass fraction, solubility, and type of dissolved gases. The higher the dissolved gas content the earlier will a cavity be formed. The same holds for the formation of a vapor bubble, only that this would occur at a lower pressure still, i.e. $p_v < p_g$.

The subambient pressure required for gaseous or vapor bubble formation represents the pressure drop or tensile stress the liquid is able to withstand before it ruptures. This Δp is a function of the maximum surface tension the liquid can tolerate. The equation connecting film pressure p, and cavity pressure $p_c = (p_g + p_v)$, and the surface tension σ, is given by

$$(p_c - p) = \sigma \left(\frac{1}{R_1} + \frac{1}{R_2} \right) \tag{8.27}$$

where R_1 and R_2 are the curvature radii of, say, an elliptical bubble geometry. Thus the formation of the cavity depends on film pressure and surface tension and thereby also on the temperatures near the zone of cavitation.

The relative amounts of gaseous and vapor components present in a cavitation bubble can be deduced from a consideration of their molar perfect gas relationship. Following Braun and Hendricks, 1984, we have for both gas and vapor mixtures

$$p_c V = \eta \mathcal{R} T$$

where $\eta = \eta_g + \eta_v$ is the molar content. For the individual components

$$p_v V = \eta_v \mathcal{R} T$$

$$p_g V = \eta_g \mathcal{R} T$$

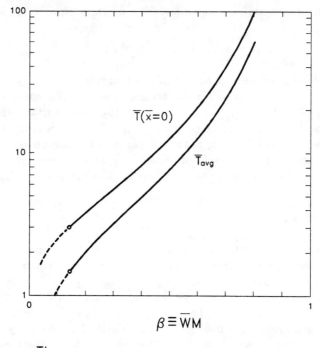

$$\overline{T} = \frac{Tk}{\mu U^2}$$

Fig. 8.28 Temperatures in melt lubrication, Bejan, 1989

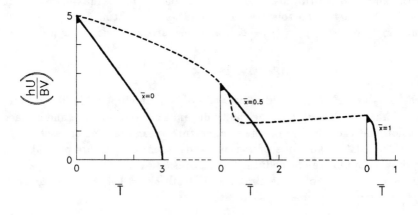

$$\overline{T} = \left(\frac{Tk}{\mu U^2}\right)$$

Fig. 8.29 Transverse temperature profiles in melt lubrication, Bejan, 1989

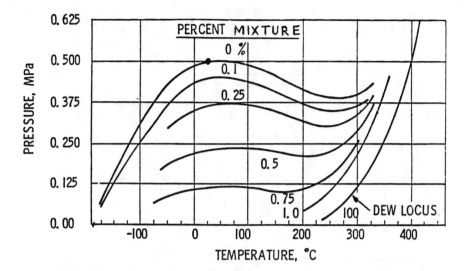

Fig. 8.30 Gaseous content in cavitation zone, Braun and Hendricks, 1984

In order to determine the ratio of the partial pressures

$$\frac{p_v}{p_g} = \frac{\eta_v}{\eta_g}$$

one has to know the molar ratio of the dissolved gases, the volatility, and concentrations of hydrocarbons in the lubricant and their respective vapor pressures.

In addition to mass transfer from the liquid to the cavity causing a change in the size of the bubble a change may also be due to a variation of p. Since the mass then remains constant we have

$$\rho V|_i = \rho V$$

For an adiabatic or isothermal transformation

$$\frac{p_{ci}}{\rho_{ci}^m} = \frac{p_c}{\rho_c^m} \qquad m = 1 \,or\, k$$

and Equ. (8.27) becomes

$$p_{ci} - p_i = 2(\sigma/\mathcal{R}_i)$$
$$p_c - p = 2(\sigma/\mathcal{R})$$
$$(8.28)$$

One can now relate the film pressure to the cavity pressure by

$$\left(\frac{p}{p_c}\right) = 1 - \frac{1}{p_c}\left(\frac{p_c}{p_{ci}}\right)^{\frac{1}{3}}\left(\frac{2\sigma}{\mathcal{R}_i}\right)$$
$$(8.29)$$

TABLE 8.2

Gas and vapor release from a synthetic oil

Component name	Total kilomols	Liquid kilomols	Vapor kilomols	K* value
		P = 0.5 MPa		
Nitrogen	0.6498347	0.5783169	0.7151496E-01	134.9110
Oxygen	0.3169256	0.2968714	0.2005186E-01	73.68898
Argon	0.1383062E-01	0.1295026E-01	0.8802979E-03	74.15950
Carbon dioxide	0.1938509E-02	0.1909867E-02	0.2864233E-04	16.36143
n-Heptadecane	99.99997	99.99995	0.2148885E-07	0.2344388E-06
Total	100.9825	100.8900	0.9247661E-01	
		P = 0.1 MPa		
Nitrogen	0.6498347	0.1150717	0.5347407	602.7195
Oxygen	0.3169256	0.8965719E-01	0.2272503	328.7454
Argon	0.1383062E-01	0.3894745E-02	0.9935413E-02	330.8623
Carbon dioxide	0.1938509E-02	0.1244915E-02	0.6935862E-03	72.26054
n-Heptadecane	99.99997	99.99995	0.7067883E-06	0.9167059E-06
Total	100.9825	100.2099	0.7726274	
		P = 0.05 MPa		
Nitrogen	0.6498347	0.5630137E-01	0.5935327	120.4198
Oxygen	0.3169256	0.4695885E-01	0.2699596	656.6794
Argon	0.1383062E-01	0.2038125E-02	0.1179245E-01	660.9146
Carbon dioxide	0.1938509E-02	0.8570638E-03	0.1081440E-02	144.1323
n-Heptadecane	99.99997	99.99995	0.1572325E-05	0.1796036E-05
Total	100.9825	100.1061	0.8763721	

*Equilibrium constant   Braun and Hendricks, 1984

$$\mu_{20°C} = 55 \; \frac{m^2}{s} \; ; \; \mu_{60°C} = 14 \; \frac{m^2}{s} \; ; \; T = 25°C$$

If the molar composition, phase diagrams, and surface tension of the lubricant are known, one can determine the radius at which gaseous cavitation will start for a given film pressure. Fig. 8.30 and Table 8.2 give the percentage of the gas mixture released from an oil solution at a given temperature and pressure. With their help one can determine what the composition of the gas mixture is in the cavity. At all pressures, the amounts of nitrogen, oxygen, argon, and CO_2 are significantly higher than of n-heptadecane (C17), that is to say that the amount of gas released is much higher than hydrocarbon vapors. As the pressure decreases the vapor-liquid equilibrium constant K for N_2, O_2, Ar and CO_2 increases markedly, indicating that the solubility and the amounts of the above-mentioned components in the liquid diminish greatly. The amounts of released n-heptadecane also increase as the pressure decreases and the composition of the cavity changes from a gaseous structure to one of hydrocarbon vapor.

As quoted in the paper by Braun and Hendricks, a ΔT is required in order to bring about the formation of a vapor bubble. This ΔT is given by

$$(T_c - T_{sat}) = \frac{2\sigma T_{sat}}{C_2 R_c h_{fg} \rho_v} \tag{8.30}$$

where C_2 is an empirical constant equal to 1.25 and R_c the effective radius of the cavity.

The associated heat flux for vapor bubble inception is then

$$Q = A p^a (T_c - T_{sat})^{\frac{b}{pc}} \tag{8.31}$$

where A, a, b, c are constants to be determined experimentally. Thus the generation of vapor in the cavity is contingent not only upon a given set of pressure and temperature, but also on the availability of a certain amount of heat energy. Given the fact that the cavitation region rarely falls to such low absolute pressures that would trigger substantial vapor bubble formation, and since such bubble generation requires a supply of heat energy to sustain it, one may conclude that most cavitation zones consist primarily of gases originally dissolved in the liquid lubricant.

Chapter 9

TIME-DEPENDENT PROCESSES

The present chapter encompasses three topics: squeeze films, transient conditions, and non-Newtonian fluids. The difference between the first two lies in that squeeze films are cyclic, whereas the second group represents non-steady processes. At first glance the inclusion of non-Newtonian fluids may seem odd, but it is sufficient to note their connection with such phenomena as retardation, transit, and relaxation times to realize the affinity of these fluids with time-dependent behavior. As in the case of steady flow where h_{min} and T_{max} constitute criteria for judging performance, so here, too, the lowest h_{min} or the highest T_{max} during a cycle offers a yardstick of satisfactory operation. However, other features of unsteady films, such as thermal inertia or hysteresis effects, are likely to overshadow the impact of the conventional criteria as applied to steady THD devices. In Chapter 7 one such consequence was seen in shaft seizure due to differential thermal expansion; in the next section another unexpected result will appear in the form of negative stiffness in THD squeeze films.

With regard to the mathematics of the unsteady THD area, the major added complexity consists in the need to include the full total derivative (DT/Dt) in the energy equation. The convective term will now read

$$\frac{DT}{Dt} = \left(\frac{\partial T}{\partial t}\right) + u\left(\frac{\partial T}{\partial x}\right) + v\left(\frac{\partial T}{\partial y}\right) + w\left(\frac{\partial T}{\partial z}\right)$$

And, of course, an added dimension, time, will appear to burden both the solution and the presentation of results.

9.1 SQUEEZE FILMS

The most elementary physical system for squeeze film action is that of a set of parallel plates approaching each other at a constant velocity.

This problem was solved in a fairly comprehensive fashion by Rohde and Ezzat, 1973, for a set of square plates including heat transfer effects into a finite thickness bearing pad. For squeeze films, the Reynolds equation with variable viscosity assumes the form

$$\frac{\partial}{\partial x}\left(F\frac{\partial p}{\partial x}\right) + \frac{\partial}{\partial z}\left(F\frac{\partial p}{\partial z}\right) = \rho V \tag{9.1}$$

where using φ as an integration variable for y, we have

$$F = \rho \int_0^{h(t)} \frac{\varphi}{\mu}\left(\varphi - \int_0^{h(t)} \frac{\varphi}{\mu}d\varphi \Big/ \int_0^{h(t)} \frac{d\varphi}{\mu}\right) d\varphi \tag{9.2}$$

The fluid film energy equation is given by

$$\rho c_p \frac{DT}{Dt} = k\nabla^2 T + \mu\left\{\left(\frac{\partial u}{\partial y}\right)^2 + \left(\frac{\partial w}{\partial y}\right)^2\right\} \tag{9.3}$$

This energy equation is subject to the initial condition

$$T(x, y, z; 0) = T_i$$

and the boundary conditions

$$\frac{\partial T(0, y, z; t)}{\partial x} = \frac{\partial T(B, y, z; t)}{\partial x} = 0 \tag{9.4a}$$

$$\frac{\partial T\left(x, -\frac{L}{2}, y; t\right)}{\partial z} = \frac{\partial T\left(x, \frac{L}{2}, y; t\right)}{\partial z} = 0 \tag{9.4b}$$

The above conditions state that heat leaves the bearing periphery by pure convection, an assumption justified by the fact that in squeeze films the fluid exit velocities are large. The boundary conditions at the fluid-solid interface for the three different thermal conditions studied are
a) Isothermal boundaries:

$$T(x, z, 0; t) = T(x, z, h(t); t) = T_i \tag{9.4c}$$

b) Adiabatic boundaries:

$$\frac{\partial T(x, z, 0; t)}{\partial y} = \frac{\partial T(x, z, h(t); t)}{\partial y} = 0 \tag{9.4d}$$

c) Complete THD solution:

$$k\frac{\partial T(x, z, h(t); t)}{\partial y} = k_s\frac{\partial T_2(x, z, 0; t)}{\partial y_2} \tag{9.4e}$$

$$T(x, z, 0; t) = T(x, z, h(t); t) \tag{9.4f}$$

The heat conduction in the solids is expressed as

$$\rho c_p \frac{\partial T_2}{\partial t} = k \nabla^2 T_2 \tag{9.5}$$

This equation is subject to the initial condition

$$T_2(x, z, y_2; 0) = T_i$$

and the boundary conditions:

$$k \frac{\partial T(x, z, h(t); t)}{\partial y} = k_s \frac{\partial T_2(x, z, 0; t)}{\partial y_2} \tag{9.6a}$$

$$-k_s \frac{\partial T_2}{\partial h} = h_{sa}(T_2 - T_a) \tag{9.6b}$$

These conditions guarantee continuity of heat flow across the fluid-solid interfaces and to the environment. The system of Equ. (9.1)–(9.6) completely describes the THD fluid film and the temperature distribution in the solids.

The normalizations used are as follows:

$$\bar{x} = \frac{x}{B} \qquad\qquad \bar{z} = \frac{z}{B}$$

$$\bar{y} = \frac{y}{h(t)} \qquad\qquad \overline{\overline{T}} = \frac{T}{T_i}$$

$$\bar{y}_2 = \frac{y_2}{d} \qquad\qquad \bar{p} = \frac{ph^2(0)}{\mu_i U_R B}$$

$$\bar{u} = \frac{u}{U_R} \qquad\qquad \bar{h}(t) = \bar{h}(0)\left(\frac{1 - \bar{t}/\bar{h}(0)}{1 - \bar{t}}\right)$$

$$\bar{v} = \frac{v}{U_R} \qquad\qquad \bar{t} = -\frac{V}{h(0)} \cdot t$$

$$\bar{w} = \frac{w}{U_R} \qquad\qquad \bar{\varphi} = \frac{\varphi}{h(t)}$$

where t denotes time, and d is used for pad thickness; U_R is some reference velocity; the subscript i refers to $t = 0$; and V is the constant squeeze velocity. With these normalizations the Reynolds and energy equations become

$$\frac{\partial}{\partial \bar{x}}\left(\overline{F}\bar{h}(t)^3 \frac{\partial \bar{p}}{\partial \bar{x}}\right) + \frac{\partial}{\partial \bar{z}}\left(\overline{F}\bar{h}(t)^3 \frac{\partial \bar{p}}{\partial \bar{z}}\right) = -\left(\frac{B}{h_0(0)}\right)\frac{V}{U_R(1 - \bar{t})} \tag{9.7}$$

$$\frac{BU_R \rho c_p}{k}\left[-\left(\frac{B}{h(0)}\right)\left(\frac{V}{U_R}\right)\frac{\partial \overline{\overline{T}}}{\partial \bar{t}^*} + \bar{u}\frac{\partial \overline{\overline{T}}}{\partial \bar{x}^*} + \bar{w}\frac{\partial \overline{\overline{T}}}{\partial \bar{z}}\right.$$

$$+ \left(\frac{B}{h(0)}\right) \frac{\overline{w}}{\overline{h}(t)(1-\bar{t})} \frac{\partial\overline{\overline{T}}}{\partial\bar{y}} \Bigg]$$

$$= \left(\frac{\partial^2\overline{\overline{T}}}{\partial\bar{x}^{*2}} + \frac{\partial^2\overline{\overline{T}}}{\partial\bar{z}^2} + \left(\frac{B}{h(0)}\right)^2 \frac{1}{\{\overline{h}(t)(1-\bar{t})\}^2} \frac{\partial^2\overline{\overline{T}}}{\partial\bar{y}^2}\right)$$

$$+ \frac{U_R^2\mu_i}{T_i k}\left(\frac{B}{h(0)}\right)^2 \frac{\overline{\mu}}{\{\overline{h}(t)(1-\bar{t})\}^2}\left\{\left(\frac{\partial\overline{u}}{\partial\bar{y}}\right)^2 + \left(\frac{\partial\overline{w}}{\partial\bar{y}}\right)^2\right\} \quad (9.8)$$

where

$$\overline{\overline{F}} = \int_0^1 \left(\frac{\bar{\varphi}^2}{\overline{\mu}} - \frac{\bar{\varphi}}{\overline{\mu}}\cdot\frac{\int_0^1 \frac{\bar{\varphi}}{\overline{\mu}}d\bar{\varphi}}{\int_0^1 \frac{d\bar{\varphi}}{\overline{\mu}}}\right)d\bar{\varphi} \quad (9.9)$$

$$\frac{\partial}{\partial\bar{x}^*} = \frac{\partial}{\partial\bar{x}} - \frac{\bar{y}}{\overline{h}(t)}\left(\frac{\partial\overline{h}(t)}{\partial\bar{x}}\right)\cdot\frac{\partial}{\partial\bar{y}}$$

$$\frac{\partial}{\partial\bar{t}^*} = \frac{\partial}{\partial\bar{t}} + \frac{\bar{y}}{\overline{h}(t)(1-\bar{t})}\cdot\frac{\partial}{\partial\bar{y}}$$

The heat conduction equation and the boundary conditions for the solids become

$$\frac{\partial\overline{\overline{T}}_2}{\partial\bar{t}} = -\frac{k_s}{\rho_s c_{p_s}}\frac{h(0)}{V}\frac{1}{B^2}\left[\frac{\partial^2\overline{\overline{T}}}{\partial\bar{x}^2} + \frac{\partial^2\overline{\overline{T}}_2}{\partial\bar{z}^2} + \left(\frac{B}{d}\right)^2\frac{\partial^2\overline{\overline{T}}_2}{\partial\bar{y}_2^2}\right] \quad (9.10)$$

$$\left(\frac{\partial\overline{\overline{T}}}{\partial\bar{y}}\right)_{\bar{y}=1} = \left(\frac{k_s}{k}\right)\left(\frac{h(0)}{d}\right)\overline{h}(t)(1-\bar{t})\left(\frac{\partial\overline{\overline{T}}_2}{\partial\bar{y}_2^2}\right)_{\bar{y}_2=0}$$

$$\frac{\partial\overline{\overline{T}}_2}{\partial\bar{x}} = -\frac{Bh_{sa}}{k_s}(\overline{\overline{T}}_a - \overline{\overline{T}}_2); \qquad \bar{x} = 0$$

$$\frac{\partial\overline{\overline{T}}_2}{\partial\bar{x}} = \frac{Bh_{sa}}{k_s}(\overline{\overline{T}}_a - \overline{\overline{T}}_2); \qquad \bar{x} = 1$$

$$\left(\frac{\partial\overline{\overline{T}}_2}{\partial\bar{z}}\right)_{z=-\frac{B}{2L}} = -\frac{Bh_{sa}}{k_s}\left(\overline{\overline{T}}_a - \overline{\overline{T}}_2\right)$$

$$\left(\frac{\partial\overline{\overline{T}}_2}{\partial\bar{y}_2}\right)_{\bar{y}=1} = \frac{Bh_{sa}}{k_s}(\overline{\overline{T}}_a - \overline{\overline{T}}_2)$$

The above equations were solved for a slider having the geometric and operational parameters listed in Table 9.1.

The formation of the velocity profiles from the inception of motion at $x = B$ is shown in Fig. 9.1. At $t = 0$ the profiles are parabolic, as in the

TABLE 9.1

Operating conditions of squeeze film

Case no.	T_i °C	T_a °C	k_s W/m-°K	-V m/sec	h_c W/m²-K	L = B m
Case 1 standard	37.7	80	15.9	.1524	175	.0254
Case 2, II				.152 .228 .304		
Case 3, III						.0254 .0382 .0508
Case 4, IV	23.9 37.7 93.4					
Case 5, V		-18 to -19				
Case 6, VI			.53 15.9 62.5			

ρ_s = 4.5 kg/m³
c_s = 129 J/(kg-K)
ρ = 0.514 kg/m³
c_p = 580 J/(kg-K)
k = .039 w/(m-K)
d = 0.0254 m
$h^{(0)}$ = 5.08 x 10⁻⁵m

Lubricant viscosity:

°C	N-s/m²
37.7	0.733 x 10⁻³
98.8	0.54 x 10⁻⁴

isothermal case. As time progresses the profiles flatten and form a core. This is due to the lower fluid viscosities near the surfaces where the dissipation is high; by contrast, at the center of the fluid film the dissipation is zero. The cross-film temperature profiles for different thermal boundary conditions are given in Fig. 9.2 at $\bar{t} = 0.5$. In the case of isothermal boundaries the temperature distribution across the film is significantly different from that given by the THD solution whereas the adiabatic case is quite close to the THD solution. The effect of high viscous dissipation at the boundaries affects strongly the THD profile. At the middle of the film,

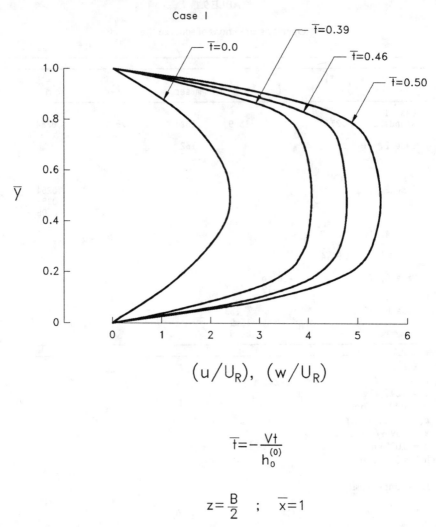

$$\bar{t}=-\frac{Vt}{h_0^{(0)}}$$

$$z=\frac{B}{2} \quad ; \quad \bar{x}=1$$

Fig. 9.1 Velocity profiles as function of time in squeeze films, Rohde and Ezzat, 1973

where the dissipation is zero, the temperature due to heat conduction in the fluid film rises above the initial value by a few percent only. The isotherms of Fig. 9.3 show the peak temperatures to occur in the fluid film adjacent to the solid interface. The large temperature gradients across the film emphasize the importance of cross-film conduction in squeeze film analyses. The temperature distribution in the solids is affected only in the immediate vicinity of the fluid film. This is due to the fact that the heat flow through the solid is governed by its thermal inertia.

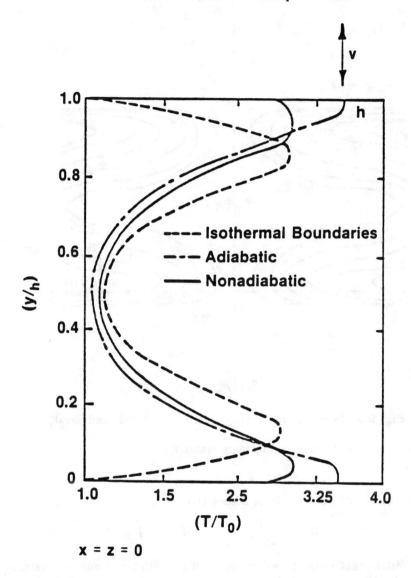

Fig. 9.2 Transverse squeeze film temperature profiles for various thermal regimes, Rohde and Ezzat, 1973

For a constant viscosity the momentum equation may be written as

$$\nabla^2 p = \frac{\mu V}{h(t)^3}$$

and the fluid film pressure expressed by

$$p = \frac{\mu V}{h(t)^3} \cdot \bar{p}$$

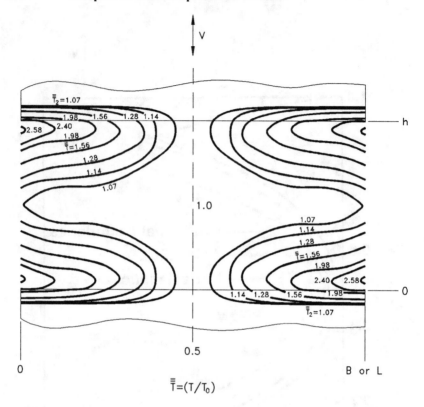

Fig. 9.3 Temperature field in squeeze films, Rohde and Ezzat, 1973

where $\bar{p}$ is the solution of Poisson's equation

$$\nabla^2 \bar{p} = -1; \quad \bar{p}|_A = 0$$

The generated fluid-film force is then given by

$$W = -\frac{\mu V}{h(t)^3} \cdot \overline{W}; \qquad \overline{W} = \int\int \bar{p}\, dA$$

It follows that when the solids approach each other at a constant velocity the load is a monotonically increasing function of time and dW/dt is positive. Numerical results for the full THD solution, however, showed that for certain input parameters the load W decreased with a decrease in h. This behavior is shown in Fig. 9.4a for different values of V; in Fig. 9.4b for different film thickness ratios; and in Fig. 9.4c for different initial viscosities.

To ascertain that this anomaly is not due to computational idiosyncrasies, Rohde and Ezzat performed a one-dimensional variable squeeze film analysis. They obtained for the $W - t$ relationship the expression

$$\left[\frac{\partial T}{\partial t}\right]_{t=0^+} = \frac{144}{\rho c_p}\frac{V^2}{h(0)^6}\left(x - \frac{B}{2}\right)^2\left(y - \frac{h(0)}{2}\right)^2 \tag{9.11}$$

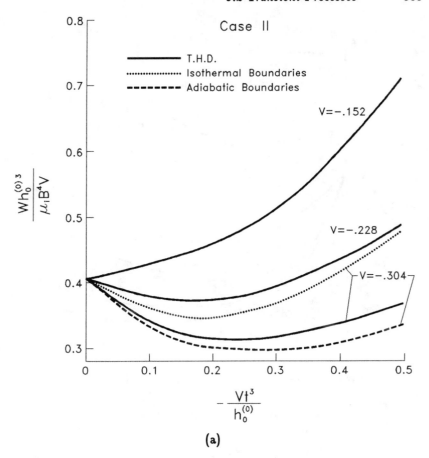

(a)

Fig. 9.4 Squeeze film forces, (a) dependence on velocity, (b) dependence on geometry, (c) dependence on T_0, Rohde and Ezzat, 1973

$$\left[\frac{dW}{dt}\right]_{t=0^+} = \frac{3V^2\mu_i B^3}{h(0)^4}(1-\delta) \tag{9.12}$$

where

$$\delta = \left(\frac{27}{25}\right)\frac{1}{\rho c_p}\left(\frac{V}{h(0)}\right)\left(\frac{B}{h(0)}\right)^2\left(\frac{\partial\mu}{\partial T}\right)_{T=T_i}$$

An examination of Equ. (9.12) shows that for some values of δ, (dW/dt) becomes indeed negative. This confirms the validity of the numerical results. One explanation of this phenomenon may be as follows. The parameter δ can be seen as an indicator of the initial rate of heat generation in the film. For high values of δ the thermal mechanism (drop in viscosity) initially overpowers the geometric hydrodynamic effects causing the fluid film forces to decrease with time. As the fluid film thickness decreases in height the situation is reversed and the geometric mechanism begins to dominate,

Case III

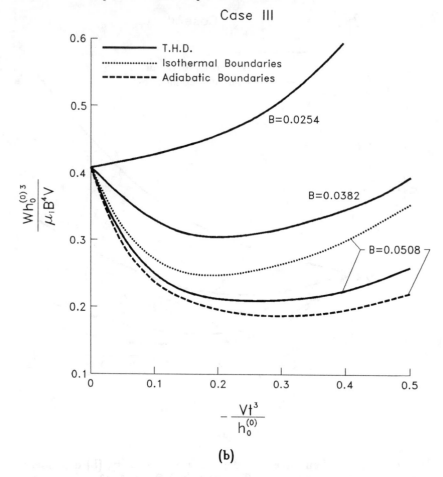

(b)

Fig. 9.4 Squeeze film forces, (a) dependence on velocity, (b) dependence on geometry, (c) dependence on T_0, Rohde and Ezzat, 1973

resulting in an increase in load. The net result is that for high values of δ the load-time function exhibits a minimum.

9.2 TRANSIENT PROCESSES

9.2.1 Thermal Transients in Sliders

There are many systems in which unsteady conditions constitute the routine mode of operation, with speed, load, temperature, etc. undergoing continuous changes. Piston rings are probably the most conspicuous example in which nearly all variables, including the composition of the lubricant (due to blowby) and the geometry of the interacting surfaces undergo radical fluctuations. Even in systems which operate in a steady mode, there

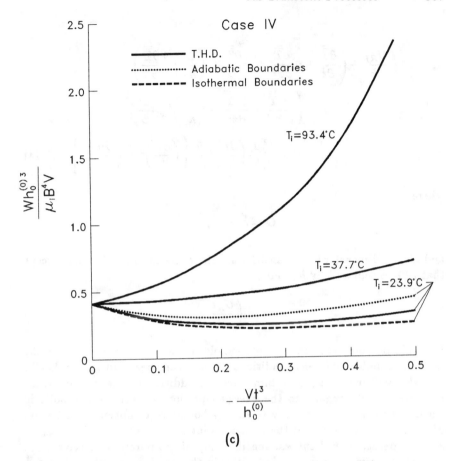

Fig. 9.4 Squeeze film forces, (a) dependence on velocity, (b) dependence on geometry, (c) dependence on T_0, Rohde and Ezzat, 1973

are frequent occasions of transient operation. This happens when a parameter is shifted from one level to another; during shut-downs and start-ups; and whenever there is any external disturbance to the system. Following such perturbations equilibrium is reached only after some settling period; in large size equipment, as in power plants of electric utilities, this may take hours, or even days. Using the same methods they did in the previous section Ezzat and Rohde, 1973, performed an analysis of unsteady operation on a slider with a linear wedge. Here the right hand side of the Reynolds equation contains the geometric parameter (dh/dx) instead of the squeeze film velocity V. In addition, the velocity U is now a function of time.

The two basic equations, then, in non-dimensional form now read

$$\frac{\partial}{\partial \bar{x}}\left(\overline{F}\bar{h}^3\frac{\partial \bar{p}}{\partial \bar{x}}\right) + \frac{\partial}{\partial \bar{y}}\left(\overline{F}\bar{h}^3\frac{\partial \bar{p}}{\partial \bar{z}}\right) = \overline{U}\frac{\partial \overline{G}}{\partial \bar{x}} \tag{9.13}$$

$$\frac{B U_R \rho c_p}{k}\left(\frac{\partial \overline{\overline{T}}}{\partial \bar{t}} + \bar{u}\frac{\partial \overline{\overline{T}}}{\partial \bar{x}^*} + \bar{w}\frac{\partial \overline{\overline{T}}}{\partial \bar{z}} + \left(\frac{B}{h_2}\right)\frac{\bar{v}}{\bar{h}}\frac{\partial \overline{\overline{T}}}{\partial \bar{y}}\right)$$

$$= \left(\frac{\partial^2 \overline{\overline{T}}}{\partial \bar{x}^{*2}} + \frac{\partial^2 \overline{\overline{T}}}{\partial \bar{z}^2} + \left(\frac{B}{h_2}\right)^2\frac{1}{\bar{h}^2}\frac{\partial^2 \overline{\overline{T}}}{\partial \bar{y}^2}\right)$$

$$+ \frac{U_R^2 \mu_i}{T_i k}\left(\frac{B}{h_2}\right)^2\frac{\bar{\mu}}{\bar{h}^2}\left\{\left(\frac{\partial \bar{u}}{\partial \bar{y}}\right)^2 + \left(\frac{\partial \bar{w}}{\partial \bar{y}}\right)^2\right\} \quad (9.14)$$

where

$$\overline{G} = \bar{h}\frac{\int_0^1 \frac{\bar{\varphi}}{\bar{\mu}}d\bar{\varphi}}{\int_0^1 \frac{d\bar{\varphi}}{\bar{\mu}}}$$

and all the dimensionless quantities are the same as in Section 9.1, except that $h(0)$ is replaced by h_2 and

$$\bar{t} = \left(\frac{U_R t}{B}\right)$$

It should be noted that the above Reynolds equation is not rigorous in the sense that under transient conditions the film thickness will vary in height and this will induce squeeze film forces in addition to the wedge-formed pressures. With regard to the energy equation its form is parabolic in time. It will thus be necessary to specify boundary conditions at all edges of the bearing as well as at the fluid-solid interfaces. At the trailing edge a zero temperature gradient was specified imposing a purely convective mode of heat transfer at that location. Within the bearing pad of thickness d, the heat transfer equation reads

$$\frac{\rho_s c_{ps} U_R B}{k_s}\cdot\frac{\partial \overline{\overline{T}}_2}{\partial \bar{t}} = \left[\frac{\partial^2 \overline{\overline{T}}_2}{\partial \bar{x}^2} + \frac{\partial^2 \overline{\overline{T}}_2}{\partial \bar{z}^2} + \left(\frac{B}{d}\right)^2\frac{\partial^2 \overline{\overline{T}}_2}{\partial \bar{y}_2^2}\right] \quad (9.15)$$

Throughout the present problem the boundary conditions are similar to those imposed on the squeeze film slider in the previous section. The dimensions and operating conditions of the slider are specified in Table 9.2.

Since time is required for the system to arrive at a thermal equilibrium (final viscosity), the final performance characteristics of the slider, too, are reached only after some time delay. Fig. 9.5a shows the time it takes the load to arrive at its equilibrium value when operating at constant speed from start-up, but with different initial temperatures; in Fig. 9.5b the behavior of W is shown when operating at the same initial temperature but different speeds. As seen, the initially high load approaches rapidly, in a few milliseconds only, its steady state value. The rate of drop is slower for high initial temperatures since the rate of heat dissipation is lower; as

TABLE 9.2

Input data for slider under
transient operating conditions

B = L = 3 in., (0.0762 m)	k = 0.075 Btu/(hr-ft-°F) (0.04 W/m-K)
d = 1 in. (0.0254 m)	c_p = 0.48 Btu/(lb-°F), 620 J/(kg-K)
h_1 = 0.004 in., (1.016 x 10⁻⁴ m)	T_a = 80°F (26.8°C)
h_2 = 0.002 in., (0.508 x 10⁻⁴ m)	h_{sa} = 100 Btu/(hr-ft²-°F), (175 in./m²-K)
ω_s = 0.28 lb/in.³, (7750 kg/m³)	Oil SAE-30
k_s = 30 Btu/(hr-ft-°F), 15.9 in./(m-K))	μ_1 = 1.53 x 10⁻⁵ Reyns at 100°F (37.8°C)
c_{ps} = 0.1 Btu/(lb-°F), (129 J/kg-K)	μ_2 = 1.13 x 10⁻⁵ Reyns at 210 °F (98.8°C)
w = 0.032 lb/in.³, (870 kg/m³)	μ_1 = (0.105 N-s/m²)
	μ_2 = (0.00775 N-s/m²)

expected, the response is faster at higher speeds. The behavior of T_{max}, shown in Fig. 9.6, is similar in that it rises rapidly to its equilibrium value, although not as fast as does the load.

The effect of the thermal boundary conditions at the fluid solid interface is shown in Fig. 9.7. The THD solution is overestimated by the assumption of isothermal boundaries and underestimated by the adiabatic boundary condition. In the case of adiabatic conditions no heat is transmitted across the fluid solid interface and a relatively hot boundary layer is formed, as exemplified by Fig. 9.8. This boundary layer has a considerable effect on the fluid film temperature profile which in turn governs the load-carrying capacity. Maintaining the bearing surfaces at the initial colder temperature has the effect of increasing the load-carrying capacity. Under both adiabatic and isothermal conditions, the nature of the transient response is the same, though under isothermal conditions equilibrium is reached earlier than with the adiabatic or THD cases.

Due to the rapid response of the fluid film to thermal transients, the effect of runner acceleration can only appear if unrealistic acceleration rates are considered. The effect of one such value is demonstrated in Fig. 9.9. The load-speed characteristic is shown for the case when the runner is accelerated and decelerated at $10,000 \, \text{ft/sec}^2$, $(3048 \, \text{m/s}^2)$. Under these academic rates of velocity change, the fluid film does not respond rapidly enough and a hysteresis-type load-speed cycling occurs. After a few oscillations a steady-state loop is reached. For realistic bearings the effect of runner acceleration or deceleration on the transient thermohydrodynmic

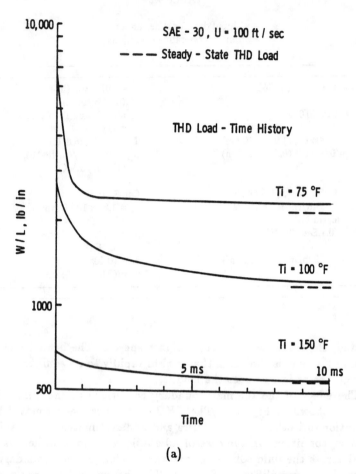

(a)

Fig. 9.5 Transients involved in reaching load equilibrium, (a) condition of constant velocity, (b) effect of various velocities, Rohde and Ezzat, 1973

performance is negligible. In effect the fluid film thermal response is almost instantaneous, a matter of milliseconds, while the bearing solids response is of the order of an hour. This is due to the large difference in the heat capacities of the fluid and solids. Since bearing performance is governed by fluid film behavior, then for all practical purposes, the bearing reaches steady state on the order of the transition time of the lubricant through the contact.

The steady state performance of a slider bearing is given in Chapter 4. In it the fluid film conduction in the direction of motion was ignored. In the present study conduction in the x direction is included and it is of interest to compare the steady-state load capacities as predicted by both solutions. Using a fictitious value for the specific heat of the solids has the effect of

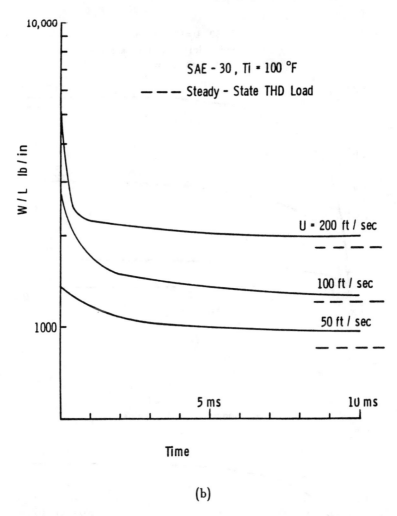

Fig. 9.5 Transients involved in reaching load equilibrium, (a) condition of constant velocity, (b) effect of various velocities, Rohde and Ezzat, 1973

increasing their thermal response, thus allowing one to present the solution in the limit as $t \to \infty$. The resulting load capacity for this calculation is, as shown in Fig. 9.10, within 5% of the steady state value obtained in Chapter 4. The corresponding temperature fields (at $t = 10\,\mathrm{ms}$ for the transient) are given in Fig. 9.11. It is seen that the fluid film temperatures exhibit an immediate response. The bearing solid, however, responds very slowly. In view of this small difference in results, heat conduction in the fluid film in the direction of motion can be said to have a negligible effect.

The above results give some idea of the time it takes to reach thermal equilibrium in the fluid film, provided everything else stays constant. As such it elucidates one element only in a syndrome of unsteady thermal in-

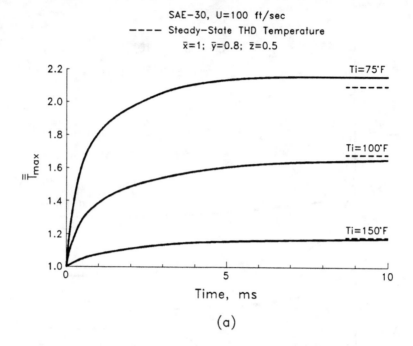

(a)

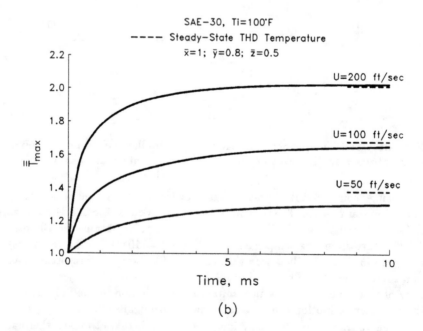

(b)

Fig. 9.6 Transients involved in reaching equilibrium T_{max}, (a) condition of constant velocity, (b) effect of various velocities, Rohde and Ezzat, 1973

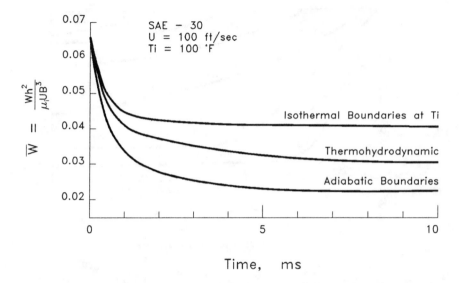

Fig. 9.7 Effect of thermal mode of operation on load capacity transients in sliders, Ezzat and Rohde, 1973

teractions and it must be stressed that isolation of such a single transient variable, with everything else kept constant, is an unrealistic condition. In the first place, as was pointed out earlier, the squeeze film effect was left out. This is still minor when compared to the fact that none of the processes illustrated in Figs. 9.5 to 9.9 can be duplicated. There is no mechanical system where the velocity starts instantaneously at some finite value whereby one could measure the time it takes for the load and temperatures to reach equilibrium; or the reverse of it, with a system running at a certain finite velocity to suddenly fill the entire gap with a lubricant at some cold T_i and measure the time it takes for temperature and load to reach equilibrium. The extreme shortness of the transient intervals obtained rests thus on an artificial construct and in reality when U undergoes acceleration or deceleration from some initial value, the time required for equilibrium will be longer. What remains valid, however, is that the fluid film will reach equilibrium much sooner than the surrounding solid surfaces, a point taken up in the next section.

9.2.2 Thermal Transients in Journal Bearings

Ettles et al, 1988, employed an en gross approach to obtain some idea of the time required to reach equilibrium due to a step change in load or speed in a journal bearing assembly. He considered the surfaces of both shaft and bearing shell, with an outer radius $2R$, to be subject to a sudden uniform temperature jump ΔT. Assuming axial symmetry the equation of

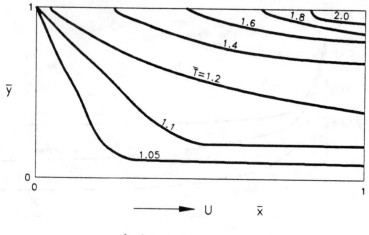

a) Adiabatic Boundaries

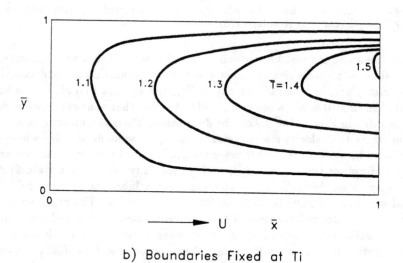

b) Boundaries Fixed at Ti

Fig. 9.8 Film temperatures in slider under transient conditions, $\bar{z} = 0.5$; $t = 10\,\text{ms}$, Ezzat and Rohde, 1973

transient heat conduction for such a case becomes

$$\frac{\partial T}{\partial t} = \frac{k}{\rho c_{ps}} \left(\frac{\partial^2 T}{\partial r^2} + \frac{1}{r} \frac{\partial T}{\partial r} \right) \tag{9.16}$$

In nondimensional form, the result is

$$\frac{\partial \overline{\overline{T}}}{\partial t^*} = \frac{\partial^2 \overline{\overline{T}}}{\partial r^{*2}} + \frac{1}{r^*} \frac{\partial \overline{\overline{T}}}{\partial r^*} \tag{9.17}$$

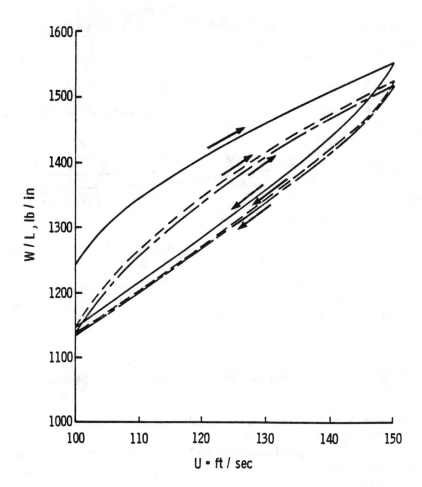

Fig. 9.9 Hysteresis effect due to acceleration-deceleration cycling, Ezzat and Rohde, 1974

$$r^* = \frac{r}{R} \qquad \overline{\overline{T}} = \frac{T}{T_0}$$

$$t^* = \frac{k}{\rho c_{ps}} \cdot \frac{t}{R^2} \equiv \text{Fourier number} \qquad (9.18)$$

To simulate the step change, the surfaces of both shaft and bearing are brought in sudden contact with a temperature differential of $T = (T - T_0)$. The surface temperatures are a function of the heat transfer coefficient h_s; if h_s is infinite, the surfaces rise by a $\Delta T = (T - T_0)$ instantaneously. For small values of h_s the time delay to equilibrium is long.

The heat transfer coefficient h_s is contained in the Nusselt number, $(h_s d/k)$, and Fig. 9.12 shows the general relation between Nu and the so-

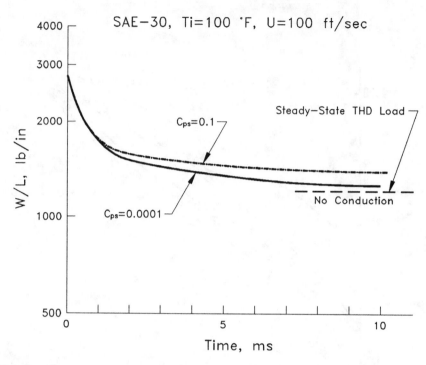

Fig. 9.10 Effect of longitudinal fluid conduction on load capacity, Ezzat and Rohde, 1973

lution of Equ. 9.16. This solution was based on an assumption of adiabatic outer bearing surface, or $(\partial T/\partial r) = 0$ at $r = 2R$.

For bearings, a proper definition of its Nusselt number must be arrived at. In general, the Nusselt number for laminar flow between two parallel walls is given by

$$Nu_f = \frac{h_s \cdot 2h}{k_f}$$

- for uniform wall temperature: $Nu = 7.54$
- for uniform heat flux: $Nu = 8.24$

Above, $2h$ is the hydraulic depth. To convert the general Nu_f to a bearing Nu the following values were used:

Clearance ratio, $2h/R = 2C/R = 2\delta \times 10^{-3}$
Lubricant conductivity, $k = 0.15 \, \text{W/m·°C}$
Solids conductivity, $k_s = 50 \, \text{W/m·°C}$
$Nu_{fa} = \frac{1}{2}(7.54 + 8.24) = 7.89$

The above gives for our case $Nu = h_s R/k = 11.85(R/C)10^{-3}$.

This value of Nu is, however, close to an upper bound of the time delay to equilibrium since the surfaces were taken to be at the maximum ΔT, whereas the temperatures throughout the film—particularly in the center—will be less than this maximum. The experimental data shown in the next

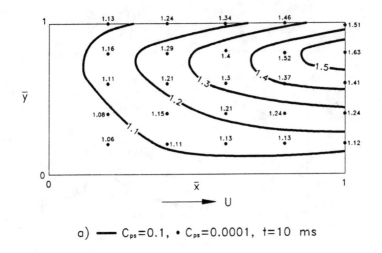

a) —— $C_{ps}=0.1$, • $C_{ps}=0.0001$, t=10 ms

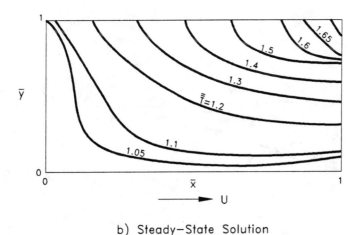

b) Steady–State Solution

Fig. 9.11 Mid-plane fluid film temperature distribution, Ezzat and Rohde, 1973

section suggest an effective value of Nu in the range of 1–5. For $Nu = 1$, the 50% response time for the surface of the shaft is $Fo = 0.29$ and for the surface of bearing, $Fo = 0.79$. Since the bearing has a greater thermal capacity than the shaft, the 50% response time for the assembly will be closer. The model used assumes the supply lubricant to be unaffected by the change in operating conditions. This will not be the case in actual bearing assemblies that include a sump. In these cases a third thermal system would have to be included.

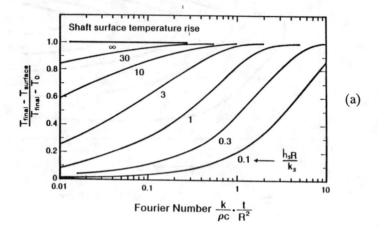

(a)

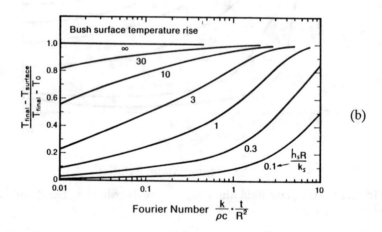

(b)

Fig. 9.12 Temperature rise of mating surfaces subject to a step ΔT, (a) shaft surface, (b) bearing surface, Ettles, 1988

9.2.3 Thermoelastic Effects

Under certain conditions a step change in load or speed is accompanied not only by a viscosity and temperature readjustment but also by thermoelastic deformations. In general, these deformations will lag the changes in film temperatures but eventually they will catch up and produce an effect. There are thus at least two aspects of transient thermal behavior, namely the time delay for the film temperatures to reach equilibrium; and thermoelastic deformations which will delay the onset of full equilibrium, even though the temperatures in the film may have levelled off. Ettles, 1982, analyzed the effects of such distortions on the bearing element shown in Fig. 9.13. It is not a very realistic model although there is at least one design called the "hydroflex" bearing which is actually supported by a solid stem. Other designs which approximate the shape of Fig. 3.19 are tilting pad bearings with large, rigid (whether deliberate or not) pivots; thrust pads sitting on a matrix of springs; journal bearing arcs loosely supported in their seats; and some face seal geometries. The cantilevered pad analyzed was assumed to be infinitely wide and both runner and pad had a finite thickness of the same order; usually, of course, the runner in thrust bearings is much more massive than the bearing pad.

Since the thermal inertia of the solids is much greater than that of the fluid film, the change in film temperature was taken from a steady state solution. Also the time scale of events is such that any squeeze film effects can be neglected; for a sudden doubling of the load any squeeze effects will attenuate to zero in less than a single time step compared with the several hundred time steps required for full thermoelastic equilibrium.

In order to evaluate the transient process vis-a-vis the equilibrium solution, steady state solutions (without distortion) were first obtained. Both the steady and transient solutions were obtained for the bearing conditions given in Table 9.3. Some typical results for $B = 0.12$ m are given in Fig. 9.14 which shows the variation in minimum film thickness and the maximum temperatures in the film and shoe against the specific load W/B. The temperature contours for this case are shown in Fig. 9.15. In the trailing part of the film these are similar to those obtained by Rohde and Oh in Chapter 7; however, in the leading part the temperature contours tend to follow the streamlines with a small thermal boundary layer formed on the portion of the pad where back flow occurs.

With the temperature throughout the film defined, the temperature in the solids at any instant later, can be found. The equation governing this transient heat conduction is the diffusion equation

$$\frac{\partial T}{\partial t} = \frac{k_s}{\rho_s c_{ps}} \left(\frac{\partial^2 T}{\partial x^2} + \frac{\partial^2 T}{\partial y^2} \right) \equiv \sigma_s \nabla^2 T \tag{9.19}$$

In classical analysis, the Fourier number $Fo = \sigma t / L^2$ emerges as the dominant parameter, where L is a characteristic length such as thickness.

TABLE 9.3

Input data for thermoelastic solution

Dimensions and heat transfer coefficients

    Pad thickness, d_s = 0.25;

    Rotor thickness, d_R = d_s; cantilever length, B_c = 0.875B; groove width, 0.15B; sliding speed U = 30 m-s^{-1}; supply temperature T_0 = 40°C; pad heat transfer coefficient h_s = 1200 W-m^{-2}-°C^{-1}; rotor heat transfer coefficient h_R = 2400 W-m^{-2}-°C^{-1}

Properties of rotor and pad material (steel)

    Density ρ_s = 7850 kg-m^{-3}

    Specific heat c_{ps} = 460 J-kg^{-1}-°C^{-1}

    Conductivity k_s = 52 W-m^{-1}-°C^{-1}

    Elastic modulus E = 207 GN-m^{-2}

    Linear coefficient of expansion, 1.1×10^{-5} °C^{-1}

Oil properties

    Density ρ = 890 kg-m^{-3}

    Specific heat c_p = 1880 J-kg^{-1}-°C^{-1}

    Conductivity k = 0.15 W-m^{-1}-°C^{-1}

    Viscosity (Pa-s) is given by the formula:

        $\mu = 2.67 \times 10^{-5} \exp\{1350/(T + 131)\}$, (T, °C)

        $\mu_{30°C}$ = 0.117 Pa-s

        $\mu_{60°C}$ = 0.0313 Pa-s

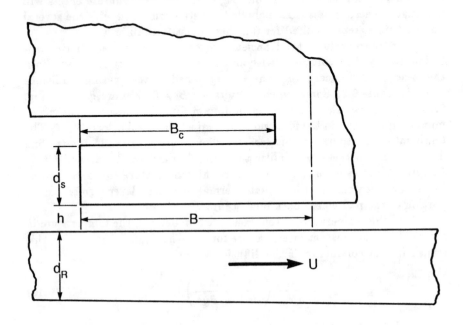

Fig. 9.13 Cantilevered thrust pad

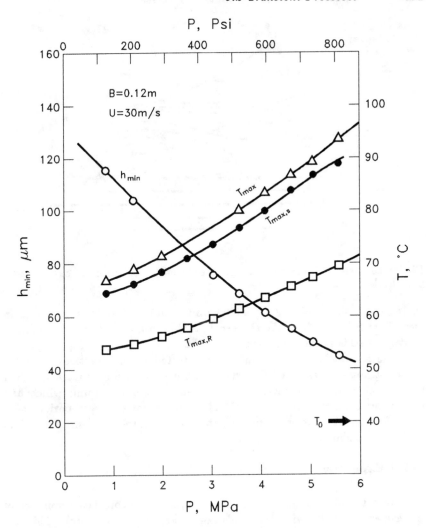

Fig. 9.14 Steady state solution of cantilevered pad, Ettles, 1982

Using this Fourier number as a parameter, Fig. 9.16 shows the change in bearing temperatures, deflection, and minimum film thickness following a sudden doubling or halving of load at constant speed. Asymptotic values are plotted where available. The instantaneous changes at $t = 0$ are shown as vertical bars. The deflection reaches a peak at $Fo \simeq 0.2$ which corresponds approximately to classical results of thermal deflection. The rotor surface temperature appears to lag slightly the pad maximum temperature.

Although the time to reach a new equilibrium condition may seem long, the initial response is rapid. For a pad with $B = 0.12$ m and 30 mm thick T_{max} in the pad will achieve 50% of its eventual increment at $Fo = 0.1$, or 6.3 sec. Fig. 9.17 shows results for a large pad, $B = 1$ m, $d = 0.25$ m. As

$T_a = 40°$

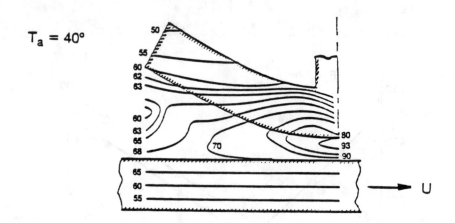

Fig. 9.15 Temperature contours for steady state conditions in cantilevered pad, Ettles, 1982

compared to the $B = 0.12\,\mathrm{m}$ case where the time transient is measured in seconds or minutes, here the time scale is that of hours. A Fourier number of at least 10 is needed for thermal equilibrium.

The above theoretical results are reflected in the practices of power plants of large electric utilities where instructions require that a specified time be allowed (often several hours) before restarting a unit which had previously been run at full load. The purpose of this rest period is to allow for the dissipation of unfavorable temperature gradients formed in the bearings shortly after stopping.

9.2.4 Experimental Correlations

An attempt was made by Ettles et al, 1988, to correlate a number of experimental results with Newton's law of heating (or cooling). This law states that the rate of temperature change of a body is proportional to the temperature difference between the body and its environment. This yields an exponential dependence of temperature on time. Fig. 9.18 shows the temperature increase at a point close to the film of a tilting pad journal bearing. These data are for a 76 mm bearing with five centrally pivoted pads, and a (C/R) ratio 0.00157. The thermocouple was located 65% downstream of the leading edge position in the hottest pad, with its junction about 1.5 mm from the film.

In the exponential expression

$$\Delta T = \Delta T_F \left[1 - e^{a\,Fo} \right] \tag{9.20}$$

the exponent a is 4.58 for Fig. 9.18a and 5.73 for Fig. 9.18b. The radius of the shaft is used as the dimension in Fo. The same type of behavior is shown for the bearing on rundown, Fig. 9.18c, where $a = 4.58$.

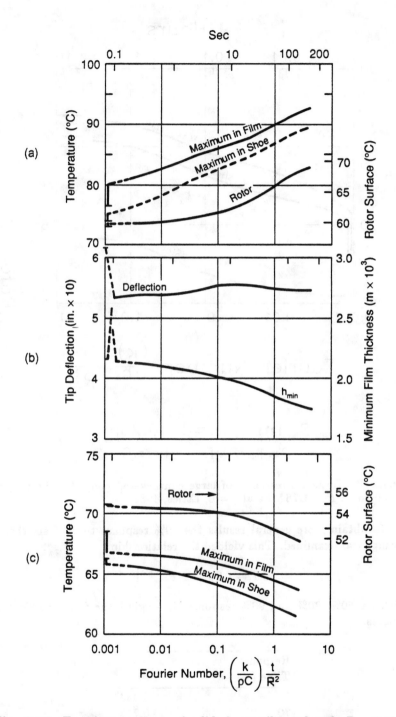

Fig. 9.16 Transient response of solids in cantilevered pad, $B = 0.12\,\text{m}$; $a,b - P = 4.64\,\text{MPa}$ at $t = 0$; $c - P = 1.37\,\text{MPa}$ at $t = 0$, Ettles, 1982

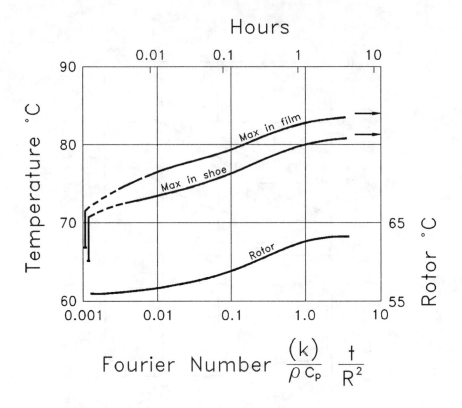

Fig. 9.17 Transient response of large cantilevered pad, $B = 1\,\text{m}$; $d_S = d_R = 0.25\,\text{m}$; $P = 1.78\,\text{MPa}$ at $t = 0$, Ettles, 1982

To obtain more general results the 50% response time of six thrust bearings was examined. This yielded the relationship

$$\Delta T = \Delta T_F \left[1 - e^{1.6\,Fo}\right]$$

which for 50%, 70% and 90% response times gives the following Fourier numbers.

Response Time (%)	Fo
50	0.433
70	0.752
90	1.44

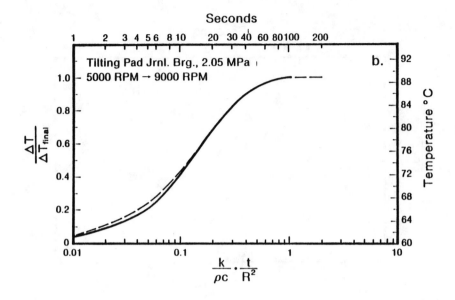

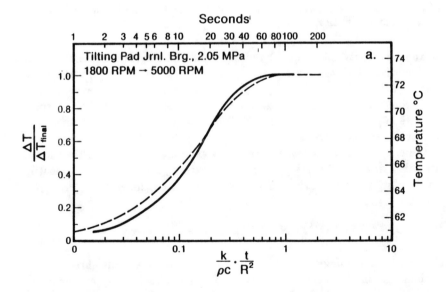

Fig. 9.18 Transient temperatures during start-up and shut-down of a thrust bearing, Ettles, 1988

The heating and cooling cycles in a journal bearing, taken from Seireg and Doshi, 1975, are shown in Fig. 9.19; the experiments were conducted

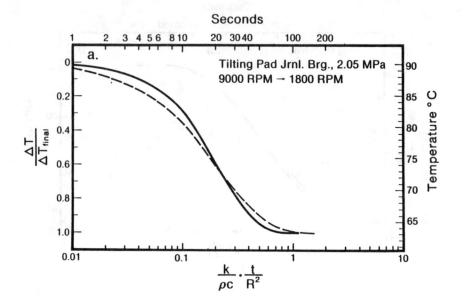

Fig. 9.18 Transient temperatures during start-up and shut- down of a thrust bearing, Ettles, 1988

at an eccentricity ratio of 0.92 and 2000 rpm. The agreement here with an exponential relationship is not as good as with thrust bearings, a fact noted also in the correlations done by Ettles. In general most of the experimental data could be matched fairly well with an exponential law of the form of Equ. (9.20). This implies that a fluid film bearing has a "time constant," and the decay time for, say, a 90% change is unaffected by the magnitude of the change. Also, in much of the data examined the cooling time for an assembly was about twice as long as for heating. This difference in heating and cooling rates decreases with a decrease in step changes.

9.3 NON-NEWTONIAN FLUIDS

From a tribological standpoint traction curves exhibit three distinct regions. In the first the strain rate and the generated heat are small, resulting in conventional lubricant behavior. In the second, non-linear region, the lubricant is subject to large strain rates and the fluid becomes non-Newtonian; here a maximum is observed after which the traction force begins to decrease. This is the start of the third, the thermal region. In this region the traction force decreases with rising speed and appreciable temperature rises occur near the exit in both the lubricant and on the solid surfaces.

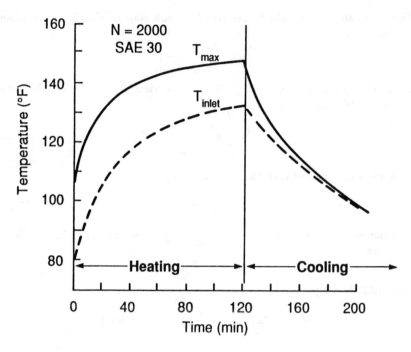

Fig. 9.19 Start-up and shut-down transients in a journal bearing, Seireg and Doshi, 1975

A viscoelastic model, the Eyring model, originally postulated for isothermal flow, represented by the so-called sinh formula

$$\mu \left(\frac{du}{dy} \right) = \tau_0 \sinh \left(\frac{\tau}{\tau_0} \right) \qquad (9.21)$$

has been shown by Conry, 1979, to be valid also in the thermal region, provided the variations of μ and τ_0 with temperature are accounted for. Moreover, the variation in the reference stress, τ_0, is not great and conforms to Eyring's prediction of being proportional to absolute temperature. The variation of shear stress, and hence traction, with temperature is thus primarily due to the effect of temperature on lubricant viscosity.

9.3.1 Traction under Thermal Conditions

Conry, 1981, analyzed the traction between two infinite rollers assuming a constant film thickness and essentially a Hertzian pressure profile. The momentum expression is given by

$$\left(\frac{\partial p}{\partial x} \right) = \frac{\partial \tau_{xy}}{\partial y} \qquad (9.22)$$

The lubricant strain rate is described by the sum of elastic and viscous components. Thus

$$\frac{\partial u}{\partial y} = \left.\frac{\partial u}{\partial y}\right|_E + \left.\frac{\partial u}{\partial y}\right|_v$$

Using the "sinh" law of Equ. (9.21) for the viscous strain rate, the total strain of the fluid at a point is

$$\frac{\partial u}{\partial y} = \left(\frac{1}{G}\frac{\partial \tau_{xy}}{\partial y}\right) + \frac{\tau_0}{\mu}\sinh\left(\frac{\tau_{xy}}{\tau_0}\right) \tag{9.23}$$

with the shear modulus of the fluid given by

$$G = G_0 + G_1 p \tag{9.24}$$

and the viscosity μ, a function of both the pressure and the absolute temperature

$$\mu = \mu_0(T)e^{\beta p}$$

Equ. (9.23) now yields

$$\frac{\partial u}{\partial y} + u\frac{\tau_{xy}G_1}{G^2}\left(\frac{\partial p}{\partial x}\right) = \frac{\tau_0}{\mu}\sinh\left(\frac{\tau_{xy}}{\tau_0}\right) \tag{9.25}$$

If the pressure distribution is a function of x alone then integrating Equ. (9.22) with respect to y yields

$$\tau_{xy} = \tau_1 + y\left(\frac{dp}{dx}\right) \tag{9.26}$$

where τ_1 is the shear stress at $y = 0$. We then obtain

$$\frac{\partial u}{\partial y} + u\frac{G_1}{G^2}\left(\tau_1 + y\frac{dp}{dx}\right)\frac{dp}{dx} = \frac{\tau_0}{\mu}\sinh\left(\frac{\tau_1 + y\dfrac{dp}{dx}}{\tau_0}\right) \tag{9.27}$$

which is the governing differential equation for $u(y)$.

Integrating (9.27) and applying the condition $u = U_1$ at $y = 0$ yields

$$u\exp\left[\frac{G_1}{G^2}y\frac{dp}{dx}\left(\tau_1 + \frac{y}{2}\frac{dp}{dx}\right)\right]$$
$$= U_1 + \int_0^y \frac{\tau_0}{\mu}\exp\left[\frac{G_1}{G^2}\varsigma\frac{dp}{dx}\left(\tau_1 + \frac{\varsigma}{2}\frac{dp}{dx}\right)\right]\sinh\left(\frac{\tau_1 + \varsigma\frac{dp}{dx}}{\tau_0}\right)d\varsigma \tag{9.28}$$

Writing for the midplace shear stress

$$\tau_m = \tau_1 + \frac{h}{2}\left(\frac{dp}{dx}\right)$$

and using $u = U_2$ at $y = h$

$$U_2 \exp\left[\frac{G_1 h}{G^2}\frac{dp}{dx}\tau_m\right] = U_1 + \int_0^y \frac{\tau_0}{\mu}\exp\left[\frac{G_1}{G^2}\varsigma\frac{dp}{dx}\right.$$
$$\left.\left(\tau_m + \left(\varsigma - \frac{h}{2}\right)\frac{dp}{dx}\right)\right]\sinh\left(\frac{\tau_m + \left(\varsigma - \dfrac{h}{2}\right)\dfrac{dp}{dx}}{\tau_0}\right)d\varsigma \qquad (9.29)$$

The value of τ_m must be so selected as to satisfy Equ. (9.29).

The problem thus far remains general and can accommodate a variation of temperature in the u direction and its effect on both τ_0 and μ. However, since τ_0 is weakly dependent on temperature, the reference stress can be assumed to be constant. The velocity gradient (the total strain rate) through the film thickness is easily obtained by back substitution into the previous equations.

Assuming conduction across the film thickness to be dominant, the energy equation assumes the form

$$\rho c_p u \frac{\partial T}{\partial x} = k\frac{\partial^2 T}{\partial y^2} + \Phi \qquad (9.30)$$

where

$$\Phi = \frac{\tau_0}{\mu}\left(\tau_m + \left(y - \frac{h}{2}\right)\frac{dp}{dx}\right)\sinh\left(\frac{\tau_m + \left(y - \dfrac{h}{2}\right)\dfrac{dp}{dx}}{\tau_0}\right)$$

The boundary conditions for Equ. (9.30) are the familiar expressions resulting from a moving heat source,

$$T_1(x) - T_1(0) = \left(\frac{1}{\pi \rho_1 c_{p1} U_1 k_1}\right)^{1/2}\int_0^x k\frac{\partial T}{\partial y}\bigg|_{y=0}\frac{d\varsigma}{(x-\varsigma)^{1/2}} \quad ; \quad y = 0$$

$$T_2(x) - T_2(0) = \left(\frac{1}{\pi \rho_2 c_{p2} U_2 k_2}\right)^{1/2}\int_0^x -k\frac{\partial T}{\partial y}\bigg|_{y=h}\frac{d\varsigma}{(x-\varsigma)^{1/2}} \quad ; \quad y = h$$

The initial conditions are that the film and surface temperatures at $x = 0$ are equal.

With τ_m obtained from Equ. (9.29) the traction force per unit width transmitted through the contact can be calculated from the integral of the midplane shear stresses over the contact area, viz.

$$\left(\frac{F_\tau}{L}\right) = \int_0^{2\pi} \tau_m dx \qquad (9.31)$$

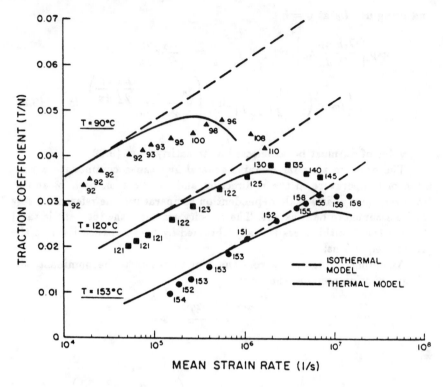

Fig. 9.20 Traction curves at different disk temperatures, $U_{roll} = 7.48\,\mathrm{m/s}$; $p_{max} = 921\,\mathrm{MPa}$, Conry, 1981

For the model having the specifics given in Table 9.4 results are shown in Fig. 9.20 for three disk temperatures; calculations were also made for an isothermal model which ignores viscous heat dissipation. The numbers next to the experimental data points are the temperatures measured at the outlet of the contact. The tractions predicted by the thermal model increase with increasing strain rate to a maximum then decrease, as observed in practice. As seen, the thermal curve starts to deviate from the isothermal curve at small values of the mean strain rate indicating that a shear heating effect is present in the second region of the traction curve even though the viscous shear heating is small. Also, the slopes of the isothermal curves tend to become parallel at the higher strain rates indicating that the traction curves approach values predicted by the sinh curve.

Figs. 9.21 – 9.23 show the temperature and shear stress distributions for the thermal model and the corresponding isothermal solutions. These are the midplane shear stress distributions for the right-hand points of the "thermal model" curves plotted in Fig. 9.20. The maximum temperatures are always at the midplane. The surface temperatures in all cases rise to a maximum at values of (x/a) between 1.4 and 1.5. The midplane shear stress

TABLE 9.4

Data for sinh traction model

Roller geometry	
Diameter	79 mm
Contact width	22.5 mm
Roller material properties	
Modulus of elasticity	2.07×10^{11} Pa
Poisson's ratio	0.30
Density	8000 kg/m^3
Specific heat	470 W-s/kgK
Thermal conductivity	38 W/mK

LVI 260 properties	
Shear modulus at atmospheric pressure	9.0×10^7 Pa
Pressure coefficient of shear modulus	3.1
Representative stress	4.5 MPa
Density	929 kg/m^3
Specific heat	2250 W-s/kgK
Thermal conductivity	0.125 W/mmK

Temperature (°C)	90	120	153
Viscosity at atmospheric pressure (Pa-s)	0.015	0.0057	0.0026
Viscosity-pressure coefficient $\beta \times 10^9$ (Pa^{-1})	20.48	16.77	13.72

distributions shown in Figs. 9.21 – 9.23 are approximately parabolic, which relates, in an average sense, to a parabolic distribution of heat along the contact. For the disk temperature of 90°C there was almost no difference between the temperatures of the two surfaces while for the disks operating at the higher temperatures, 120°C and 153°C, the slower moving surface had the higher temperatures.

9.3.2 Elasto-Plastic Model

In the work of Berthe et al, 1980, the investigation of non-Newtonian fluids is carried into the region of elasto-plastic models. In the authors' approach the hydrodynamic aspects are subordinated to thermal and non-linear effects as reflected, for example, in the assumptions of constant pressure in the contact zone. This simplification immediately leads to

$$\frac{\partial \tau}{\partial y} = \frac{\partial p}{\partial x} = 0 \quad \longrightarrow \quad \tau = \tau(x)$$

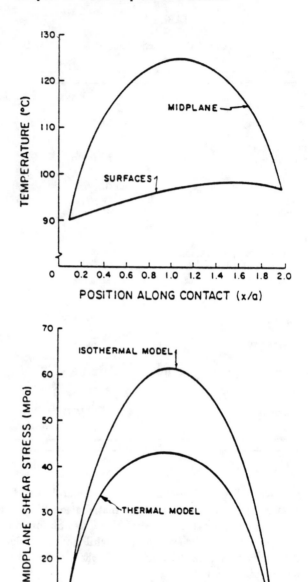

Fig. 9.21   Temperature and midplane shear stress in traction drive, $T_{disc} = 90°C$; $\partial u/\partial y = 8.24 \cdot 10^5\,s^{-1}$; other data as in Fig. 9.20, Conry, 1981

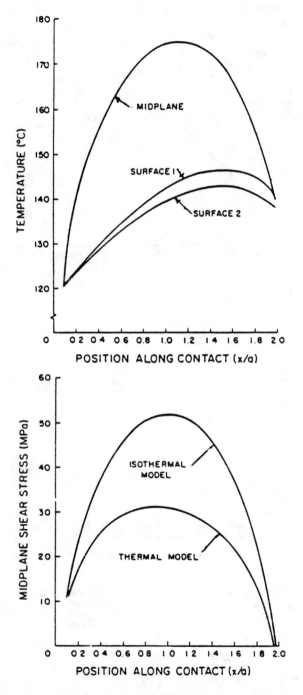

Fig. 9.22 Temperature and midplane shear stress at $T_{disc} = 120°C$; $\partial u/\partial y = 7.63 \cdot 10^6 \, \text{s}^{-1}$; other data as in Fig. 9.20, Conry, 1981

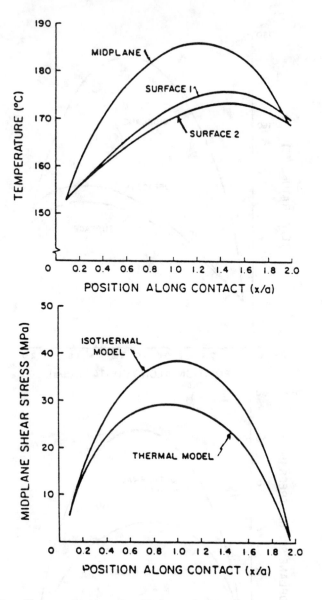

Fig. 9.23   Temperature and midplane shear stress at $T_{disc} = 153°C$; $\partial u/\partial y = 1.45 \cdot 10^7 \, \mathrm{s^{-1}}$; other data as in Fig. 9.20, Conry, 1981

expressed in the form

$$\left(\frac{\partial u}{\partial y}\right) = f\left[\left(\tau, \frac{d\tau}{dt}\right), T(y)\right] \tag{9.32}$$

Furthermore, it is postulated that the slope of the traction curve can be expressed by a rheological μ_R which is a function of the equilibrium viscosity μ, the transit time in the contact, t, and a relaxation time Λ. The rheological viscosity would then assume the form

$$\mu = \mu_R(T,p)\left[1 - e^{-(t/\Lambda)}\right] \tag{9.33}$$

For the linear part of the traction curve we have

$$\left(\frac{\partial u}{\partial y}\right) = \frac{\tau}{\mu_t} = F_1(\tau) \tag{9.34}$$

where μ_t is the transient viscosity. For the thermal portion, the familiar sinh relation is used

$$\left(\frac{\partial u}{\partial y}\right) = \left(\frac{\tau_0}{\mu_t}\right)\sinh\left(\frac{\tau}{\tau_0}\right) = F_2(\tau) \tag{9.35}$$

where τ_0 is the characteristic shear stress still to be defined. A purely elasto-plastic behavior of the fluid would read

$$\left(\frac{\partial u}{\partial y}\right) = \frac{U}{G}\left(\frac{\partial \tau}{\partial x}\right) + F_3(\tau) \tag{9.36a}$$

where G is the shear modulus, $F_3(\tau)$ is a plastic function of shear rate expressed as follows

$$F_3(\tau) = \frac{1}{A}\left(\frac{\tau}{\tau_e}\right)^{1/N} \tag{9.36b}$$

τ_e is the yield stress at zero shear rate, and A and N are parameters to be defined.

In the energy equation, convection is ignored and only conduction in the transverse direction is considered. Thus the energy equation becomes simply

$$k\frac{\partial^2 T}{\partial y^2} + \Phi = 0 \tag{9.37a}$$

The authors show that for the rheological fluids considered here the dissipation reduces itself to

$$\Phi = \tau F_i(\tau) \quad , \qquad i = 1,2,3 \tag{9.37b}$$

where F_i is one of the three coefficients defined previously. For the temperature rise in the moving solids the familiar Jaeger integrals are used, viz.

$$\Delta T_1(x) = \frac{1}{\sqrt{\pi \rho_1 c_{p1} k_1 U_1}}\int_{-b}^{x}\frac{-k\left(\dfrac{\partial T}{\partial y}\right)(\xi)d\xi}{\sqrt{x - \xi}}$$

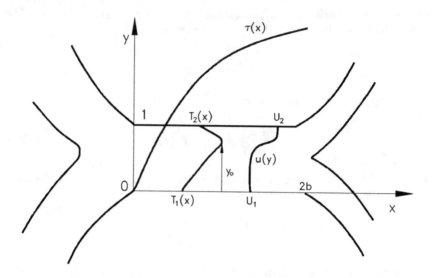

Fig. 9.24 Quantities of interest in rheological model

for solid no. 1; and likewise for solid no. 2.

The approach to the solution of a configuration shown in Fig. 9.24 is to calculate at each x the shear stress, surface temperatures T_1 and T_2, and the temperature and velocity profiles $T(y)$ and $u(y)$ so that the four boundary conditions on T and u on each surface are satisfied. The shear stress is then integrated over the elliptical surface to obtain the traction force and friction coefficient. In this integration, the rheological parameters μ_R, τ_0 and G are allowed to vary with temperature. The two main items of interest sought are the influence of surface temperature differentials on the shear stress and temperature profiles; and the influence of the rheological parameters on the individual performance items, particularly traction.

Results were obtained for the following conditions:

- Pressure (GPa) :

$$p_1 = 0.57, \quad p_2 = 0.71, \quad p_3 = 0.91, \quad p_4 = 1.14$$

- Mean rolling speed $\overline{U}$ (m/sec): $\left(\dfrac{U_1 + U_2}{2}\right),$

$$\overline{U}_1 = 21, \quad \overline{U}_2 = 32, \quad \overline{U}_3 = 46, \quad \overline{U}_4 = 61$$

- Fluid temperature, °C:

$$T_1 = 24, \quad T_2 = 49, \quad T_3 = 71, \quad T_4 = 105, \quad T_5 = 120$$

The linear viscous model used was

$$\mu_R = 2.33 \cdot 10^{-2} \exp \left| 1.33 \cdot 10^{-8} p + 3800 \left(\frac{1}{T} - \frac{1}{313}\right) \right| \quad (\text{Pa} \cdot \text{s})$$

$$\Lambda = 1.04 \cdot 10^{-7} \exp \left| 1.03 \cdot 10^{-8} p \right| \quad \text{(sec)}$$

$$\tau_0 = 3 \cdot 10^6 \exp \left| 585 \left(\frac{1}{T} - \frac{1}{340} \right) \right| \quad \text{(Pa)}$$

Fig. 9.25 gives the locus of $\bar{y}$ and $\bar{\tau}$ for various temperature differentials $(T_1 - T_2)$. If the temperature differences are not high, the figure shows that the point of maximum temperature is located approximately in the middle of the film. As this is usually the present case, $\bar{y}_{max}$ (where $T = T_{max}$) can be kept equal to 0.5 so that only one surface temperature needs to be calculated.

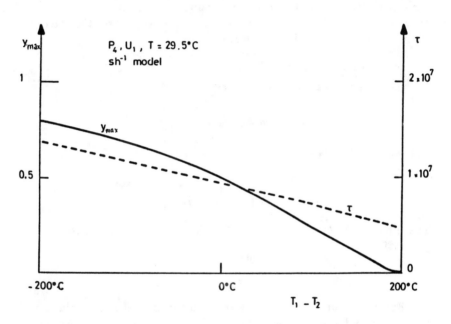

Fig. 9.25 Location of T_{max} and value of τ, Berthe et al, 1980

Viscous model
The equilibrium viscosity of the fluid is taken as a function of the mean pressure and temperature. A relaxation time or a shear modulus defined from low shear rate experiments is used, namely

$$\mu_R = 2.33 \cdot 10^{-2} \exp \left| 1.33 \cdot 10^{-8} p + 3800 \left(\frac{1}{T} - \frac{1}{313} \right) \right| \quad \text{(Pa \cdot s)}$$

$$G = 1.2 \cdot 10^5 \exp \left| 2.5 \cdot 10^{-9} p + 3800 \left(\frac{1}{T} - \frac{1}{313} \right) \right| \quad \text{(Pa)}$$

$$\Lambda = 1.98 \cdot 10^{-7} \exp |1.78 \cdot 10^{-8} p| \quad (\text{sec})$$

Non-linear viscous model

Fig. 9.26 gives traction curves obtained with the non-linear viscous model where τ_0 is assumed to be expressed by

$$\tau_0 = C_0 \exp \left| C_T \left(\frac{1}{T} - \frac{1}{340} \right) \right|$$

where C_0 and C_T are yet to be defined. Curve 1 in Fig. 9.26 is obtained with $C_0 = 6 \cdot 10^6$ Pa and $C_T = 10^3 \, °\text{K}$ and is taken as reference. Curves 2 and 3 are obtained by varying only C_0; when C_0 increases the traction curve tends to the limit of curve 4 obtained with a linear viscous model. Curve 6 shows that an increase of the parameter C_T causes the traction to fall more rapidly with sliding velocity.

Elasto-Plastic Model

Here the yield stress is written as

$$\tau_e = C_0 \exp \left| C_T \left(\frac{1}{T} - \frac{1}{340} \right) \right|$$

and the limiting plastic shear stress, deduced from Equ. (9.36), is

$$\tau_p = \tau_e \left\{ 1 + \left[A \left(\frac{\partial u}{\partial y} \right) \right]^N \right\}$$

The influence of A, C_0, C_T, and N is investigated about a standard set given by

$$A_0 = 2.5 \cdot 10^{-6} \, \text{s}, \qquad C_{00} = 4.5 \cdot 10^6 \, \text{Pa}$$

$$C_{T_0} = 4230 \, °\text{K} \qquad N_0 = 1$$

Fig. 9.27, curve 1, gives the traction for the reference values, while curves 2, 3 and 4 are respectively for $2A_0$, $(A_0/2)$ and $C_{T_0}/2$. An increase of A increases the traction coefficient and the slope of the traction curve. An increase of C_T causes the traction to fall more rapidly with sliding velocity. Fig. 9.28 gives additional data for traction in an elasto-plastic model. Fig. 9.29 shows the variation of shear stress along the contact and of the transverse temperatures and velocities for all three regimes. These curves were obtained with the same traction coefficients. The shear stress profiles are very different for the elasto-plastic model. The temperature rise with this model is zero at the entry to the contact zone. The plastic zone itself begins at the center of the film and expands as the flow progresses downstream. For the varying conditions listed, surface ΔT's range from $0°$ to $30°$ and temperature differences between surface and center of the film vary from $0°$ to $150°$. Maximum shear rate at the center of the film can be much greater than the mean shear rate $(U_2 - U_1)/h$.

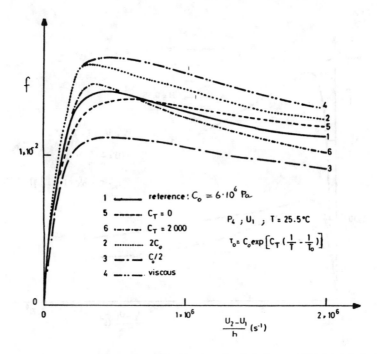

Fig. 9.26 Traction curves for non-Newtonian lubricant, Berthe et al, 1980

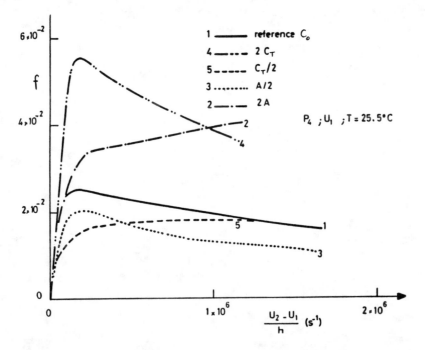

Fig. 9.27 Traction curves for non-linear model, Berthe et al, 1980

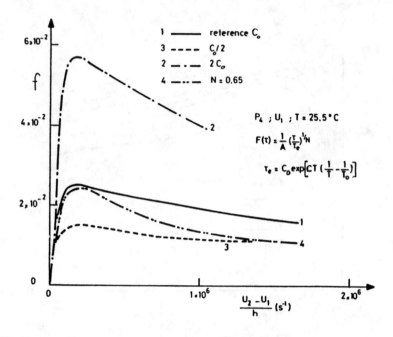

Fig. 9.28 Traction curves for elasto-plastic model, Berthe et al, 1980

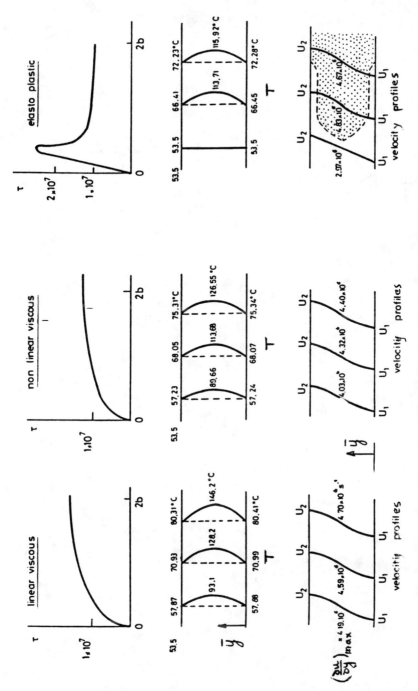

Fig. 9.29 Transverse mapping of film for three traction models, Berthe et al, 1980

Chapter 10

INSTRUMENTATION TECHNIQUES

The preceding nine chapters have been subdivided into two parts. One dealt with aspects which from a conceptual viewpoint are more or less defined, while the second considered topics which still lack a solid technical foundation. However, even the first group contains some disturbing anomalies, such as heat transfer solutions which yield lower load capacities and higher temperatures than do adiabatic cases, or the fact that no analysis yielded temperature variations in the axial direction, $T(z)$. The first example seems to violate the first law of thermodynamics; the second contradicts a substantial body of experimental data. Thus verification of the theoretical results, remains here, too, an important objective. The main function, however, of thermal fluid film experiments is that of guidance. This emphasis on the nature of the required experimental work pertains in particular to the technically immature disciplines, discussed in the second part of the book such as turbulence, two-phase flow, rheodynamic models, etc. Proper parametric correlations and boundary conditions have to be extracted from carefully planned experiments to provide inputs and constraints for analyses of thermal fluid films. While far from resolving problems of turbulence or two-phase flow in their general context such experimentally formulated tribological models would provide a framework for what is permissible and valid in the field or tribology. The formulation of conceptual foundations for advanced subjects in tribology constitutes, then, the core of a projected experimental program in thermal fluid films.

Such an experimental program represents a challenge not only as to the kind of tests required—but also as to how to perform them. Chapter 1 discussed to some degree the far reaching effects that variable viscosity has on any parametric variation, whether theoretical or experimental. These complications are due to the fact that a variation in any operational or geometrical quantity immediately produces a reorientation of the temperature and viscosity fields. Another complicating element is the sheer technical difficulty in obtaining the kind of measurements required in thermal fluid films. Since the clearances are of the order of mils, and often of microns, it is very difficult to measure the strains, stresses, and temperatures across the film. Even to obtain the surface temperatures of runner and bearing, or of the mixing inlet temperature, is a difficult and often uncertain task. Lastly, there is the problem of the universality of the obtained results. Usually experiments are conducted in test rigs which are specific to the particular investigation. Since the peripheral equipment and the mode of assembly have a considerable impact on the heat flow pattern, it is in most instances difficult to generalize the results to cases where the environment is different. While load capacity is affected but little by such diversities, thermal effects in fluid films are very sensitive to these and other environmental factors.

It seems, therefore, that any successful program of investigating the thermal behavior of fluid films is predicated on the availability of advanced, sophisticated instrumentation capable of recording the kind of quantities required in tribological measurements. New techniques and new concepts in instrumentation would have to be resorted to if the goals of a discerning experimental program are to be realized. The present chapter, therefore, reviews some of the advanced instrumentation techniques that may prove useful in the conduct of tribological experiments. The purpose of the descriptions is not to delve into the theory and practice of the various measurement techniques which embrace a great variety of physical and chemical principles and whose elucidation would be prohibitive in the present volume. Techniques which seem particularly promising will be discussed in terms of their adaptability for tribological experiments as recorded by those who used them either in demonstration or in actual application. Even if some may still be in their infancy, they are included in the expectation that, upon further development, a number of them will prove suitable for tribological research.

10.1 FILM THICKNESS MEASUREMENTS

A number of standard devices exist for measuring film thickness, most of them of the reluctance and capacitance types. The greatest need for advancement is in the area of measuring very small film thicknesses, small even by tribological standards. Thus the primary emphasis in the following paragraphs will be on instrumentation showing promise of accurately recording film thicknesses in the micron and submicron ranges.

10.1.1 Microtransducers

Special transducers which are themselves small in size and capable of differentiating between extremely small dimensions have been developed by a number of investigators. The following description is taken from van

Leeuwen et al, 1987, who applied them to EHD contacts. Both film thickness and temperature measurements are possible with the use of vapor deposited thin layer transducers. Since chemical films develop at elevated temperatures, typically over 80°C, the temperatures should be kept low. Also at high sliding speeds a chemical film may form, which sets an upper limit to speed.

Fig. 10.1 gives a schematic drawing of a thin film microtransducer. Generally, for vapor deposited transducers, investigators favor SiO_x as an insulating layer. However, its electrical, mechanical, and thermal properties differ substantially from steel, thus affecting the temperature reading. A material with steel-like properties like Al_2O_3 may be preferred. Titanium is used for temperature transducers because it has a very low pressure sensitivity. Since capacitance transducers require only that the material be a good conductor, the same materials can be used as for the temperature transducer. The only difference is in the geometry used.

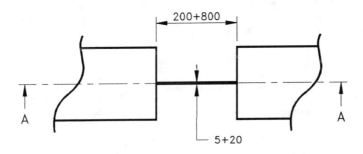

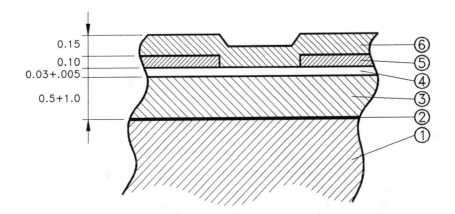

Fig. 10.1 Schematic drawing of a thin film transducer, (1) substrate, (2) adhesive layer, (3) insulating layer, (4) transducer, (5) conductor, (6) protective layer (optional). All dimensions in μm

The electrical resistance can be arbitrary, but the conductor pattern should have a much lower resistance. To keep the parasitic capacitance low, and to avoid pinholes, the conductor pattern area is kept small. The terminals of the conductor pattern are connected to a print board by 20 μm diameter gold wires. The transducers can be made by vapor deposition followed by laser beam cutting, or by a photolithographic process. Photolithography is a very versatile technique. All layers are deposited in one sequence, without breaking the vacuum of the evaporation jar. The layers are partially removed by selective etching using a photomask. What remains are the conductor and transducer patterns. Photolithography can accomplish the installation of many transducers on the same substrate. Flexible photomasks will easily accommodate to the substrate curvature. Fig. 10.2 shows an example of the geometry of the transducers used in the van Leeuven experiments.

Section A–A

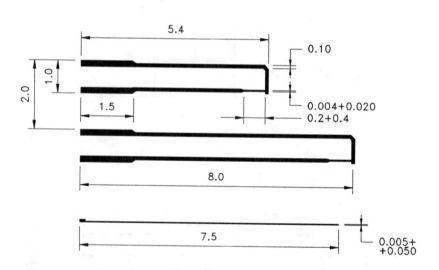

Fig. 10.2 Transducer geometries. Section A-A of Fig. 10.1. Upper two: temperature transducers. Lower two: capacitance transducers. All dimensions are in μm

In most cases a Wheatstone bridge, or a compensation circuit is used when small changes in Ohmic resistance are to be measured. A constant electric current of 1 mA flows through the transducer, which corresponds to a certain voltage drop. The difference with an externally adjustable reference voltage is amplified by a broadband instrumentation amplifier (50×). High quality, low noise operational amplifiers are used, resulting in a large bandwidth and high slew rate. To allow high gain for maximum sensitivity of the dynamic signal, the static component of the signal—due to resistance changes caused by variations in the operating temperature—is nulled out by adjusting the reference voltage. The absolute temperature can now be deduced from the change in reference voltage.

10.1.2 Optical Interferometry

In using optical interferometry to measure oil film thickness, the main disadvantage is, of course, that one of the mating surfaces must be transparent. On the other hand, extremely small dimensions can be recorded with this method, some, as will be seen below, of the order of less than one wavelength of light.

The standard use of optical interferometry is here exemplified by the tests conducted by Robinson and Cameron, 1975. With oil as the fluid, each fringe represented in these experiments a change in film thickness of 0.184 μm. The disadvantage of using monochromatic light is that it only gives *changes* of film thickness. To obtain absolute values in the 0.1 – 1.0 mils range two or more colors have to be used. An alternate approach is to use fringes of equal chromatic order which requires a spectrometer with a line-by-line picture. The main consideration in designing such a system is that the fringe visibility should be maximum. This comes about when the two interfering rays are of the same intensity. This requires that the transparent member be given a semi-reflective coating on the running surface.

In choosing the transparent member a glass runner can be used because it is more nearly isothermal than the bearing; and the differences in thermal properties of glass compared with those of steel are not serious. As a bearing material glass is poor but when it is given a semi-reflecting coating—in the present case chromium —the system behaves like an oil-chromium rather than oil-glass system.

Originally thrust pads having a surface of tin babbitt were polished using a ground glass lapping plate. Although a sufficiently reflective surface was obtained, babbitt was found to be an unsuitable interferometric surface. Stainless steel pads were substituted. These were lapped and polished to an overall flatness of 0.13 μm. The reflectivity of the pads and of the chromium semi-reflecting layer had to be matched. For the pad, reflectivities were measured in air and oil by comparing them with known reflectivities of a silver plate, using a constant light source and a photocell.

For good visibility the intensity of the two interfering beams must be similar. In Fig. 10.3 an incident ray of unit intensity strikes a semi-reflecting layer. One beam is reflected at the chromium-oil interface with an intensity of R_{cr}. Of the remainder some is absorbed and some is transmitted with an intensity of T_{cr}. At the steel oil surface, this is reflected with an intensity of $(R_{steel}T_{cr})$. The part that continues past the semi-reflecting chromium will have an intensity $(R_{steel}T_{cr}^2)$ and will interfere with the first reflected

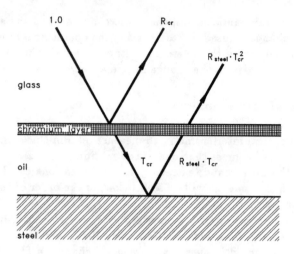

Fig. 10.3 Intensities of interference rays

ray. For their intensities to be the same R_{cr} must equal $(R_{steel}T_{cr}^2)$.

The test setup used by Robinson and Cameron is shown in Fig. 10.4. Their interest was in measuring film thickness in flat and tilting pad thrust bearings. The beam splitter was a 50 mm square glass plate with a 50/50 vacuum deposited layer of TiO_2. The reflectivities they measured were:

Surface	In air	In oil
Tin babbitt	0.25	0.20
Steel tilting pad	0.30	0.26
Steel fixed pad	0.56	0.48

Fig. 10.5 shows typical results obtained for a thick film while Fig. 10.6 shows results with an EHD spike near the trailing edge of the pad.

An improved version of optical interferometry is described by Spikes and Guangteng, 1987, who measured films thinner than the conventional limit of one quarter the wavelength of visible light. A wedge of spacer layer material is sputtered onto the chromium-plated glass disc of an optical elasto-hydrodynamic point contact. As the disc rotates, the spacer layer thickness varies between zero and one wavelength of light whereas the elastohydrodynamic film thickness is constant. From the variation in the position of interference figures around the circumference of the glass disc, accurate measurements below $10^{-3}\,\mu m$ (1/40th wavelength of light) have been achieved. The principle of this method is shown in Fig. 10.7. A flat layer of transparent silica or alumina is deposited on top of a semi-reflecting chromium layer on the transparent glass disc surface. In conventional interferometry, neglecting any phase change effects, destructive interference

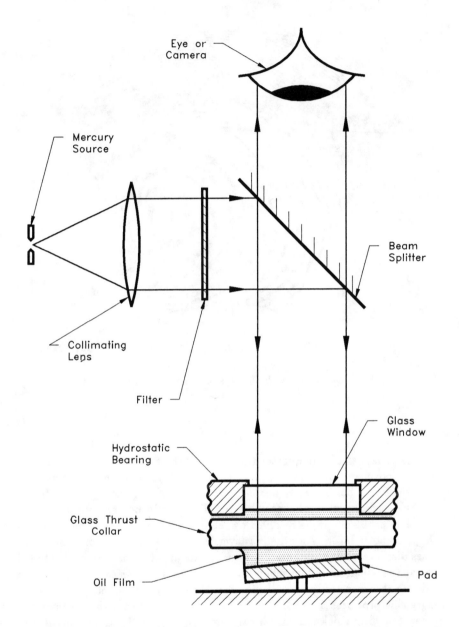

Fig. 10.4 Instrumentation setup for optical interferometry, Robinson and Cameron, 1975

occurs when

$$h = \left(n + \frac{1}{2}\right)\frac{\lambda}{2}, \qquad n = 0, 1, 2, 3$$

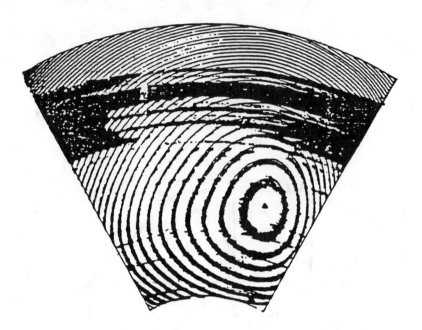

Fig. 10.5 Sample interferogram, Robinson and Cameron, 1975

If a spacer layer of thickness h_s is present, then interference occurs at

$$h = \left(n + \frac{1}{2}\right)\frac{\lambda}{2} - h_s \qquad (10.1)$$

that is at less than one quarter the wavelength of light. One disadvantage of the above approach is that a given spacer layer can only measure one film thickness below $(\lambda/4)$ i.e. $h = (\lambda/4 - h_s)$. To measure a range of thicknesses, a number of spacer layered discs would be needed.

The problem can be resolved by the use of a variable thickness, wedge-shaped spacer layer. An alumina layer is sputtered on a chromium-plated glass disc so that its thickness varies from about 0.05 μm at one edge to 0.1 μm at the other, as shown schematically in Fig. 10.8. As the disc rotates, a varying thickness of spacer layer passes through the contact. Correspondingly this produces a varying interference fringe pattern, which consists of a constant film thickness plus a varying spacer layer. It is then possible to relate fringes, seen at different times at the microscope position, to their corresponding angular positions on the moving track and the applicable h_s.

The distance the fringes move depends on the oil film thickness, as shown in Fig. 10.9. With the spacer layer wedge angle small, thin oil films can give measurable fringe pattern movements. For a disc with 0.2 μm thickness at one edge, there is approximately 0.02 μm change of spacer layer thickness for every centimeter across the disc diameter. Hence a 0.02

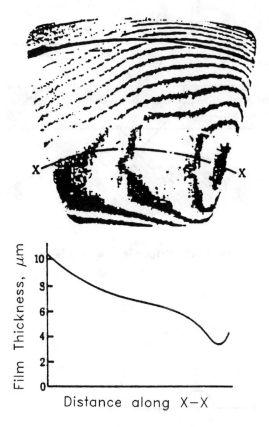

Fig. 10.6 Interferogram for EHD Contact, Robinson and Cameron, 1975

μm thickness of oil film will move fringes by up to one centimeter from their position when no oil film is present.

In the Spikes and Guangteng experiments, white light was used to give chromatic interferometry. The following two recording methods were investigated.

a) Video Recording Method

Initial work used continuous white light and a video recorder attached to the microscope. Short recordings of the static fringe color at 5° positions round the disc were made to map the alumina wedge thickness profile. Recordings were then made of the varying fringe colors per revolution at constant speed. This was done at a range of different disc speeds. The video recordings were analyzed to determine the position around the disc tracks at which interference colors occurred for each speed. Film thickness for each speed was then calculated from the movement of the fringe color

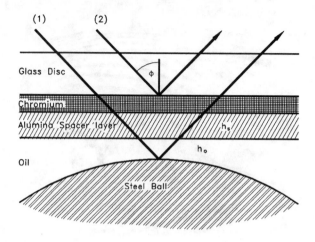

Fig. 10.7 Principle of spacer layer

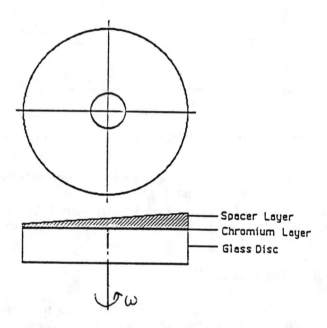

Fig. 10.8 Variable thickness spacer layer

boundaries. In practice more than one such boundary was usually available on a disc at any given time so the movements of all the boundaries could be averaged.

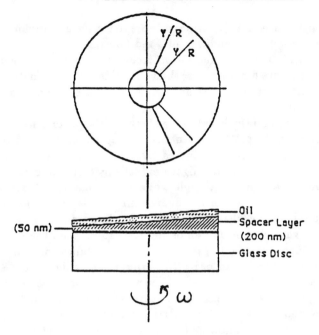

Fig. 10.9 Effect of oil film on position of color fringes

b) Camera Recording Method

A disadvantage of the video recorder method is that it is only possible to view one image at a time. It is thus difficult to precisely define at what point colors change in a sequence of images and to compare colors from one test speed with another. To overcome this limitation a camera recording method with flash tube as a light source was used. As the disc rotates it first triggers the flash unit, which exposes a quarter frame of film in a camera attached to the microscope, and then activates a stepper motor to wind the film on. This produces a series of contact images covering one revolution. In practice the images are arranged to trigger at a cosine spacing rather than equiangular spacings around the disc so as to correspond to equal steps in spacer-layer thickness. The displacement of fringes at different speeds was determined by laying strips of images from halfway around a disc on a light box and moving them laterally so as to align them with a corresponding strip taken from the static contact.

10.1.3 Laser Fluorescence Technique

Mineral oil lubricants produce fluorescence when illuminated with ultraviolet light. A similar fluorescence is produced with a blue light source which has a longer wavelength. A He-Cd laser can then be used directly, eliminating the need for additional mirrors to generate ultraviolet light with low output power. The addition of a red fluorescent dye further enhances oil fluorescence. The laser fluorescent technique may be a more convenient method to measure film thickness than optical interference method, because unlike optical interference the fluorescent method does not require a high finish of the surfaces.

To measure film thickness change between the piston rings and a transparent cylinder, Ting, 1980, used the optical arrangement shown in Fig. 10.10. The oil film was illuminated by a blue light beam issuing from a He-Cd laser head. The visible light produced by oil fluorescence could then be observed at a convenient angle while a filter blocked off scattered blue light. Since the intensity of the fluorescent light is proportional to the illuminated oil volume, the change of oil film thickness could be determined by using a light detector and an oscilloscope to measure the light intensity changes. Quantitative oil film thickness was obtained once a calibration curve was established. In order to perform the calibration, piston rings with one or two machined steps of known dimensions can be used to establish the light intensity versus oil film thickness relationship. Calibration is best performed with the piston stationary so that the piston ring will be in contact with the bore, allowing the latter to serve as a reference for piston ring step depth.

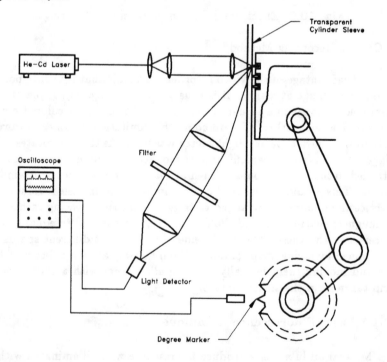

Fig. 10.10 Instrumentation setup for piston ring measurements, Ting, 1980

Fig. 10.11 shows signals of light intensity change, due to the passage of piston and piston rings through the laser beam path. Of the two signals simultaneously displayed, the upper trace is the fluorescent light intensity signal picked up some distance below the top dead center where the oil film thickness is measured. The lower saw tooth trace is a crank angle position signal generated by a marker attached to the crankshaft (Fig. 10.10), with the peak to peak distance corresponding to a crank angle of 10°. With this signal, the relationship between the top light intensity trace and the crank angle can be established. In this manner oil distribution during an upward stroke motion may be qualitatively constructed as shown in Fig. 10.12. Improved trace resolution, using a smaller light spot size, is likely to show whether the piston rings operate under flooded or cavitating conditions. By installing a device which moves the laser beam synchronously with the piston, the film thickness can be measured for the complete cycle.

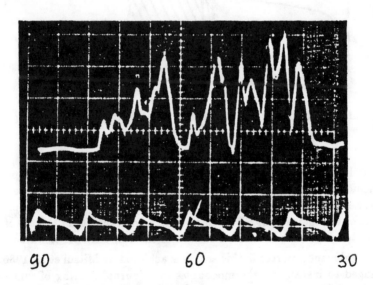

Fig. 10.11 Fluorescent light output (upper curve) and crank angle degree marker (lower curve), Ting, 1980

10.2 TEMPERATURE MEASUREMENTS

10.2.1 Special Thermocouples

Unlike film thickness, temperature measurements do not present problems of scale or size since they usually fall within the moderate temperature spectrum. But they still do have in common with film thickness the fact that access to desired locations represents a problem of size. Thus, for example, while transverse temperature variations may be of the order of 10°

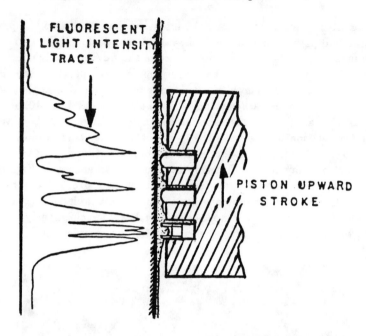

Fig. 10.12 Schematic for measuring film thickness in piston ring, Ting, 1980

or 100°F and thus easily measured, access to individual locations presents a formidable problem. Tied to access is the problem of non-intrusion or non-interference by the instrument. Both of the above add up to a problem of size of the instrument. Thus one of the advancements in the art of measuring temperatures is miniaturization of the conventional thermocouple.

Significant progress in this area was achieved by Mitsui et al, 1986, who managed to install 144 thermocouples in a journal bearing of dimensions $D \times L \times R/C = 100\,\text{mm} \times 70\,\text{mm} \times 10^3$. The mounting of the thermocouples is represented schematically in Fig. 10.13. 0.24 mm wire thermocouples were implanted by soldering within 0.2 mm from the surface. The output of journal surface temperatures was measured through slip rings.

To measure more accurately temperatures on the journal surface the device must be sufficiently sensitive to detect temperature variations even if they involve only a thin layer of the surface. Moreover, it must not induce any alterations in the surface examined and the measurement must not be influenced by normal temperature gradients as a result of the thermic resistance between the measurement point and the shaft surface. To adapt to these requirements Andrisano, 1988, used a thermocouple whose hot junction was made not by direct welding, but with an intermediate metal, deposited in a thin layer through a chemical deposition process.

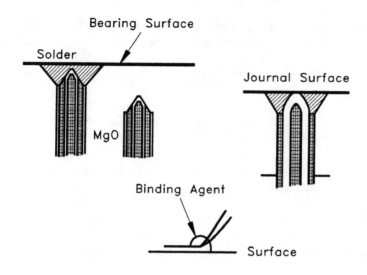

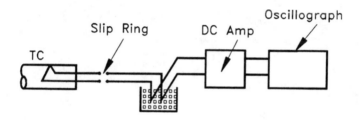

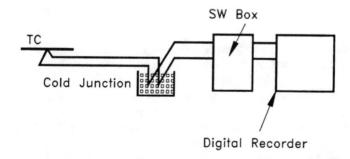

Fig. 10.13 Miniaturized thermocouple, Mitsui et al, 1986

A cylindrical-sinusoidal wave of the type

$$T = T_w \sin \left[\left(\frac{2\pi x}{L_w} \right) + \left(\frac{2\pi v t}{L_w} \right) \right]$$

having an amplitude T_w, wavelength L_w and which travels at velocity v, has to be attenuated and dephased, at a depth y from the surface, according to

$$T_y = T_w e^{-\alpha(\cos \beta)y} \left[\sin \left(\frac{2\pi x}{L_w} \right) + \left(\frac{2\pi v t}{L_w} \right) + \Psi_w \right] \qquad (10.2)$$

$$\alpha = \left(\frac{2\pi}{kL_w} \right)^{1/2} \left[v^2 + \left(\frac{2\pi k}{L_w} \right)^2 \right]$$

$$\beta = \frac{1}{2} \tan^{-1} \left(\frac{v L_w}{2\pi k} \right)$$

$$\Psi_w = \alpha(\sin \beta)y$$

Fig. 10.14 plots the maximum value of (T_y/T_w) versus depth y for different journal velocities. In order to keep T_y nearly equal T_w, y must be of the order of a micron. At this depth the phase displacement Ψ_w is negligible.

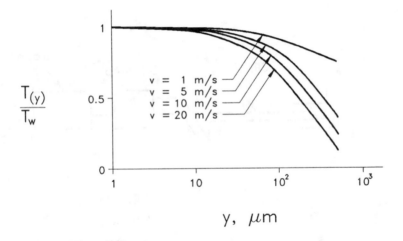

$$\frac{T_{(y)}}{T_w}$$

v = 1 m/s
v = 5 m/s
v = 10 m/s
v = 20 m/s

y, μm

Fig. 10.14 Journal temperatures as function of radial depth and surface speed; $(\sigma = 12 \times 10^{-6}\,\text{m}^2/\text{s})$, Andrisano, 1988

The mounting of the thermocouple in the journal is illustrated in Fig. 10.15. Comparable with the overall dimensions of the wire the radial hole housing the thermocouple must be as small as possible because the hole can alter the value of the measured temperature, especially near the exit. Chromel-alumel alloy thermocouple wires of 0.3 mm diameter were used, one of which was covered with a thin insulating sheath. The two wires were embedded in a 1.5 mm diameter radial hole until they protruded a few millimeters beyond the ground surface. The following operations were then performed:

a) a fluid acrylic resin suspension charged with nickel powder less than 20% in volume was cast through the shaft axial hole. The purpose of the

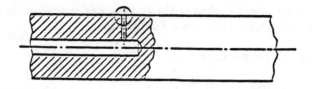

0.003 mm covering thickness

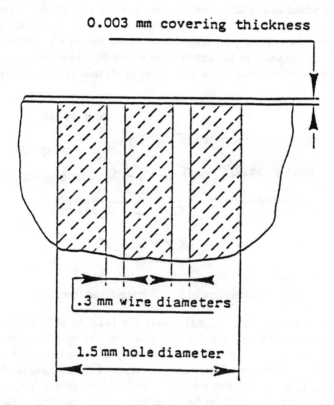

.3 mm wire diameters

1.5 mm hole diameter

Fig. 10.15 Mounting of thermocouple in shaft, Andrisano, 1988

powder was to activate, for the metallization process, the non-metallic surface where the wires emerged, and at the same time prevent short-circuiting them below the surface;

b) hardening and stabilizing of the acrylic resin, in order to prevent shrinkage during the experiments;

c) resetting of the original cylindrical surface and removing the excess resin and the protruding wires;

d) pickling with with HCl acid and surface activation with a $PdCl_2$ chloridic solution;

e) chemical nickeling at ambient temperature in a bath with: $NiCl_2 \cdot 6H_2O = 0.5\,M$; ammonium citrate $= 0.4\,M$; $NaH_2PO_2 = 1.0\,M$:

NaOH(pH = 9.8): in 30 minutes the surface was covered with a continuous, regular Ni-P alloy layer some 3 μm thick.

10.2.2 Bisignal Transducer

At the 1988 Thermal Workshop Kannel described a bisignal transducer capable of measuring simultaneously temperature and pressure. The vapor deposited sandwich transducer is shown in Fig. 10.16. It consists of a titanium temperature transducer coated on top of a manganin pressure transducer. As the bisignal transducer passes through the interface, the change in resistance due to temperature of the titanium and the change in resistance due to contact pressures of the manganin are sensed simultaneously.

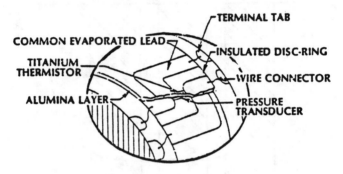

Fig. 10.16 Construction of a p-T transducer, Kannel, TW 1988

Construction of the bisignal transducer consists of first sputtering a layer of alumina (Al_2O_3) about 1 μm thick over a fairly wide portion of the surface of the disk. This coating is followed by a vapor-deposited manganin coating. The dimensions of the "active" region of the manganin transducer (center region in Fig. 10.16), where the bulk of the resistance change takes place, are about 50 × 200 μm by 1 μm thick. This coating is followed by another sputtered layer of alumina and finally the vapor deposited titanium element. The dimensions of the titanium element are similar to those of the manganin transducer.

Several techniques were tried for establishing the temperature sensitivity of the titanium element. The most reproducible method was to direct a stream of oil, at a preset temperature, onto a transducer. The titanium element was evaporated onto a glass slide and had an active region about 6.6 mm by 0.25 mm. The oil jet was directed at the active region of the transducer and the transducer output was read on an oscilloscope set at a very slow sweep speed (2 sec/div). In a typical calibration curve a rise of 8°C produced 1 mV on the bridge circuit.

During the course of the experiments, it became clear that the output of the vapor-deposited titanium was influenced by pressure. When the disks, with a thin sheet of insulating paper between them, were statically loaded, under ambient temperature conditions the titanium transducer produced an output that was:

a) Opposite in sign to that produced by a high temperature.

b) Equivalent to 44% of that produced by the pressure transducer.

To compensate for the pressure effect in measuring the temperature distribution, two techniques were employed. The first involved compensating the temperature signal after the experiments had been performed. This approach naturally does not produce immediate readings during the experiments. The second technique dynamically corrected the temperature signal with a modified pressure signal. Since the output of the titanium transducer is reduced by the influence of the pressure, inserting the pressure transducer on the opposite leg of the bridge adds a voltage proportional to pressure. A resistance network can be used to produce the level of pressure signal required for correcting the temperature trace. A schematic diagram of such a circuit is shown in Fig. 10.17. To balance the bridge, the resistance in the leg containing the temperature transducer, R_T, must equal the resistance of the leg containing the pressure transducer, R_P. For this condition,

$$R_T = R_2 + \frac{R_P R_1}{R_P + R_1} \tag{10.3}$$

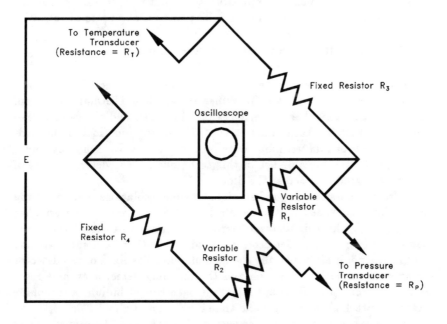

Fig. 10.17 Pressure compensating circuit, Kannel, TW 1988

If it is assumed that the pressure coefficient of resistivity of the temperature transducer is constant, then

$$\frac{dR_T}{R_T} = C\frac{dR_P}{R_P} \tag{10.4}$$

Here, C is a constant relating the pressure coefficient of resistivity of the titanium to the pressure coefficient of resistivity of the manganin. By differentiating Equ. (10.3), and combining with Equ. (10.4), there results

$$\frac{R_1}{R_P} = \left[\left(\frac{R_P}{CR_T}\right)^{\frac{1}{2}} - 1\right]^{-1} \tag{10.5}$$

If R_T, R_P, and C are known, equations (10.5) and (10.3) can be used to determine R_1 and R_2 in the circuit of Fig. 10.17.

For the transducer used by Kannel $R_T = 1595\,\Omega$ and $R_P = 925\,\Omega$. From tests it was found that $C = 0.44$. Using equations (10.3) and (10.5),

$$R_1 = 6247\,\Omega$$

$$R_2 = 805\,\Omega$$

In order to measure both temperatures and pressures, two separate circuits are required. For temperature measurements, the self-compensating circuit has to be used, whereas a "normal" Wheatstone bridge circuit is sufficient for pressure measurements. Typical pressure and temperature traces are shown in Fig. 10.18 for cases of pure rolling and for slip up to about 20%.

10.2.3 Infrared Radiation Emission Technique

The most advanced of the temperature measuring techniques is that of infrared emission presented by Winer at the 1988 Thermal Workshop. This is based on a number of previous works by Winer and his co-workers— Nagaraj et al, 1978, Ausherman et al, 1976, Winer and Kool, 1980. Additional details of this technique were provided at the Workshop by Lauer who discussed the possible use of this technique for the determination of temperature gradients across the film.

Infrared emission has good spatial and time resolutions, but one of the surfaces must be transparent to be infrared. The best choice of material has been sapphire, which simulates a bearing material both thermally and mechanically. Two infrared detectors are used, one with a fixed spot detector and the other scanning optically the field of view. The fixed optics detector has an area resolution of 38 μm and the scanning detector an area resolution of about 1 μm. Both are liquid nitrogen cooled Indium Antimonide detectors with high sensitivity and time resolution of about 8 μs.

The experimental contact consisted of a rotating 31.8 mm diameter chrome steel ball. The fluid was a naphthenic base oil having a peak in the

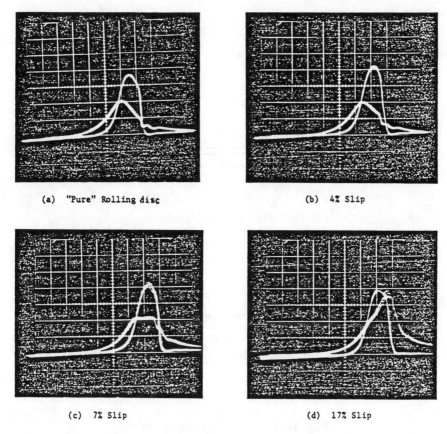

(a) "Pure" Rolling disc   (b) 4% Slip

(c) 7% Slip   (d) 17% Slip

Fig. 10.18 Experimental output using the p-T transducers, $P = 0.7\,\text{GPa}$; $N = 28,000\,\text{rpm}$; oil temperature – 35°C, Kannel, TW 1988

emission spectra at 3.4 μm, and significant emission in the band 3.0 to 3.7 μm. The ball surface, on the other hand, emits as a grey body. Fig. 10.19 shows these surfaces along with the spectral characteristics of the narrow and wide band filters. The detector with the sapphire surface will only respond to radiation in the band 1.8 to 5.5 μm. The narrow band filter has been chosen so that essentially all of the radiation emitted by the oil film is transmitted along with a portion of the grey body ball radiation. The wide band filter, on the other hand, transmits only radiation from the ball surface. When using one filter or the other, the influence of the filter was assigned to the detector rather than the radiation source, i.e., the detector calibration curve depended on the filter used, as well as on both the temperature and spectral characteristics of the radiation source.

Fig. 10.20 shows the radiation contribution and the signal attenuation before passing through the filter. The total radiation N is

$$N = \eta_b N_b + \eta_F N_F + \eta_S N_S + \eta_0 N_0 \qquad (10.6)$$

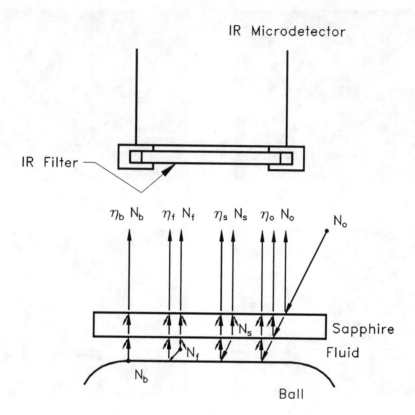

Fig. 10.19 Interferometry principle

where the attenuation factors include losses due to reflection and absorption. Above, and subsequently, the following nomenclature is used:

N_{BB} = black body radiation
N_b = ball radiation received
N_F = lubricant radiation received
N_0 = ambient radiation received
N_S = sapphire radiation received
$N_{\lambda BB}$ = monochromatic black body radiation
$N_{\lambda F}$ = monochromatic lubricant radiation received
$\eta_{b,F,O,S}$ = attenuation factor for ball, lubricant,
                ambient, and sapphire radiation, respectively
α' = absorption coefficient
ϵ_b = ball emissivity
τ = transmissivity of fluid
τ_s = transmissivity of sapphire
$\rho_{1,2}$ = Fresnel reflections of upper and lower sapphire surface
ρ = ball surface reflectivity

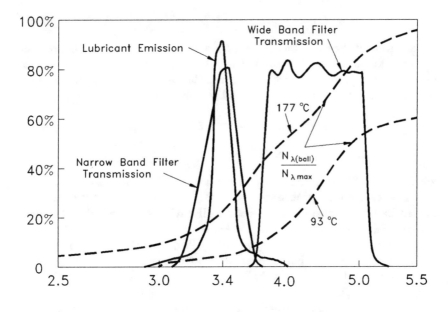

Wavelength − λ (μm)

Fig. 10.20 Spectral characteristics of lubricant contact, Winer et al, TW 1988

There are four contributions to the radiant energy collected by the micro-detector: (1) from the ball surface, N_b, (2) from the lubricant film, N_F, (3) from the sapphire, N_S, and (4) from reflected ambient radiation, N_0. The sapphire contribution $\mu_S N_S$ would be less than 2% of the ball contribution and contribution (4) can be accounted for knowing the optical properties of the system and ambient temperature.

The radiation emitted by the ball surface N_b is only a function of ball emissivity ϵ_b and ball surface temperature T_b since the ball is a grey body. Therefore, from the detector response, the sum of contributions (1) and (2) at any point in the contact can be determined. The technique for separating the two components consists of using filters which take advantage of the different spectral characteristics of the two contributions as shown in Fig. 10.19.

The pertinent equations as derived by Winer and his co-workers are grouped below:

$$\eta_b = \tau_S(1 - \rho_1)(1 - \rho_2)\tau(h, T_F) = \eta_s \tau(h, T_F) \qquad (10.7)$$

$$\eta_0 = \rho_1 + \tau_s^2(1 - \rho_1)^2 \rho_2 + \rho_b \eta_S^2 \tau(2h, T_F) \qquad (10.8)$$

$$N_F = [\epsilon_b \epsilon_F(h, T_F) + \rho_b \epsilon_F(2h, T_F)] N_{BB}(T_F) \qquad (10.11)$$

$$N_{\lambda F} = \epsilon_F N_{\lambda BB} + \rho_b \tau \epsilon_F N_{\lambda BB} \qquad (10.9)$$

$$\tau = 1 - \epsilon_F = e^{-\alpha'(\lambda, T_F)h} \tag{10.10}$$

Substitution of the contributions N_b, N_F and N_0 into equation (10.6) results in two unknowns ϵ_F and τ required to determine temperature from a given radiation measurement. These values must be obtained through calibration. Suitable models for the emissivity and transmissivity are

$$\epsilon(h, T_F) = 1 - e^{-\alpha h e^{\beta T}} \tag{10.12}$$

$$\tau(h, T_F) = e^{-\gamma h e^{\delta T}} \tag{10.13}$$

where T is in °K and α, β, γ, δ are constants to be determined.

The calibration procedure consists of submerging a stationary unloaded ball in a constant temperature oil bath. With the film thickness h determined by geometry and T_F known Equs. (10.12) and (10.13) are used to find the constants α, β, γ, and δ. Two balls of different emissivity are needed for this purpose. With the above formulation, the wide band filter data directly yields the ball surface temperature, since in this case $\epsilon_F = 0$ and $\tau = 1$.

In the tests conducted by Winer et al data were taken with a reservoir temperature of 40°C, a sliding speed varying from 0.35 to 12.7 m/s, and normal loads of 67 N to 215 N. Fig. 10.21a shows the ball surface temperature where, due to symmetry, only one half of the map is shown. Along with the boundary of the Hertzian contact zone, isotherms are shown for increments of 10°C. As seen, from a reservoir temperature of 40°C, the ball temperature increases to a maximum of 117°C downstream of the contact center. Fig. 10.21b shows a similar plot for the lubricant temperature which is seen to peak at 120°C before it enters the contact zone.

10.3 OTHER MEASUREMENTS

New instrumentation techniques were presented and discussed at the 1988 Thermal Workshop which although not aimed directly at measuring thermal variables, would yet provide important information in any experimental thermal program. Two such areas are measurement of wear and flow visualization.

10.3.1 Implanted Radiation Method

Studies with a proton beam on carbon and nitrogen targets demonstrated that 7_{Be} (Beryllium) is produced with a large cross section. Coupled with the development of superconducting heavy ion cyclotrons, this allows for a reverse nuclear physics kinematic process, namely fragmentation of carbon or nitrogen beams on light mass targets. The consequence of this is that the nuclear fragments are made at the beam velocity and can then escape from the thin target in the direction of the primary accelerated beam.

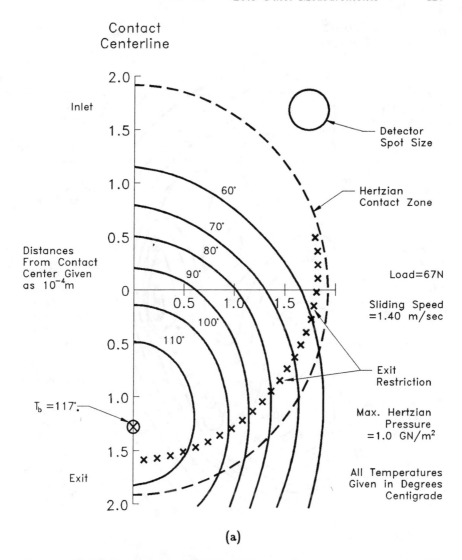

Contact
Centerline

(a)

Fig. 10.21 Temperatures in EHD contact, (a) temperatures on the ball surface, (b) average oil film temperatures, Winer et al, TW 1988

The above phenomena have led to the development of radioactive ion beams for nuclear physics use.

The implanted radiation method discussed below was presented by Mallory TW, 1988, much of it based on a 1988 paper by the author and his co-workers. It was indicated that for wear studies, a secondary beam of 7_{Be} which has a half-life of 53 days and a single detectable ray would be ideal. All other radioisotopes produced in the fragmentation of the nitrogen or carbon beam projectiles have half-lives that are short compared to 7_{Be}.

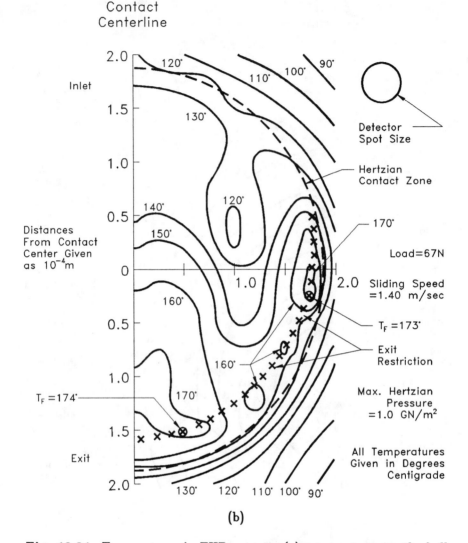

Fig. 10.21 Temperatures in EHD contact, (a) temperatures on the ball surface, (b) average oil film temperatures, Winer et al, TW 1988

The high velocity 7_{Be} ions also allow target implantation to take place in air. A suitable energy degrader has to be built, so that particle depositions at the front surface of the test sample is obtained. In the degrading process, the primary beam is stopped in the degraders and does not reach the the test sample. This protects the wear sample from radiation damage. Fig. 10.22 shows how an activated part would be monitored. The part is first activated in a cyclotron facility, then installed in the laboratory or an engine. The irradiated part emits low intensity gamma rays whose signature

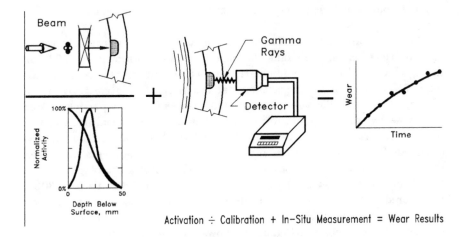

Fig. 10.22 Principle of irradiation emission, Mallory et al, TW 1988

is monitored as the part is worn, the wear being inversely proportional to the level of measured radiation. It was indicated that the radiation level of the activated part would not present a health hazard.

Implantation of 7_{Be} for wear analysis represents a significant advancement over other techniques. The principal advantage, for example, over presently available surface layer activation is that 7_{Be} implantation does not require a host metal. This means that a wide class of materials including ceramics, graphite composites, plastics and rubber compounds can be studied. Another unique feature of this method is that only the 7_{Be} is implanted and no nuclear reactions take place in the wear sample. This eliminates any background radiation. As wear is measured in situ, the technique can be used to monitor wear rates in engines during operation. This flexibility, coupled with the ability to measure wear rates in the $10^{-7}\,\mu m$ per hour range, offers great promise in the testing of materials.

10.3.2 Flow Visualization

In general, flow visualization employs a coherent source of light to illuminate a fluid seeded with light-reflecting particles. In most cases, this yields only a qualitative picture of the flow pattern. The present method, developed by Braun and associates, TW 1988, aims at quantifying the recorded image. In essence, very high optical magnification is aimed at small areas, which are then stitched together to reproduce the overall flow field.

A test setup would consist of three components:

a) the laser and ancillary optics,
b) the front end image processing installation, and

c) the computer-based image processing software and hardware.

In this setup the beam of an argon-ion laser is passed through a system of cylindrical lenses. This fans the beam into a curtain of light, 0.025 in. thick, which projects perpendicularly onto the longitudinal axis of the test section. The laser beam is positioned at the location of interest, by means of a set of mirrors. The thin plane of light ensures that only particles which have true two-dimensional motion are followed. The optical front end of the TV camera is located in line with the lateral optical port on the side of the shaft and positioned in such a way that it can survey a portion of the pocket. A succession of consecutive positions of a set of particles is thus brought into view. The images are recorded at intervals of $1/30$ sec and stored. The digitization procedure records the consecutive radial and tangential positions of the particles, which are then converted to vectors of the particles' trajectories. Fig. 10.23 shows, on a scale of 26:1, the result of combining a series of thirty such images.

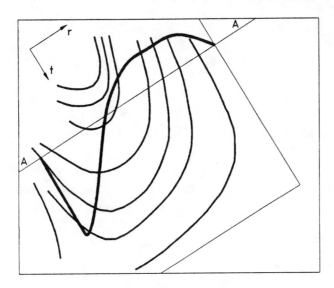

Superimposed Tangential
Velocity Profiles

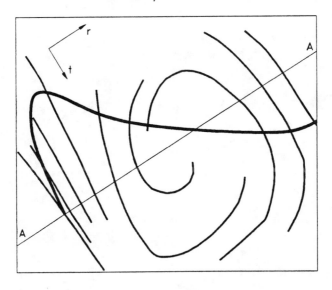

Fig. 10.23 Trajectories in two observation windows, Braun, TW 1988

Chapter 11

THE NEEDS OF TECHNOLOGY

While the previous chapters have delved into the conceptual and methodological aspects of THD fluid films, the ultimate destiny of the analyses is to serve the technologies utilizing this discipline. This chapter, therefore, can be visualized as approaching the THD subject from the opposite direction, that of the user. First the needs and requirements of the more relevant technologies will be briefly scanned, followed by the evaluation and recommendations of the Thermal Workshop as to the importance of the various branches of THD knowledge, the extent of discovered technical gaps, and the priorities to be assigned to future projects in this area.

Clearly the needs and applications to be cited will not parallel exactly the technical subtleties of thermal analysis. Moreover, they may seem quite remote from the likelihood of immediate fulfillment. They do, however, represent reality and they tell of the requirements in the field for which all theoretical effort is but the tool. So, if nothing else, they provide the direction THD is to follow if it is to remain a viable engineering discipline. The spokesmen for these advanced needs were personnel from the U.S. Air Force and U.S. Army Aviation Systems Command.[1] As was the case in the fifties and sixties when the space age has given tribology a potent boost after years of obscurity—so it is again the space vehicle that provides the impetus for THD research with such challenges as ceramic bearings, cryogenic seals, solid lubricants, and other innovations, all involving extremes in the thermal environment. Together with the subsequent analytical summation these articulated needs provide an all-encompassing blueprint for the future course of THD scientific effort.

The preceding ten chapters could not possibly be summarized under a single set of conclusions. If attempted, they would certainly miss a num-

---

[1] Bobby D. McConnel, AFWAL/MLBT, and Robert C. Bill, U.S. Army ASC.

ber of key concerns, as well as oversimplify, if not actually distort, those presented. Perhaps then, the best way to characterize the summation presented at the end of this chapter is that in terms of their importance to the THD field, these items were seen by the Thermal Workshop as priorities in need of special attention. These conclusions were arrived at in the course of selected group meetings, in which the participants deliberated on the merits and importance of each of these items. The topics selected for inclusion in the list of recommendations thus passed the scrutiny of the two-day open discussions, Committee evaluation, and final plenum approval. They represent, therefore, the technical consensus of the Thermal Workshop.

11.1 ADVANCED TRIBOSYSTEMS

The various systems that rely to a large extent on technological innovation are characterized by the fact that they span extremes in the temperature range. On the one hand, they impinge upon the cryogenic region and, on the other hand, require seals and bearings operating close to 2000°F and higher. The range of temperatures, speeds, and stresses under which some common tribological devices have to operate is shown in Fig. 11.1; stresses and speeds, for example, are seen to range over four orders of magnitude. Thus the challenge facing the engineer is not only that of variable temperature and its importance to performance, but one of reliability and integrity of these devices, involving as they do complex chemical and physical phenomena associated with extremes of the temperature spectrum. Thus in terms of its application to modern technology there is a new dimension to the THD problem which, as discussed in the preceding chapters, will have to come from the basic physical sciences. Until such results are available, the tribologist would have, as in the field of turbulence or two-phase flow, manage as best he can with the experimental and analytical tools available to him from his own field.

The most common area requiring innovative tribological input is probably the family of compact, highly efficient turbomachines employed in aircraft and space vehicles. These usually demand high performance components in two areas: reliable materials and lubricants and system-integrated bearings, seals and dampers which yield optimum rotordynamic characteristics and stability. Hot fluid pumps, for example, or heat storage systems may require operation up to $1,000°C$. In addition, unconventional fluids such as LO_2, LH_2, LCH_4, or NH_3 may be involved, tempting the engineer to use the process fluids as lubricants. Conventional demands are severely put to the test by such advanced prime movers as the compound cycle engine, or the stoichiometric combustion engine. The first one, shown schematically in Fig. 11.2, combines a highly supercharged lightweight diesel engine with a turbo machine, producing a high power density unit with severe demands on its bearings, seals and reduction gears. The piston rings in a diesel operating at $1,000°F$ are expected to have sliding speeds two to three times the

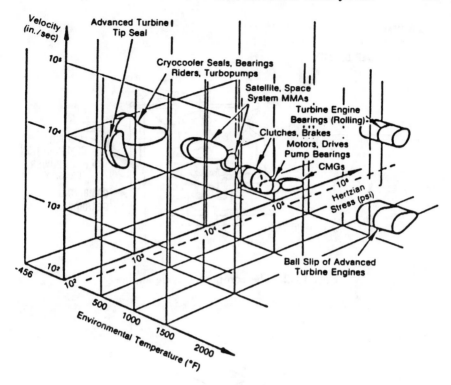

Fig. 11.1 Range of operational conditions of some tribological devices, Dill, 1989

present standards. The technological progress of the stoichiometric engine is partly portrayed in Fig. 11.3 and, as indicated, evolutionary progress is not likely to be sufficient for success in this area; rather a breakthrough in new materials, and also in methods of analysis, will be required to achieve tangible results.

But perhaps the most conspicuous of the new tribo-engines are those using cryogenic fluids. The crucial role played by liquid oxygen and liquid hydrogen turbo pumps in propulsion systems of the shuttle and other space vehicles are probably the most familiar of these devices. But cryogens are also used in orbit maneuvering subsystems; in pumps and turbo-alternators in dynamic power conversion units; in pointing and control mechanisms; and, as shown in Fig. 11.4, in refrigeration units for control of detection systems, infrared sensors, and optical instrumentation. They put extreme demands on the tribological components such as five-year continuous service without replacement; 95% reliability; sealing at 400,000 to 600,000 rpm; the use of helium fluids; and operation at $10°$K. In applications to space Rankine, Brayton, and Stirling cycles such systems must be able to withstand the shock loads and g forces during launch; 7 to 10 years multistop life cycles; and temperatures up to $1,500°$C.

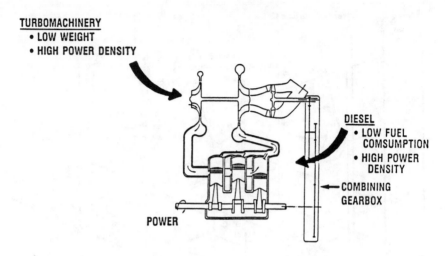

Fig. 11.2 The compound cycle engine

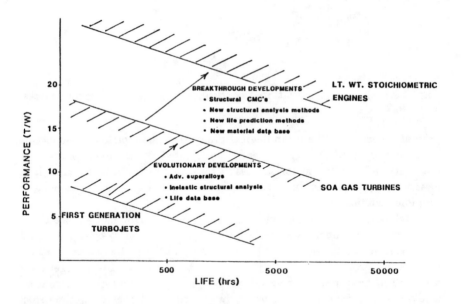

Fig. 11.3 Goals and means in the development of a stoichiometric engine, McConnell, TW 1988

Even with regard to more immediate applications such as transmissions on advanced rotor crafts, temperatures up to 450°F and non-metallic rolling element bearings may be required, the ultimate aim being a system that dispenses altogether with lubricant supply. Table 11.1 gives some pa-

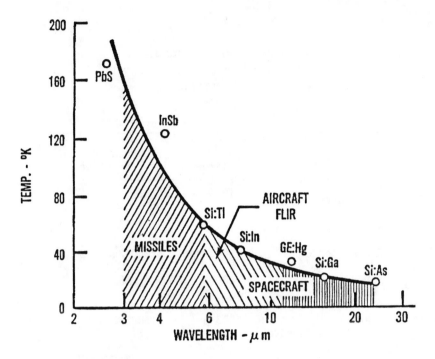

Fig. 11.4 Relation between temperature and material detection capability, McConnell, TW 1988

rameters for a 1, 000 hp projected helicopter engine in which a comparison is made of the present design capabilities versus projected requirements. Not only conventional performance is to be raised to more demanding levels but additional features such as noise reduction and ability to resume operation after prolonged shut-downs are stressed as parts of the specifications of the new systems.

Table 11.2 gives a list of some of the tribological devices which await development in the short range and over the next one or two decades. Coatings, lubricants, and ceramic materials stand out among the technologies that call for urgent development. Simultaneously the need is emphasized for proper modelling of these tribosystems in order to arrive at viable analytical results, based mostly on finite element methods. In the area of cryogenics the need for bearings and seals capable of using liquid hydrogen and liquid oxygen must, in addition to proper routine performance, also satisfy the following operational requirements.

- Low friction start-stop capability
- Avoidance of high speed rubs
- Whirl free operation
- Allowance for thermal expansion
- Minimal power loss and temperature rise

TABLE 11.1

Comparison of design parameters
from 1000 hp helicopter engine study with
current design practice

Parameter	Target	Current design practice
$P/P_{compressor}$	10.6	2-3
Comp. ratio	14-17	7.5
Diesel speed	6000 rpm	1800-2500 rpm
Top ring reversal temperature	800°F	<450°F
Brake mean effective pressure	393 psi	150-250 psi
Mean piston speed	3000 fpm	<2000 fpm
lbm/hp	0.432	5-10
hp/in^3	5.7	0.5-1

11.2 GAPS AND NEEDS

The demands on the advanced engineering systems and components
discussed above call for studies of a number of THD topics many of which
still lack adequate conceptual formulations. Thus, in addition to theoretical
efforts, basic experimental investigations are needed in many of the areas
listed below.
- Turbulence effects up to Reynolds numbers of 100,000
- Interaction of a rotationally high Reynolds number flow with pressure-
  driven high Reynolds numbers in seals and hydrostatic bearings
- Problems with fluid bearing designs which result in
  - unacceptable power losses
  - excessive temperature rise and vaporization
  - unpredictable dynamic response
- More adequate bearing design codes which would
  - include the variation of pertinent variables

TABLE 11.2
Tribological requirements of the future

	Presently applied state of art	Art input	> 10 yrs. ahead (research topics)
Lubricants	MIL L-23699	Mini-lube; high temperature lubes; grease lube	Dry lubricants; unlubricated
Bearings	DVM steels; stacked balls, ball/roller; lundberg palmgren life prediction	Ceramic rolling elements; spherical roller, high-speed tapered roller	Magnetic bearings; wide use of ceramics; prob. life prediction for ceramics
Gears	Carb/Nit 9310; involute tooth forms; scoring, scuffing, bending fatigue life prediction	High hot hardness steels, ion implant. & plating, near net forged; high contact ratio, non involute forms (conformal), zero kinematic error; finite element modelling, dynamic analysis	Dual alloy light weight gears, advanced coatings and surface treatments; full transient load structural model
Configurations	Planetary	Self aligning bearingless Planetary; Split Torque	Variable speed; direct turb. drive; electromagnetic systems
Clutches	Sprague; roller ramp	Spring	
Shafts	Steel forgings	Composite (organic matrix)	Metal matrix composites
Housings	Cast Mg	Composite	
Analysis	Principles of mech. design; component level struct. analysis	Dynamic finite element analysis of all important structures	Comprehensive transmission modeller

- use of proper boundary conditions
- yield results of reasonable generality
- Cryogenic fluids
  - Include in the analysis the variable properties such as density, viscosity, etc.
  - Establish the unique flow characteristics of these fluids
  - Two-phase behavior
- Stoichiometric combustion
  - Reaction flow and turbulence
  - Soot formation
  - Radiation modeling
- 4000°F environment
  - Stable long-life materials

- Transient heat transfer
- High temperature engine sensors/actuators
- High temperature research instrumentation
- High temperature composite materials
  - Materials behavior
  - Fabrication techniques
  - Durability modeling (damage/failure analysis)
- Boundary condition in THD analysis
  - Contradiction between experiments which show $T(z) \neq const.$ and analyses which yield $T(z) = const.$
  - Runner temperature. Analyze the validity of the Jaeger integral model which yields essentially a constant T_R; if constant, what is this constant value?
  - Reason for drop in T after h_{min}
  - Reason why heat transfer solutions often yield higher temperatures and lower loads than do adiabatic cases

11.3 CONCLUSIONS AND RECOMMENDATIONS OF THE THERMAL WORKSHOP

At the end of the Thermal Workshop the participants were organized into four study groups each dealing with a range of THD subjects (see Appendix A). The sub-divisions embraced the following four generic areas:

A. Boundary conditions in THD analysis
B. Non-conventional operating conditions (turbulence, EHD, etc.)
C. Two-phase regimes and cavitation
D. Experiments and instrumentation

These study groups together with the follow-up plenum session assigned the following priorities for future THD study:

Priority Projects Selected by Group A

1. Develop criteria for two-dimensional versus three-dimensional solutions
2. THD effects on stability
3. Thermo-elastic deformations and effects on dynamics
4. Experimental three-dimensional data for temperature distribution
5. Inlet mixing temperatures
6. Heat transfer in pads
7. Mechanism of journal seizure
8. Formulation of correct boundary conditions

Priority Projects Selected by Group B

1. Resolution of discrepancy between theory and experiment for values of spring and damping coefficients

2. Measurement and visualization of turbulent film
3. Combined effects of Couette and Poiseuille turbulent flows
4. Heat transfer through the laminar sublayer in fluid films
5. Three-dimensional solution of Navier-Stokes equation for turbulent flows
6. High Reynolds number flows
7. Heat dissipation in EHD
8. Fluid rheology in EHD
9. Transient thermal characteristics
10. Heat dissipation in dry contacts

Priority Projects Selected by Group C

1. Design codes for two-phase flow
2. Experimental verification of Item 1
3. Physics of cavitation
   - Nucleation
   - Boundary conditions
   - Surface energy and surface tension
   - Time scale
4. Mechanism of melt lubrication
5. Hydrodynamics of very thin films
6. Mechanism of two-phase flow
   - Phase distribution
   - Boundary conditions
   - Turbulence effects

Priority Projects Selected by Group D

1. Techniques for measuring three-dimensional temperature distribution in a non-intrusive way
2. Surface characteristics as a function of time
3. Determination of heat transfer coefficients
4. EHD parameters under cryogenic conditions
5. Mechanism of film collapse
6. Effects of cooling on wear
7. Two-phase flow and onset of instability
8. Measurement of mixing inlet temperatures
9. In situ rheological measurements
10. Residual stress measurements

APPENDIX A

WORKSHOP SPECIFICS

<u>ANNOUNCEMENT</u>

WORKSHOP ON THERMAL PROBLEMS IN TRIBOLOGY

GEORGIA INSTITUTE OF TECHNOLOGY
ATLANTA, GEORGIA

May 19 and 20, 1988

<u>Introduction</u>

A workshop on the multidisciplinary aspects of thermal problems in tribology took place at the George W. Woodruff School of Mechanical Engineering at Georgia Institute of Technology on May 19 and 20, 1988. The project received the endorsement of the Research Committee of Tribology of the ASME and was sponsored by the National Science Foundation Tribology Program and the Wright Aeronautical Labs of the U.S. Air Force (B.D. McConnell, Program Manager, AFWL/MLBT). The motivation, scope, objectives and organization of the workshop are detailed below.

<u>Motivation</u>

There is a need to alleviate the stalemate prevailing in the field of solving thermal problems in tribology. While a most impressive body of solutions has accumulated for isothermal cases which can be easily extracted from books, papers, and design manuals, no such solutions exist for non-isothermal cases. Even when in-house specific thermohydrodynamic (THD) solutions are attempted the analyst is often faced with unresolved conceptual difficulties. Yet no bearing or seal is isothermal and variable viscosity must be considered for at least three important reasons:

- Ordinary performance parameters such as power consumption, film thickness, and flow depend on a knowledge of the temperature field.
- Thermal effects are important in the evaluation of film thickness and traction in concentrated contacts.
- Knowledge of the maximum bearing temperature is a basic criterion of bearing, seal, or gear failure.

The state of THD engineering solutions is not due to a lack of effort, but rather the result of scattered efforts and the absence of a conceptual engineering framework. To date, the more numerous the papers and the more intricate the analyses, the more complex and more obscure the subject becomes. Moreover, there is little quantitative or qualitative agreement among the published results. A concerted effort is therefore needed to identify the causes of this inverted relationship between effort invested and the results obtained. One objective of this workshop would be to propose new directions for research in the field of thermal phenomena in lubrication.

Scope

The identification of the sources of confusion in the field of THD lubrication and the formulation of steps to clarify that confusion constitute the two main goals of the Workshop. To make clear what the objectives may be one can cite for illustrative purposes the following major needs:

a. Much of the disarray prevailing in the THD field is due to the large number of variables present. It was the task of the Workshop to evaluate their relative importance and to set up a hierarchy of relevance of these various parameters with regard to engineering THD solutions.

b. Much of the difficulties can be traced to lack of generality of the systems analyzed in the available literature. The Workshop attempted to arrive at several generic models or prototypes which would provide for the formulation of sufficiently generalized THD solutions.

c. Basic experiments are needed to answer a number of conceptual uncertainties. Proposed experiments were formulated and discussed at the Workshop.

d. Much of the available data from THD tests is unsatisfactory because vital measurements were not made, in many cases due to the lack of proper instrumentation. A case in point is measurement of temperature profiles across films only microns thick. The Workshop addressed the issue of instrumentation requirements for this field.

Format of Workshop

A two and one-half day Workshop was proposed with the attendance of 100–150 representatives from industry, universities, and government. Run under the joint chairmanship of Professor Ward O. Winer of Georgia Tech and Dr. Hooshang Heshmat of MTI, it took take place May 18, 19, 20, 1988 on the campus of the Georgia Institute of Technology. Introductory position papers were delivered to stimulate and focus discussion. The primary deliberations consisted of working sessions with open critical discussion leading to consensus or a spectrum of judgements on the research needs in the thermal aspects of tribology. The presentations and discussion group reports were collected and organized by an editor of the Workshop document.

The suggested Workshop format was as follows:

a. Each day had morning and afternoon sessions attended by all delegates.

b. The sessions were organized by topics, some of which were as follows:

  1) Thermal aspects of rigid films,
     • Mixing temperatures,
     • Time varying films,
     • Cavitation phenomena

    2) Critical experiments needed

    3) Instrumentation for thermal experiments

    4) Special topics

c. Experts from other fields were invited to participate in discussions involving non-tribological areas such as turbulence, heat transfer, measurement techniques, etc.

Workshop Steering Committee

Ward O. Winer, Chair
Regents' Professor
School of Mechanical Engineering
Georgia Institute of Technology
Atlanta, Georgia 30332-0405

F. F. Ling
William Howard Hart Professor
Mechanical Engineering
Rensselaer Polytechnic Institute
Troy, New York 12818

Hooshang Heshmat
Senior Program Manager
Mechanical Technology, Inc.
968 Albany-Shaker Road
Latham, New York 12110

D. F. Hays
Department Head
Fluid Mechanics Department
General Motors Research
Warren, Michigan 48090

AGENDA

WORKSHOP
ON
THERMAL PROBLEMS IN TRIBOLOGY

Sponsored by
Wright Aeronautical Labs, U.S. Air Force
and
National Science Foundation, Tribology Program

Atlanta, Georgia
May 18-20, 1988

WEDNESDAY MAY 18

TIME	TOPIC
12:00 - 6:00 PM	Registration
8:30 - 9:30 PM	Address by:
	Larry Fehrenbacher, AT&T, Inc.
	Robert Bill, AVSCOM, NASA-LRC

THURSDAY, MAY 19

SESSION	TIME	TOPIC
	8:30 - 9:00 AM	Introductory Remarks by
		W. Winer, Presiding Chairman
		H. Heshmat, Organizing Chairman
1a	9:00 - 10:00 AM	Temperatures in Conventional
		Hydrodynamic Films (THL)
1b	10:30 - 11:15 AM	THL (continued)
2	11:15 AM - 12:15 PM	Mixing Inlet Temperature
3	1:30 - 2:30 PM	Cavitation Zone Temperatures
4	2:30 - 3:30 PM	Two-Phase Films
5	4:00 - 5:00 PM	Time Varying Films

FRIDAY, MAY 20

SESSION	TIME	TOPIC
6	8:00 - 9:30 AM	Thermal Phenomena Under Turbulent Conditions
7	9:30 - 10:30 AM	Elastohydrodynamic (EHD) Films
8	10:45 - 11:45 AM	Experiments Needed
9	11:45 AM - 12:45 PM	Instrumentation for Tribological Experiments
10	2:00 - 2:45 PM	Special Topics
	2:45 - 4:00 PM	Concluding Session; Sum-Up

Each session consisted of 50 minutes discussion followed by 5 to 10 minutes sum-up. A graphical overview of the Workshop proceedings is given in charts A–1 and A–2.

LIST OF SESSION CHAIRMEN

Workshop Topics (Sessions)	Session Chairman
Rigid Films (THL)	Professor A. Z. Szeri University of Pittsburgh
Mixing Temperatures	Professor C. M. Ettles Rensselaer Polytechnic Institute
Time Varying Films	Professor W. A. Gross University of New Mexico
Cavitation Phenomena	Mr. D. E. Brewe NASA Lewis Research Center
EHD	Professor Harmen Blok University Delft, The Netherlands
Two-Phase Flow Phenomena	Professor W. F. Hughes Carnegie Mellon University
Turbulence	Professor J. A. Tichy Rensselaer Polytechnic Institute

THE WORKSHOP AT A GLANCE

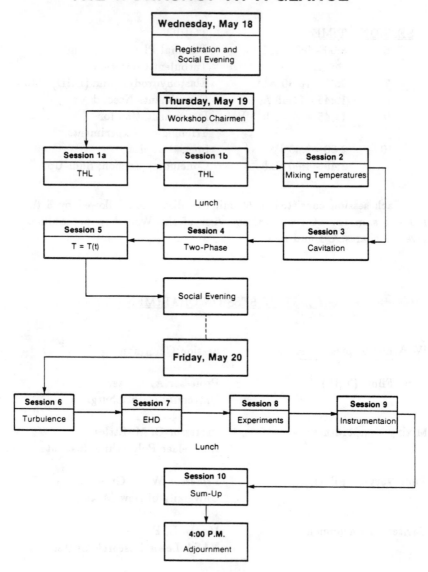

Fig. A-1 The Workshop at a glance

Experiments Needed   Dr. J. W. Kannel
                                          Battelle Memorial Institute

Instrumentation for Thermal   Dr. V. King
Experiments   Memory Tech. Group Lab 3M

WORKSHOP SUMMATION

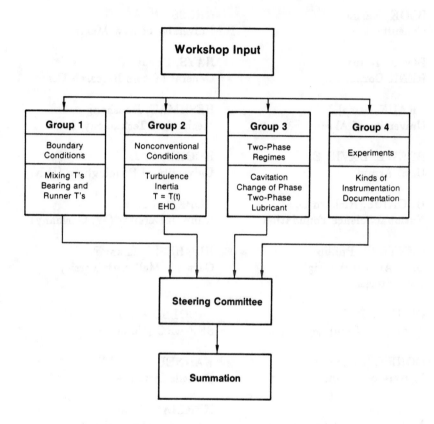

Fig. A–2 Workshop summation

PARTICIPANTS
In The
1988 THERMAL WORKSHOP

ADVANI, S.D.
Michell Bearings - Vickers (U.K.)

BALL, Dr. J. H.
Waukesha Bearings Corporation

BASU, Prithwish
EG&G Sealol

BEJAN, Adrian
Duke University

BILL, Robert C.
U. S. Army

BLOK, Harmen
Consultant

BOCK, Klaus L.
RENK Corporation

BRAUN, Minel J.
University of Akron

BREWE, DAVID E.
U. S. Army

BUCKIUS, Richard O.
National Science Foundation

CENTERS, Phillip
U. S. Air Force Wright
Laboratories

CHIU, Y. P.
Torrington Company

CONRY, Thomas F.
University of Illinois

DILL, Jim
Mechanical Technology, Inc.

DREW, Donald
Rensselaer Polytechnic Institute

ETTLES, Chris M.
Rensselaer Polytechnic Institute

FEHRENBACHER, Larry L.
Technology Assessment and
Transfer Inc.

FILLON, Michel
Universite de Poitiers (France)

GENTLE, C. R.
Trent Polytechnic

GREEN, I.
Georgia Institute of Technology

GROSS, W. A.
University of New Mexico

HAYS, Donald F.
General Motors Research Center

HESHMAT, Hooshang
Mechanical Technology, Inc.

HOOKE, C. J.
University of Birmingham (U.K.)

HOPF, Guido
Ruhr-Universitat (W. Germany)

HUGHES, William F.
Carnegie Mellon University

JACOBSON, Bo
SKF-ERC (Holland)

KANNEL, Jerrold W.
Battelle Institute

KERIBAR, Rifat
Integral Technologies, Inc.

KETTLEBOROUGH, C. F.
Texas A&M University

KHONSARI, Michael M.
University of Pittsburgh

KING, Vincent W.
3M Company

LARSEN-BASSE, J.
Georgia Institute of Technology

LAUER, James L.
Rensslelaer Polytechnic Institute

LING, Frederick F.
Columbia University

MALLORY, Merrit L.
NSCL - Michigan State University

MARTIN, Fred A.
F. A. Martin (U.K.)

McCONNELL, Bobby D.
Wright-Patterson AFB

MECKLENBURG, Karl
Wright-Patterson AFB

MOYER, Charles A.
The Timken Company

NICHOLS, Fred A.
Argonne National Lab.

PARANJPE, Rohit
General Motors Corporation

PASCAL, Marie Therese
EDF-DER (France)

PINKUS, Oscar
Sigma Tribology Consultants

PRASAD, Vishwanath
Columbia University

PROCTOR, Margaret
NASA Lewis Research Center

RAJAGOPAL, K. R.
University of Pittsburgh

SALANT, R.
Georgia Institute of Technology

SCHOCK, Harold J.
Michigan State University

SPAUSCHUS, Hans O.
GTRI

STUDZINSKI, Andrew
Reliance Electric Company

SZERI, Andras A.
University of Pittsburgh

TANAKA, Masato
University of Tokyo (Japan)

TICHY, John
Rensselaer Polytechnic Institute

TÖNDER, Kristian
University of Trondheim (Norway)

WEISINGER, Fred
Kingsbury, Inc.

WINER, Ward O.
Georgia Institute of Technology

APPENDIX B

THERMOPHYSICAL
PROPERTIES

TABLE B.1

Physical properties of some solids

Metals and alloys	w (lb/ft³) (68°F)	c_p (Btu/lb-°F) (68°F)	k (Btu/hr-ft-°F) (68°F)	(212°F)	σ (ft²/hr) (68°F)
Aluminum	169	0.214	118	119	3.665
Brass (70%Cu,30%Zn)	532	0.092	64	74	1.322
Copper	559	0.0915	223	219	4.353
Iron	493	0.108	42	39	0.785
cast (C ~ 4%)	454	0.10	30		0.666
wrought (C < 0.5%)	490	0.11	34	33	0.634
Lead	710	0.031	20	19.3	0.924
Magnesium	109	0.242	99	97	3.762
Molybdenum	638	0.060	71	68	2.074
Nickel	556	0.1065	52	48	0.882
Silver	657	0.056	242	240	6.601
Steel, mild, 1% C	487	0.113	25	25	0.452
Stainless steel(18Cr,8Ni)	488	0.11	9.4	10	0.172
Tin	456	0.054	37	34	1.505
Zinc	446	0.092	64.8	63	1.591

TABLE B.2

Thermal conductivities of solids

	Substance	Temperature °F	k
Metals:	Aluminum	32	117
	Aluminum	212	119
	Antimony	32	10.6
	Antimony	212	9.7
	Cadmium	64	53.7
	Cadmium	212	52.2
	Copper	32	226
	Copper	212	222
	Iron	64	39.0
	Iron	212	36.6
	Iron, wrought	64	34.9
	Iron, wrought	212	34.6
	Iron, cast	32	29.0
	Iron, cast	212	28.0
	Steel, mild	32	36
	Steel, mild	212	33
	Steel, 1% C	64	26.2
	Steel, 1% C	212	25.9
	Lead	32	20.0
	Lead	212	19.8
	Magnesium	32-212	92.0
	Silver	32	244
	Silver	212	240
	Tin	32	36
	Tin	212	34
	Tungsten	68	92.5
	Zinc	32	65
	Zinc	212	62
Alloys:	Babbitt	68	13.7
	Brass, 90-10	32	59
	Brass, 90-10	212	68
	Brass, 60-40	32	61
	Brass, 60-40	212	69
	Bronze	68	33.6
	Bronze	210	40
	Constantan, 60 Cu; 40 Ni	64	13.1
	Constantan, 60 Cu; 40 Ni	212	15.5
	Manganin, 84 Cu; 4 Ni; 12 Mn	64	12.8
	Manganin, 84 Cu; 4 Ni; 12 Mn	212	15.2

k = Btu/hr-ft-°F

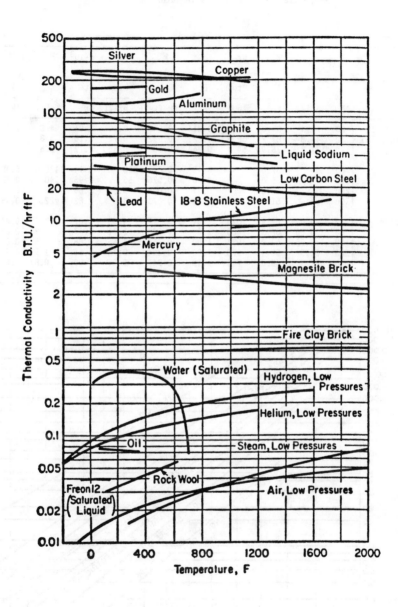

Fig. B.1 Thermal conductivity of various substances, Rosenhow and Choi, 1961

TABLE B.3

Properties of some liquids in saturated state

	T	p	w	h_{fg}	c_p	k	μ	$\sigma \cdot 10^3$	Pr	$\alpha \cdot 10^3$
			lb	Btu	Btu	Btu	lbm	ft²		1
	°F	psia	ft³	lb	lbm-°F	hr-ft-°F	hr-ft	hr		°R
Freon-12	-20	15.3	93.0	71.8	0.214	0.040	0.90	2.01	4.8	1.03
(CCl_2F_2)	32	44.8	87.2	66.6	0.223	0.042	0.72	2.16	3.8	1.72
	120	171.8	75.9	54.0	0.244	0.039	0.56	2.12	3.5	
Oil	60		57.0		0.43	0.077	210	3.14	1170	0.38
	100		56.0		0.46	0.076	55	2.95	340	0.39
	200		54.0		0.51	0.074	9.0	2.69	62	0.42
	300		51.8		0.54	0.073	3.0	2.62	22	0.45
Water	32	0.089	62.4	1076	1.01	0.319	4.33	5.07	13.4	-0.04
	60	0.26	62.3	1060	1.00	0.340	2.71	5.47	7.9	0.08
	100	0.95	62.0	1037	1.00	0.364	1.65	5.88	4.5	0.2
	200	11.5	60.1	978	1.00	0.394	0.74	6.55	1.9	0.4
	500	681	49.0	714	1.19	0.349	0.26	5.99	0.87	1.0

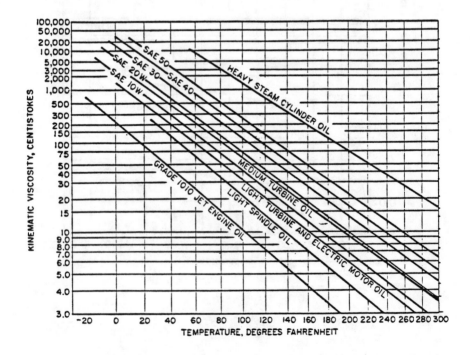

Fig. B.2 Viscosity of common petroleum oils

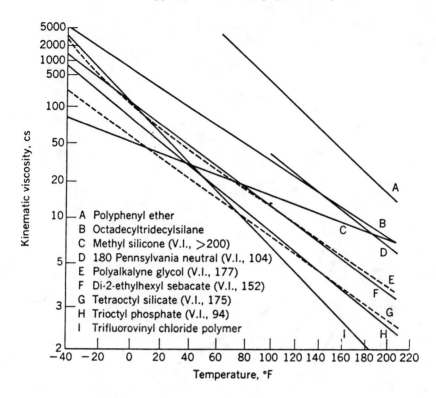

Fig. B.3 Viscosity of synthetic lubricants, Bisson and Anderson, 1964

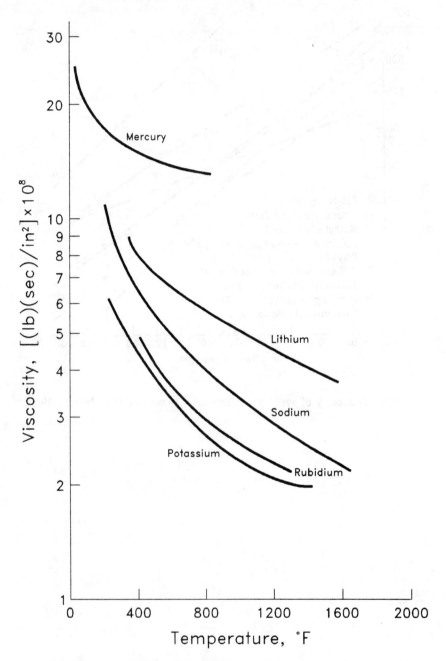

Fig. B.4 Viscosity of liquid metals, Bisson and Anderson, 1964

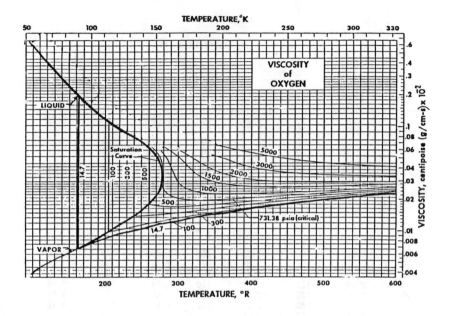

Fig. B.5 Viscosity of oxygen

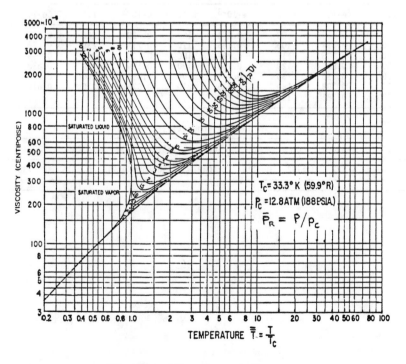

Fig. B.6 Viscosity of hydrogen

TABLE B.4

Properties of liquid kerosene

a. Viscosity ($N\text{-}s/m^2$)

T,°C	$\mu \cdot 10^3$	T,°C	$\mu \cdot 10^3$	T,°C	$\mu \cdot 10^3$	T,°C	$\mu \cdot 10^3$
-50	11.5	-15	3.14	40	1.08	180	0.296
-45	9.04	-10	2.75	60	0.832	200	0.262
-40	7.26	- 5	2.42	80	0.664	220	0.234
-35	5.96	0	2.15	100	0.545	240	0.211
-30	4.98	5	1.92	120	0.457	260	0.191
-25	4.22	10	1.73	140	0.390	280	0.174
-20	3.62	20	1.49	160	0.338	300	0.159

b. Heat Capacity (kJ/kg-deg)

T,°C	c_p	T,°C	c_p	T,°C	c_p	T,°C	c_p
20	2.00	90	2.33	160	2.68	220	3.00
30	2.04	100	2.38	170	2.73	230	3.05
40	2.09	110	2.43	180	2.79	240	3.11
50	2.14	120	2.48	190	2.84	250	3.16
60	2.18	130	2.53	200	2.89	260	3.21
70	2.23	140	2.58	210	2.94	270	3.26
80	2.28	150	2.63				

c. Density (kg/m^3)

T,°C	$e \cdot 10^{-3}$	t,°C	$e \cdot 10^{-3}$	t,°C	$e \cdot 10^{-3}$	t,°C	$e \cdot 10^{-3}$
20	0.819	90	0.774	160	0.720	220	0.668
30	0.814	100	0.766	170	0.711	230	0.658
40	0.808	110	0.759	180	0.703	240	0.649
50	0.801	120	0.751	190	0.694	250	0.638
60	0.795	130	0.744	200	0.685	260	0.628
70	0.788	140	0.736	210	0.676	270	0.618
80	0.781	150	0.728				

d. Thermal conductivity

T,°C	-50	0	50	100	150	200	250	300
W/m-deg	0.127	0.1192	0.1114	0.1042	0.0965	0.0891	0.0816	0.0738

TABLE B.5

Pproperties of some gases at atmospheric pressure

Gas	°F	w lb $\overline{ft^3}$	c_p Btu $\overline{lb\text{-}°F}$	k Btu $\overline{hr\text{-}ft\text{-}°F}$	μ lbm $\overline{hr\text{-}ft}$	ν ft^2 $\overline{hr}$	σ ft^2 $\overline{hr}$	Pr
Dry air	0	0.0863	0.239	0.013	0.040	0.457	0.633	0.722
	60	0.0763	0.240	0.015	0.043	0.56	0.799	0.712
	100	0.0709	0.241	0.016	0.046	0.648	0.919	0.706
	200	0.0601	0.242	0.018	0.052	0.864	1.25	0.694
	500	0.0413	0.248	0.025	0.068	1.63	2.40	0.680
Carbon dioxide	0	0.13	0.184	0.0076	0.0316	0.241	0.313	0.77
	100	0.11	0.203	0.010	0.0378	0.352	0.455	0.77
	500	0.063	0.247	0.012	0.060	0.957	1.27	0.75
Helium	-200	0.021	1.24	0.054	0.030	1.44	2.05	0.70
	0	0.012	1.24	0.078	0.044	3.68	5.29	0.70
	200	0.0083	1.24	0.098	0.056	6.71	9.49	0.71
Hydrogen	0	0.0060	3.39	0.094	0.0194	3.20	4.62	0.70
	200	0.0042	3.44	0.122	0.0249	5.94	8.45	0.69
Steam	212	0.0372	0.45	0.0145	0.0314	0.842	0.870	0.96
	500	0.0258	0.47	0.0228	0.0455	1.76	1.88	0.94

TABLE B.6

Properties of oxygen, O_2

Property		English	Metric units
Boiling point		-297.4°F	-183.0°C
Freezing point		-361°F	-218.3°C
Liquid specific weight		9.53 lb/gal at -297.4°F	1.14 g/ml at -183°C
Critical pressure		737 psia	5.09×10^6 N/m2 at -118.4°C
Critical temperature		-181.1°F	-118.4°C
Vapor pressure		37 psia at -280°F	2.55×10^5 N/m^2 at -173°C
		167 psia at -240°F	1.15×10^6 N/m^2 at -151°C
		492 psia at -200°F	3.39×10^6 N/m^2 at -129°C
		615 psia at -190°C	4.24×10^6 N/m^2 at -123°C
Viscosity	Kinematic	0.17 centistokes at -297.4°F	------------------
	Absolute	------------------	0.19 centipoises at -183.0°C

TABLE B.7

Properties of methane, CH_4

• Molecular weight	16.03
• Color	colorless
• Freezing (melting) point	-292.2°F (-184.0°C)
• Boiling point	-258.7°F (-161.5°C)
• Density at -263.2°F (-164.0°C)	25.91 lb/ft³ (0.415 g/cm³)
• Critical temperature	360.5°F (182.5°C)
Critical pressure	673.1 psi (45.8 atm)
Critical density	10.11 lb/ft³ (0.162 g/cm³)
• Heat of fusion at -296.5°F (-182.5°C)	26.15 Btu/lb (14.53 cal/g)
Heat of vaporization at -254.2°F (-159.0°C)	284.54 Btu/lb (138.08 cal/g)

• Heat capacity			
°F	°C	Btu/lb-°F	cal/g-°C
-175.0	-115.0	0.4502	0.4502
-112.0	- 80.0	0.5038	0.5038
-101.2	- 74.0	0.4979	0.4979
14.0 to 393.0°F (-10.0 to 200.0°C)		0.5931	0.5931

• Heat of formation	33,012.0 Btu/lb-mole (18,340.0 cal/g-mole)
• Heat of combustion	383,310.0 Btu/lb-mole (212,950.0 cal/g-mole)

• Viscosity			
°F	°C	lb/ft-sec x 10^{-3}	centipoise
-294.9	-181.6	0.0023	0.0035
-109.3	- 78.5	0.0051	0.0076
32.2	0.0	0.0069	0.0103
62.6	17.0	0.0072	0.0108
212.0	100.0	0.0091	0.0135

TABLE B.8

Properties of cryogenic liquids

Liquid	Freezing point, °F	Boiling point, °F	Liquid density lb/ft³, at boiling point	Liquid viscosity Reyns·10^{10}	
				Boiling point	-320°F
Helium	-458*	-452	7.6	7	------
Hydrogen	-434	-423	4.4	19	------
Nitrogen	-346	-320	50.1	130	230
Fluorine	-360	-306	94.0	372	489
Argon	-309	-303	87.4	------	------
Oxygen	-361	-297	71.2	274	333
Methane	-299	-258	25.8	------	------

* At pressure of 26 atm.

TABLE B.9
Physical properties of oxygen, O_2

• Molecular weight	32.00
• Color	colorless gas, slightly blue liquid
• Freezing (melting) point	-362.0°F (-218.9°C)
• Boiling point	-297.4°F (-183.0°C)

• Density

°F	°C	lb/ft^3	g/cm^3
-361.9	-218.8	85.65	1.372 (solid)
-296.9	-182.7	71.20	1.140
-196.0	- 90.1	0.2766	0.00443 (gas at saturation)

• Critical temperature	-181.8°F (-118.8°C)
Critical pressure	730.4 psi (49.7 atm)
Critical specific weight	26.8 lb/ft^3 (0.430 g/cm^3)
• Triple point temperature	-361.903°F (-218.835°C)
Triple point density, solid	85.64 lb/ft^3 (1.372 g/cm^3)
• Coefficient of thermal expansion at	
-422.7°F (-252.6°C)	0.00087/°F (0.00157/°C)

• Heat of fusion at	
14.696 psi (1.0 atm) at	
-361.79°F (-218.77°C)	5.979 Btu/lb (3.32 cal/g)
• Heat of vaporization	91.62 Btu/lb (1,628.8 cal/g-mole)

• Heat capacity, solid, crystalline form

°F	°C	Btu/lb-°F	cal/g-mole-°C
-399.6	-239.8	0.2420	7.73
-391.5	-253.3	0.2850	9.12
-383.6	-230.9	0.3353	10.73
-377.0	-227.3	0.3444	10.02
-366.6	-221.5	0.3447	11.03
-365.8	-221.0	0.3456	11.06

Heat capacity, liquid

°F	°C	Btu/lb-°F	cal/g-mole-°C
-357.2	-216.8	0.3988	12.76
-350.0	-212.2	0.3972	12.71
-341.7	-207.6	0.3972	12.71
-332.5	-202.5	0.3991	12.77
-324.8	-198.2	0.4016	12.85
-318.1	-194.5	0.4009	12.83
-310.4	-190.2	0.4025	12.88
-304.1	-186.7	0.4034	12.91
-297.1	-182.8	0.4059	12.99

Heat capacity, gas, constant
pressure at 14.696 psi (1.0 atm)

°F	°C	Btu/lb-°F	cal/g-°C
-100.0	- 73.2	0.21838	0.21838
80.0	26.8	0.21971	0.21971
260.0	126.8	0.22492	0.22492
800.0	426.8	0.24626	0.24626
1700.0	926.8	0.26632	0.26632
2240.0	1226.8	0.27291	0.27291
3140.0	1726.8	0.28198	0.28198
4040.0	2226.8	0.29098	0.29098

TABLE B.10
Physical properties of hydrogen, H_2

• Molecular weight		2.01418	
• Color		colorless gas, transparent liquid	
• Odor		odorless	
• Freezing (melting) point		-434.4°F (-259.1°C)	
• Boiling point		-423.0°F (-252.7°C)	

• Density

°F	°C	lb/ft³	g/cm³
-432.8	-258.2	4.71	0.0755
-429.2	-256.2	4.61	0.0738
-425.6	-254.2	4.48	0.0718
-422.0	-252.2	4.34	0.0695
-418.4	-250.2	4.18	0.0670
-414.8	-248.2	3.99	0.0640
-411.2	-246.2	3.77	0.0604
-407.6	-244.2	3.50	0.0560
-404.0	-242.2	3.14	0.0503
-400.4	-240.2	2.22	0.0356

• Critical temperature	-399.8°F (-239.9°C)
Critical pressure	188.11 psi (12.80 atm)
Critical specific weight	1.95 lb/ft³ (0.031 g/cm³)
• Triple point temperature	-434.6°F (-259.3°C)
Triple point pressure	1.044 psi (54.0 mm-Hg)

• Heat of vaporization, liquid, para hydrogen

°F	°C	Btu/ft³	cal/liter
-427.4	-255.2	898.83	7,998.093
-423.8	-253.2	860.16	7,654.089
-420.2	-251.2	804.12	7,155.283
-416.6	-249.2	744.19	6,622.077
-413.0	-247.2	674.61	6,002.870
-409.4	-245.2	599.22	5,332.062

• Heat capacity, solid, normal hydrogen

°F	°C	Btu/lb-mole-°F	cal/g-mole-°C
-441.8	-263.2	0.58	0.58
-440.0	-262.2	0.76	0.76
-438.2	-261.2	0.95	0.95
-436.4	-260.2	1.16	1.16
-434.6	-259.2	1.37	1.37

Heat capacity, saturated liquid, normal hydrogen

-434.6	-259.2	3.31	3.31
-432.8	-258.2	3.46	3.46
-431.0	-257.2	3.63	3.63
-429.2	-256.2	3.83	3.83
-427.4	-255.2	4.04	4.04
-425.6	-254.2	4.27	4.27
-423.8	-253.2	4.50	4.50

Heat capacity, saturated liquid, para hydrogen

-426.8	-254.9	4.18	4.18
-422.1	-252.3	4.71	4.71
-418.9	-250.5	5.33	5.33
-414.8	-248.2	6.03	6.03
-410.1	-245.6	7.85	7.85
-402.2	-241.2	14.56	14.56

APPENDIX C

BIBLIOGRAPHY

1. Abdel-Latif, L.A., Peeken, H., and Benner, J., "THD Analysis of Thrust Bearings with Circular Pads Running on Bubbly Oil," J. Trib. Trans. ASME, Vol. 107, Oct. 1985, pp. 527–537.

2. Abdel-Latif, L.A., "Analysis of Heavily Loaded Tilted Pads Thrust Bearings with Large Dimensions Under TEHD Conditions," J. Lub. Tech. Trans. ASME, Vol. 110, July 1988, pp. 467–476.

3. Aggarwal, B.B. and Wilson, W.R.D., "Improved Thermal Reynolds Equations," TET*, 1980, pp. 152–161.

4. Andrisano, A.O., "An Experimental Investigation on the Rotating Journal Surface Temperature Distribution in a Full Circular Bearing," Journal of Tribology, ASME Trans., Vol. 110, October 1988, pp. 638–645.

5. Archard, J.F. and Rowntree, R.A., "Metallurgical Phase Changes in The Wear of Steels: A Critical Assessment," TET, 1980, pp. 285–297.

6. Arwas, E.G., "Thrust Bearing Program," 1957 Symposium on Hydrodynamic Bearings, Gen. Electric Report No. 59GL26, August 1959.

7. Ausherman, V.K., Nagaraj, H.S., Sanborn, D.M. and Winer, W.O., "Infrared Temperature Mapping in Elastohydrodynamic Lubrication," J. Lub. Tech. Trans. ASME, Vol. 98, April 1976, pp. 236–243.

8. Balkinds, A.V., "Heat Flow in the Oil-Film of Hydrodynamic Bearings," Russ. Eng. Journal, Vol. 53, August 1973, pp. 20–24.

9. Barrett, L.E. and Branagan, L.A., "Cavitation Modeling for Analysis of Fluid Film Journal Bearings," 1988 STLE Annual Meeting Symposium on Cavitation Fundamentals, May 9–13, Cleveland, Ohio.

10. Barwell, F.T. and Lingard, S., "The Thermal Equilibrium of Plain Journal Bearings," TET, 1980, pp. 24–32.

11. Bejan, A., "The Process of Melting by Rolling Contact," Inst. of Heat and Mass Transfer, Vol. 31, No. 11, 1988, pp. 2273–2283.

12. Bejan, A., "The Fundamentals of Sliding Contact Melting and Friction," ASME Journal of Heat Transfer, Vol. 111, Feb. 1989, pp. 13–20.

13. Berthe, D., Houpet, L. and Flamand, L., "Thermal Effects in EHD Contacts for Different Rheological Behaviour of the Lubricant," TET, 1980, pp. 241–250.

14. Bhushan, Bharat and Cook, N.H., "Temperatures in Sliding," J. Lub. Tech. Trans ASME, October 1973, p. 535.

---

* Here and subsequently TET refers to Thermal Effects in Tribology, Proceedings of the 6th Leeds-Lyon Symposium on Tribology, Sept. 1979, Publ. by Mech. Eng. Publs. Ltd., London, 1980.

15. Biorkland, I.S. and Kays, W.M., "Heat Transfer Between Concentric Rotating Cylinders," ASME Paper No. 58-A-99, 1958.

16. Bisson, E.E. and Anderson, W.J., "Advanced Bearing Technology," NASA SP-38, 1964.

17. Blahey, A.G. and Schneider, G.E., "A Numerical Solution of the Elastohydrodynamic Lubrication of Elliptical Contacts with Thermal Effects," Proc. 13th Leeds Lyon Symposium on Tribology, Elsevier, 1987.

18. Boncompain, R. and Frene, J., "Thermohydrodynamic Analysis of a Finite Journal Bearing's Static and Dynamic Characteristics," TET, 1980, pp. 33–41.

19. Boncompain, R. and Frene, J., "Modification de la portance due a l'effet thermique dans les paliers lisses—determination des cartes de temperatures," Mec. Mat. Electr., No. 347–348, 1978, pp. 465–474.

20. Boncompain, R., Fillon, M. and Frene, J., "Analysis of Thermal Effects in Hydrodynamic Bearings," J. Trib. Trans. ASME, Vol. 108, 1986, pp. 219–224.

21. Booser, E.R., Ryan, F.D. and Linkinhoker, C.L., "Maximum Temperature for Hydrodynamic Bearings Under Steady Load," Lubrication Engineering, July 1970, pp. 226–235.

22. Booser, E.R. and Wilcock, D.F., "Temperature Fade in Journal Bearing Exit Regions," STLE Preprint 87-TC-1A-1, October 1987.

23. Bowen, E.R. and Medwell, J.O.A., "Thermohydrodynamic Analysis of Journal Bearings Operating Under Turbulent Conditions," Wear, Vol. 51, 1978, pp. 345–353.

24. Braun, M.J., Batur, C., Ida, N., Rose,B., Hendricks, R.C. and Mullen, R.L., "A Non-Invasive Laser-Based Method in Flow Visualization and Evaluation in Bearings," Paper C 188 Inst. Mech. Engrg. Proceedings, Vol. 1, 1987, pp. 37–45.

25. Braun, M.J., Wheeler, R.L., III and Hendricks, R.C., "Thermal Shaft Effects on Load-Carrying Capacity of a Fully Coupled, Variable-Properties Cryogenic Journal Bearing," ASLE Preprint No. 86-TC-6B-1, October 1986.

26. Braun, M.J., Wheeler, R.L., III and Hendricks, R.D., "A Fully Coupled Variable Properties Thermohydraulic Model for a Cryogenic Hydrostatic Journal Bearing," ASME Paper 86-Trib-55, 1986.

27. Braun, M. and Hendricks, R.C., "An Experimental Investigation of the Vaporous/Gaseous Cavity Characteristics of an Eccentric Journal Bearing," ASLE Trans., Vol. 27, No. 1, 1984, pp. 1–14.

28. Braun, M., Mullen, R. and Hendricks, R.C., "An Analysis of Temperature Effect in a Finite Journal Bearing with Spatial Tilt and Viscous

Dissipation," ASLE Trans., Vol. 27, No. 4, 1984, pp. 405–412.

29. Brüggemann, H. and Kollmann, F.G., "A Numerical Solution of the Thermal Elastohydrodynamic Lubrication in an Elliptical Contact," ASME Paper 81-Lub-29, 1981.

30. Brugier, D. and Pascal, M.T., "Influence of Elastic Deformations of Turbo-Generator Tilting Pad Bearings on the Static Behavior and on the Dynamic Coefficients in Different Designs," J. Trib. Trans. ASME, Vol. 111, April 1989, pp. 364–371.

31. Bruin, S., "Temperature Distributions in Couette Flow With and Without Additional Pressure Gradient," Int'l. J. Heat Transfer, Vol. 15, 1972, pp. 341–349.

32. Burr, A.H., "The Effect of Shaft Rotation on Bearing Temperatures," ASLE Trans., 1959, pp. 235–241.

33. Burton, R.A., "Thermally Coupled Reynolds Flows; The Poiseuille Component," STLE Paper 89-TC-1b-1, October 1989.

34. Burton, R.A., "High Speed Seal Flows with Temperature Sensitive Viscosity," STLE Trans. Vol. 33, 1990, pp. 48–52.

35. Cameron, A., "The Viscosity Wedge," Trans. ASLE, 1958, 1, 248.

36. Cameron, A., Basic Lubrication Theory, Wiley, 1967, pp. 101–103.

37. Cann, P.M. and Spikes, H.A., "Determination of the Shear Stresses of Lubricants in Elastohydrodynamic Contacts," STLE Annual Meeting, Cleveland, 1988.

38. Cann, P.M. and Spikes, H.A., "The Influence of Lubricant on Temperature Generated in Elasto-Hydrodynamic Contacts," 14th Leeds-Lyon Symposium, "Interface Dynamics," September 1987.

39. Capitao, T.W., "Performance Characteristics of Tilting Pad Thrust Bearings at High Operating Speeds," ASME Paper, 75-Lubs-7, 1975.

40. Castelli, V. and Malanoski, S.B., "Method for Solution of Lubrication Problems with Temperature and Elasticity Effects: Application to Sector Tilting-Pad Bearings," J. Lub. Tech. Trans. ASME, Vol. 91, No. 4, 1969, pp. 634–640.

41. Charnes, A. and Saibel, E., "On the Solution of the Reynolds Equation for Slider-Bearing Lubrication–II: The Viscosity as a Function of the Pressure," ASME Trans. 75, 1953.

42. Charnes, A., Osterle, F. and Saibel, E., "On The Energy Equation for Fluid Film Lubrication," Proc. of Roy. Soc., Series A, Vol. 214, 1952, pp. 133–136.

43. Chen, C.M., "Contribution of Fluid Film Inertia to the Thermohydrodynamic Lubrication of Sector-Pad Bearings," ASME Paper 76-Lub-D,

1976.

44. Cheng, H.S. and Sternlicht, B., "A Numerical Solution for the Pressure, Temperature and Film Thickness Between Two Infinitely Long, Lubricated Rolling and Sliding Cylinders, Under Heavy Loads," J. Basic Eng., Trans. ASME, September 1965, pp. 695–707.

45. Cheng, C.M. and Dareing, D.W., "The Contribution of Fluid Film Inertia to the Thermohydrodynamic Lubrication of Sector-Pad Thrust Bearings," ASME J. Lub. Tech., Vol. 98, 1976, 125–136.

46. Cheng, H.L., A Refined Solution to the Thermal-Elastohydrodynamic Lubrication of Rolling and Sliding Cylinders," ASLE Trans., Vol. 8, 1965, pp. 297–410.

47. Cheng, Xilin, and Bian, Yongquin, "Temperature Distribution of Piston Rings (in Chinese)," Neiranji Xuebao/Transactions of CSICE (Chinese Society for Internal Combustion Engines) Vol. 3, No. 3, 1985, pp. 257–264.

48. Chien, H.L. and Cheng, H.S., "Thermohydrodynamic Performance of Smooth Plane Sliders," Proc. of the JSLE International Tribology Conference, Tokyo, 1985, pp. 559–564.

49. Clayton, Wilkie, "Temperature Distribution in the Bush of a Journal Bearing," Engineering, Vol. 166, p. 49, July 16, 1948.

50. Cole, J.A., "An Experimental Investigation of Temperature Effects in Journal Bearings," Proc. Convention on Lubr. and Wear, Inst. Mech. Engrs., London, 1957, pp. 111–117.

51. Colin, F., Floquet, A. and Play, D., "Thermal Contact Simulation in 2-D and 3-D Mechanisms," J. Lub. Tech. Trans. ASME, Vol. 110, April 1988, pp. 247–252.

52. Conry, T.F., "Thermal Effects on Traction in EHD Lubrication," J. Trib. Trans. ASME, Vol. 103, October 1981, pp. 533–538.

53. Conry, T.F., Johnson, K.L. and Owen, S., "Viscosity in the Thermal Regime of EHD Traction," TET, 1980, pp. 219–227.

54. Constantinescu, V.N., "Note on the Influence of the Heat Transfer, etc.," ASME Paper No. 68-Lub-11, 1968.

55. Constantinescu, V.N., "Basic Relationships in Turbulent Lubrication and Their Extension to Include Thermal Effects," J. Lub. Tech. Trans. ASME, Vol. 95, 1973, pp. 147–154.

56. Conway-Jones, J.M. and Leopard, A.J., "Plain Bearing Damage," Proc. 4th Turbomachinery Symposium, Texas A&M University, College Station, Texas, Oct. 14–16,1975, pp. 55–63.

57. Cope, W., "The Hydrodynamic Theory of Film Lubrication," Proc. Royal Soc. (London), A197, 1949, pp. 201–216.

58. Cowking, E., "Thermohydrodynamic Analysis of Multi-Arc Journal Bearings," Tribol. Int'l., 14–15, 1981, pp. 217–223.

59. Craighead, I., Dowson, D., Sharp, R.S. and Taylor, C.M., "The Influence of Thermal Effects and Shaft Misalignment on the Dynamic Behaviour of Fluid Film Bearings," TET, 1980, pp. 47–55.

60. Cuenca, R.M. and Raynor, S., "Thermal Effects on a Cylindrical Hydrostatic Bearing with Central Fluid Supply," ASLE Trans., Vol. 7, 1964, pp. 304–310.

61. Currie, I.G., Brockley, C.A. and Dvorak, F.A., "Thermal Wedge Lubrication of Parallel Surface Thrust Bearings," ASME Paper No. 65-Lub-S2, 1965.

62. Czeguhn, K., "Variability with Time of Bearing Clearing Due to Temperature and Pressure," Proc. Inst. Mech. Eng., Vol. 181, Pt. 30, 1966–67.

63. Dakshina, Murthy H.B., "Thermal Instability of Plastohydrodynamic Lubricant Film at Work Zones," TET, 1980, pp. 273–280.

64. DeChudbury, P. and Barth, E.W., "A Comparison of Film Temperatures and Oil Discharge Temperatures for a Tilting Pad Journal Bearing," ASME Paper 80-CZ/Lub-37, 1980.

65. DeChudbury, P. and Masters, D.A., "Performance Tests of Five-Shoe Tilting-Pad Journal Bearing," ASLE Trans., Vol. 27, No. 1, 1984, pp. 61–66.

66. Derdouri, A. and Carreau, P.J., "Non-Newtonian and Thermal Effects in Journal Bearings," STLE Transactions, Vol. 32, 1989, pp. 161–169.

67. Dill, J.F., "Extreme Measures; Tribological Testing in Hostile Environment," Mechanical Engineering, April, 1989, pp. 83–88.

68. Di Pasquantonio, F. and Sala, P., "Influence of Thermal Field on the Resistance Law in Turbulent Bearing-Lubrication Theory," J. Tribol., Trans. ASME, Vol. 106, 1984, pp. 368–376.

69. Dong, Z. and Shi-zuh, W., "Numerical Solution for the Thermo-elastohydrodynamic Problem in Elliptical Contacts," J. of Trib. ASME Vol. 106, April, 1984, pp. 246–254.

70. Dow, T.A. and Kannel, J.W., "Evaluation of Rolling/Sliding EHD Temperatures," TET, 1980, pp. 228–240.

71. Dow, T.A., Kannel, J.W. and Bupara, S.W., "A Hydrodynamic Lubrication Theory for Strip Rolling Including Thermal Effects," ASME Paper No. 74-Lub-8, 1974.

72. Dowson, D., Economou, P.N., Parsons, B. and Ruddy, B., "The Influence of Thermal Distortions of a Piston Upon Piston Ring Lubrication,"

Proceedings of the 6th Leeds-Lyon Symposium on Tribology, published by Mech. Engrg. Publ., 1979.

73. Dowson and March, "A Thermodynamic Analysis at Low Pressure," Proc. Inst. Mech. Eng., Vol. 181, Pt.30, 1966, pp. 117–126.

74. Dowson, D. and Hudson, J.D., "Thermohydrodynamic Analysis of the Infinite Slider-Bearing: Part I, The Plane-Inclined Slider-Bearing: Part II, The Parallel-Surface Bearing," Proc. Inst. Mech. Eng., 3–4, 1964, pp. 34–51.

75. Dowson, D. and March, C.N.A., "A Thermohydrodynamic Analysis of Journal Bearings," Proc. Inst. Mech. Eng., Vol. 181, Pt. 30, 1966–67, p. 117.

76. Dowson, D., Hudson, J.D., Hunter, E. and March, C.N., "An Experimental Investigation of the Thermal Equilibrium of Steadily Loaded Journal Bearings," Proc. Inst. Mech. Eng., Vol. 181, Pt. 38, 1966–67, p. 70.

77. Dowson, D., "A Generalized Reynolds Equation for Fluid Film Lubrication," Int. Jour. Mechanical Sciences, Pergamon Press, Vol. 4, 1962, pp. 159–170.

78. El-Ariny, A. and Aziz, A., "A Numerical Solution of Entrance Region Heat Transfer in Plane Couette Flow," ASME J. Heat Transfer, 1976, pp. 427–431.

79. Elrod, H.G. and Burgdorfer, A., "Refinement in the Theory of Gas Lubricated Journal Bearings of Infinite Length," Proc. Int. Symposium on Gas Lubricated Journal Bearings, Washinton DC, Oct., 1959.

80. Elrod, H.G. and Chu, T.Y., "Inertia and Energy Effects in the Developing Gas Film Between Two Parallel Flat Plates," J. Lub. Tech. Trans. ASME, October 1973, pp. 524–534.

81. Elrod, H.G. and Brewe, E., "Thermohydrodynamic Analysis for Laminar Lubricating Films," Proc. 13th Leeds-Lyon Symp. on Tribology, Sept. 1986, pp. 443–450.

82. Elrod, H.G. and Ng, C.W., "A Theory for Turbulent Fluid Films and Its Application to Bearings," J. Lub. Tech. Trans. ASME, July 1967, pp. 346–362.

83. Ettles, C.M., "Three Dimensional Computation of Thrust Bearings," Proc. 13th Leeds Lyon Symposium on Tribology, Elsevier, 1987, pp. 95–104.

84. Ettles, C.M. and Advani, S., "The Control of Thermal and Elastic Effects in Thrust Bearings," TET, 1980, pp. 105–116.

85. Ettles, C.M., "Three Dimensional Computation of Thrust Bearings," Proceedings of the 13th Leeds-Lyon Symposium on Tribology, held

in Bodington Hall, The University of Leeds, England, Paper IV(i), September 8–12, 1986, pp. 95–104.

86. Ettles, C.M., "Development of a Generalized Computer Analysis for Sector Shaped Tilting Pad Thrust Bearings," ASLE Paper 75AM-8A-2, May 1975.

87. Ettles, C.M., "The Thermal Control of Friction at High Sliding Speeds," ASME Paper 85-Trib-37, 1985.

88. Ettles, C.M., "Transient Thermoelastic Effects in Fluid Film Bearings," Wear, Vol. 79, 1982, pp. 53–71.

89. Ettles, C.M., "The Analysis and Performance of Pivoted Pad Journal Bearings Considering Thermal and Elastic Effects," J. Lub. Tech., Vol. 102, April 1980, pp. 182–192.

90. Ettles, C.M., "Solutions for Flow in a Bearing Groove," Proc. Inst. Mech. Eng., Vol. 182, Pt. 3N, 1967–68, pp. 120–131.

91. Ettles, C.M. and Cameron, A., "The Action of the Parallel Surface," Br. Proc. Conf. Lubrication and Wear, IHE, London, 1966.

92. Ettles, C.M. and Cameron, A., "Thermal and Elastic Distortions in Thrust Bearings," Inst. Mech. Eng. Proc., May 1963.

93. Ettles, C.M. and Cameron, A., "Considerations of Flow Across a Bearing Groove", J. Lub. Tech. Trans. ASME, January 1968, pp. 313–319.

94. Ettles, C.M., "Hot Oil Carry Over in Thrust Bearings," Proc. Inst. Eng., Vol. 184, Pt. 3L, 1969–70, pp. 75–81.

95. Ettles, C.M., Heshmat, H., and Brockwell, K., "Elapsed Time for the Decay of Thermal Transients in Fluid Film Bearing Assemblies," 15th Leeds-Lyon Symposium on Tribology, September 1988.

96. Ezzat, H. and Seireg, A., "Thermohydrodynamic Performance of Conical Journal Bearings," First World Conf. in Industrial Tribol., New Delhi, Paper H13, 1972, pp. H131–H1311.

97. Ezzat, H. and Rohde, S., "A Study of the Thermodynamic Performance of Finite Slider Bearings," J. Lub. Tech. Trans. ASME, Vol. 95, No. 3, July 1973, pp. 298–307.

98. Ezzat, H.A. and Rohde, S.M., "Thermal Transients in Finite Slider Bearings," J. Lub. Tech. Trans. ASME, July 1974, pp. 315–321.

99. Felder, E., "Description and Easy Characterization of Thermal Effects in Viscous Lubricant Films," TET, 1980, pp. 147–151.

100. Feng, N.S. and Hahn, E J., "Density and Viscosity Models for Two-Phase Homogeneous Hydrodynamic Damper Fluids," ASLE Transactions, Vol. 29, No. 3, 1986, p. 361.

101. Ferron, J., Frene, J. and Boncompain, R., "A Study of the Thermohydrodynamic Performance of a Plain Journal Bearing Comparison Between Theory and Experiments," J. Lub. Tech. Trans. ASME, Vol. 105, 1983, pp. 442–428.

102. Fillon, N., Frene, J. and Boncompain, R., "Etude Experimentale de L'effect Thermique dans les paliers a patins oscillants," Congres International de Tribologie EUROTRIB 85, Lyon, 4–1.6, 1985.

103. Floquet, A., "Comparison Between Analytical and Numerical Methods in Heat Transfer Analysis in Tribology," TET, 1980, pp. 162–174.

104. Fogg, A., "Fluid Film Lubrication of Parallel Thrust Surfaces," Proc. Inst. Mech. Eng., Vol. 155, 1946, p. 49.

105. Furuhama, S. and Suzuki, H., "Temperature Distribution of Piston Rings and Piston in High Speed Diesel Engine," Bulletin of the JSME, Vol. 22, No. 174, December 1979, pp. 1788–1795.

106. Galakhov, M.A. and Kovalev, V.P., "Thickness of the Lubricant Film at the Contact Surface of Bodies of Different Temperatures," Russ. Eng. Journal, Vol. 56, November 1966, pp. 38–39.

107. Gardner, W.W., "Tilting Pad Thrust Bearing Tests–Influence of Pivot Location," J. Trib., Trans. ASME, Vol. 110, No. 4, October 1988, pp. 609–613.

108. Gecim, B., "Steady Temperature in a Rotating Cylinder Subject to Surface Healing and Convective Cooling," ASME Paper 83-Lub-8, 1983.

109. Gecim, B. and Winer, W.O., "Transient Temperatures in the Vicinity of an Asperity Contact," J. Lub. Tech. Trans. ASME, July 1985, pp. 333–342.

110. Georges, J.M., Martin, J.M., and Brison, J.F., "Contact Temperatures in Boundary Lubricated Interface. Relation to Anti-Wear Additives Action," TET, 1980, pp. 307–315.

111. Georgopoulos, E. and Burton, R.A., "Thermal Convection in Short Bearing Films," J. Lub. Tech. Trans. ASME, January 1980, pp. 119–123.

112. Gero, L.R. and Ettles, C.M., "A Three Dimensional Thermohydrodynamic Finite Element Scheme for Fluid Film Bearings," STLE Trib. Trans., Vol. 31, No. 2, pp. 182–191.

113. Gethin, D.T. and Medwell, J.O., "An Experimental Investigation Into the Thermohydrodynamic Behaviour of a High Speed Cylindrical Bore Journal Bearing," J. Trib. Trans. ASME, Vol. 107, 1985, pp. 538–543.

114. Gethin, D.T., "An Application of the Finite Element Method to the Thermohydrodynamic Analysis of a Thin Film Cylindrical Bore Bearing Running at High Sliding Speed," J. Lub. Tech. Trans. ASME, Vol.

109, April 1987, pp. 283-9.

115. Gethin, D.T., "Predictive Models for the Design of Profile Bore Bearings," J. Trib. Trans. ASME, Vol. 112, No. 1, January 1990, pp. 156-164.

116. Giordano, M. and Boudet, R., "Thermohydrodynamic Flow of Piezo Viscous Fluid Between Two Parallel Discs," TET, 1980, pp. 127-132.

117. Gould, P., "Parallel Surface Squeeze Films: The Effect of the Variation of Viscosity with Temperature," J. Lub. Tech. Trans. ASME, July 1967, pp. 375-380.

118. Gould, P., "Flow Instability Induced by Viscosity Variation in High Pressure Two-Dimensional Laminar Flow Between Parallel Plates," J. Lub. Tech. Trans. ASME, October 1971, pp. 465-469.

119. Guilinger, W.H. and Saibel, E.A., "The Effect of Heat Conductance on Slider-Bearing Characteristics," Trans. ASME, May 1958.

120. Hagg, A., "Heat Effects in Lubricating Films," Trans. ASME, Vol. 66, A72-A76, 1944.

121. Hahn, E.J. and Kettleborough, C.F., "The Effects of Thermal Expansion of Infinitely Wide Slider Bearing-Free Thermal Expansion," J. Lub. Tech. Trans. ASME, January 1968, pp. 233-239.

122. Hahn, E.J. and Kettleborough, C.F., "Thermal Effects in Slider Bearings," I. Mech. E., Proceedings, 1968-69, Vol. 183, Part I.

123. Hahn, E.J. and Kettleborough, C.F., "Thermal Effects in Slider Bearings," Proc. Inst. Mech. Eng., Vol. 181, Pt. 3B, 1966-67, pp. 55-62.

124. Hahn, E.J. and Kettleborough, C.F., "Solutions for the Pressure and Temperature in an Infinite Slider Bearing of Arbitrary Profile," J. Lub. Tech. Trans. ASME, October 1967, pp. 445-452.

125. Hakansson, B., "The Journal Bearing Considering Variable Viscosity," Transactions of Chalmers University of Technology, Gothenburg, Sweden, No. 298, 1965.

126. Hamilton, J., Burr, A. and Ocvirk, F., "Experimental Investigation of a Method of Predicting Heat Dissipation and Temperature of Plain Journal Bearings," Cornell University Progress Report No. 14 to NACA, Ithaca, NY, October 1955.

127. Hamrock, B.J. and Dowson, D., "Isothermal EHD Lubrication of Point Contact—Part III," J. Lubr. Tech. Trans. ASME, Vol. 99 No. 2, 1977, pp. 264-276.

128. Hansen, P.K. and Lund, J.E., "An Analytical Study of the Heat Balance for a Journal Bearing," AGARD Conference Proceedings, AGARD-CP-323, 1982, 25-1 to 25-9.

129. Hashimoto, H. and Wada, C., "Turbulent Lubrication of Tilting Pad Thrust Bearings with Thermal and Elastic Deformations," J. Trib. Trans. ASME, Vol. 107, No. 1, January 1985, pp. 82–90.

130. Hausenblas, H., "Die nichtisotherme laminare Stroemung, etc.," Eng. Archiv. XVIII Band, 1950.

131. Heckelman, D.D. and Ettles, C.M. "Viscous and Inertial Pressure Effects at the Inlet to a Bearing Film," STLE Trans., Vol. 31, No. 1, 1–5, 1988.

132. Heshmat, H. and Pinkus, O., "Performance of Starved Journal Bearings with Oil Ring Lubrication," J. Trib. Trans. ASME, Vol. 107, No. 1, January 1985, pp. 23–31.

133. Heshmat, H. and Pinkus, O., "Thrust Bearing Behavior under Oil Starvation Conditions" (in Russian), USSR Academy of Sciences, Trenye, Iznosysmazochnye Materialy, Vol. 3, Part 1, Publ. by FAN, Tashkent, USSR, 1985, pp. 31–46.

134. Heshmat, H. and Pinkus, O., "Mixing Inlet Temperatures in Hydrodynamic Bearings," J. Trib. Trans. ASME, Vol. 108, 1986, pp. 231–248.

135. Heshmat, H., Artiles, A. and Pinkus, O., "Parametric Study and Optimization of Starved Thrust Bearings," Leeds-Lyon Symposium, September 1986, pp. 105–112.

136. Heshmat, H., "On the Mechanism of Operation of Flat Land Bearings," 14th Leeds-Lyon Symposium on Tribology, 8–11 September 1987.

137. Hirn, G.A., "Sur Les Principaux Phenomenes que Presentent les Frottements Mediats, et Sur les Diverses Manieres Employees au Graissage des Machines," Bull Soc. Ind. de Mulhouse, 1954, XXVI, pp. 188–277.

138. Hiruma, M., Furuhama, S. and Mochizuki, N., "Tribology of Piston-Rod Seal Under High Temperature and Oilless Conditions," Lubrication Engineering, Vol. 38, No. 2, February 1982, pp. 104–110.

139. Hopf, G. and Schüler, D., "Investigations on Large Turbine Bearings Working under Transitional Conditions between Laminar and Turbulent Flow," J. Trib. Trans. ASME, Vol. 111, October 1989, pp. 628–634.

140. Hornbeck, R., "Numerical Marching Techniques for Fluid Flows with Heat Transfer," NASA Publication SP-297, 1973.

141. Huebner, K.H., "Solution for the Pressure and Temperature in Thrust Bearings Operating in the Thermohydrodynamic Turbulent Regime," J. Lub. Tech. Trans. ASME, Vol. 96, January 1974, pp. 58–68.

142. Huebner, K.H., Application of Finite Element Methods to Thermohydrodynamic Lubrication," Int. J. Num. Methods Eng., Vol. 8, pp. 139–165.

143. Huebner, K., "A Three-Dimensional Thermohydrodynamic Analysis of Sector Thrust Bearings," ASLE Trans., Vol. 17, No. 1, 1973, pp. 62–73.

144. Huffenus, J.P. and Khaletzky, D., "Theoretical Study of Heat Transfer in Thrust Bearings and Hydraulic Machines, Application to the Cooling of the Oil Film," TET, 1980, pp. 117–126.

145. Hughes, W.F. and Chao, N.H.,"Phase Change in Liquid Face Seals II— Isothermal and Adiabatic Boundaries with Real Fluids," ASME Paper 79–Lub–4, 1979.

146. Hughes, W.E. and Osterle, N., "Temperature Effects in Journal Bearing Lubrication," Trans. ASLE Vol. 1, pp. 210–216.

147. Hunter, W.B. and Zienkiewicz, O.C., "Effect of Temperature Variations Across the Lubricant Films in the Theory of Hydrodynamic Lubrication," J. Mech. Eng. Sci., Vol. 2, No. 1, 1960, pp. 52–58.

148. Jaeger, J.C., "Moving Sources of Heat and the Temperature at Sliding Contacts," Proc. Royal Society, New South Wales, Vol. 76, 1943, pp. 203–224.

149. Jain, P.K., Jain, S.C., and Kar, Subir, "Influence of Pressure Gradient of Temperature Distribution in an Incompressible Viscous Couette Flow," Indian Institute of Technology, September 1976.

150. Jarchow, F. and Theissen, J., "Thermohydrodynamicshe Naherungsrechnung fur Gleitlager mit erhöten Warmeentzug," Konstruktion 30, H. 4, 1978, pp. 155–161.

151. Jeng, M.C., Zhou, G.R. and Szeri, A.Z., "A Thermohydrodynamic Solution of Pivoted Thrust Pads, Part I: Theory; Part II: Static Loading; Part III: Linearized Force Coefficient," J. Trib. Trans. ASME, 108, 1986, pp. 195–218.

152. Kacou, A., Rajagopal, K.R. and Szeri, A.Z., "A Thermohydrodynamic Analysis of Journal Bearings Lubricated by a Non-Newtonian Fluid," J. Lub. Tech. Trans. ASME, Vol. 110, July 1988.

153. Kacou, A., Rajagopal, K.R. and Szeri, A.Z., "A Thermohydrodynamic Analysis of Journal Bearings Lubricated by a Non-Newtonian Fluid," J. Trib., Trans. ASME, Vol. 110, July 1988, pp. 414–420.

154. Kahlert, W., Der Einfluss Der Tragheitskrafte bei der hydrodynamischen Schmiermitteltheorie, Ingenr.-Arch., Band XVI, 1948, pp. 321–342.

155. Kanarachos, A., "Ein Beitrag zur thermoelastohydrodynamischen Analyse von Gleitlagern," Konstruktion 29, H. 3, pp. 101–106.

156. Kaniewski, W., "Randbedinungen des Schmierfilmes," Wear, Vol. 45, 1977, pp. 113–125.

157. Kannel, J.W., Zugaro, F.F. and Dow, T.A., "A Method for Measuring Surface Temperature Between Rolling/Sliding Steel Cylinders," J. Lub. Tech. Trans., ASME, Vol. 100, January 1978, pp. 110–114.

158. Kawaike, K., Okano, K. and Furukawa, Y., "Performance of a Large Thrust Bearing with Minimized Thermal Distortion," ASLE Trans., Vol. 22, No. 2, 1979, pp. 125–134.

159. Kays, W. and Bjorklund, I., "Heat Transfer from a Rotating Cylinder With or Without Crossflow," ASME Trans., 1958, pp. 70–78.

160. Kennedy, F.E., "Surface Temperature in Sliding Systems—A Finite Element Analysis," ASME Paper No. 80 C2Lub-28, 1980.

161. Kennedy, J.S., Sinha, Prawal and Rodkiewicz, C.M., "Thermal Effects in Externally Pressurized Conical Bearings with Variable Viscosity," J. Lub. Tech. Trans. ASME, Vol. 110, April 1988, pp. 201–211.

162. Khonsari, M. M. and Beaman, J., "Thermohydrodynamic Analysis of Laminar Incompressible Journal Bearings," ASLE Trans., Vol. 29, No. 2, 1986, pp. 141–150.

163. Khonsari, M.M., "A Review of Thermal Effects in Hydrodynamic Bearings. Part I: Slider/Thrust Bearings; Part II: Journal Bearings," ASLE Trans., Vol. 30, 1987, pp. 19–20.

164. Khonsari, M.M. and Esfahanian, V., "Thermohydrodynamic Analysis of Solid-Liquid Lubricated Journal Bearings," J. Lub. Tech. Trans. ASME, Vol. 110, April 1988, pp. 367–374.

165. Khonsari, M.M. and Kim, H.J., "On Thermally Induced Seizure in Journal Bearings," J. Trib. Trans. ASME, Vol. 111, October 1989, pp. 661–667.

166. Kim, K., Tanaka, M. and Hori, Y., "A Three-Dimensional Analysis of Thermohydrodynamic Performance of Sector-Shaped, Tilting-Pad Thrust Bearing," J. Lub. Tech. Trans. ASME, Vol. 105, 1983, pp. 406–413.

167. Kingsbury, E.P., "The Heat of Adsorption of a Boundary Lubricant," ASLE Annual Conf., April 1959.

168. Kingsbury, A., "Heat Effects in Lubricating Films," Mech. Eng., Vol. 55, 1933, pp. 685–688.

169. Knight, J.D., "Prediction of Temperatures in Tilting Pad Journal Bearings," STLE Trans. Vol. 33, 1990, pp. 185-192.

170. Knight, J.D. and Barrett, L.E., "Analysis of Axially Grooved Journal Bearings with Heat Transfer Effects," ASLE Trans., Vol. 30, No. 3, pp. 316–323.

171. Knight, J. and Barrett, L., "An Approximate Solution Technique for

Multilobe Journal Bearings Including Thermal Effects with Comparison to Experiment," ASME Trans., Vol. 26, 1983, pp. 502–508.

172. Knight, J., Barrett, L. and Cronan, R., "The Effects of Supply Pressure on the Operating Characteristics of Two-Axial Groove Journal Bearings," ASLE Trans., Vol. 28, No. 3, 1985, pp. 336–342.

173. Kuhn, E.C., "Load Capacity and Losses in the Infinite Slider Bearings with a Transversely Forced-Cooled-Lubricant Film," ASME Paper No. 68 LC-18, 1968.

174. Kuhn, E.C., "Load Capacity and Losses in the Infinite Slider Bearing with a Transversely Forced-Cooled Lubricant Film," ASLE Trans., Vol. 12, 1969, pp. 135–139.

175. Kumar, S. and Sachidanandra, S.C., "Effects of Viscosity Variation and Surface Roughness in Short Journal Bearings," Wear, Vol. 52, 1979, pp. 341–346.

176. Langheim, R. and Bartz, W.J., "The Significance of the Effective Viscosity in Unstationary Loaded Bearings," 1981 ASLE/ASME Lubr. Conf.

177. Launder, B.E. and Leschziner, M., "Flow in Finite Width Thrust Bearings Including Inertial Effects, Part I: Laminar Flow; Part II: Turbulent Flow," J. Lub. Tech. Trans. ASME, Vol. 100, July 1978, pp. 330–334.

178. Li, Chin-Hsiu, "The Influence of Variable Density and Viscosity on Flow Transition Between Two Concentric Rotating Cylinders," J. Lub. Tech. Trans. ASME, Vol. 100, April 1978, pp. 261–270.

179. Ling, F.F. and Rice, J.S., "Surface Temperature with Temperature-Dependent Thermal Properties," ASLE Trans., Vol. 9, 1966, pp. 195–201.

180. Lumley, J.L., "Turbulence Modeling," J. of Appl. Mech., Trans. ASME, Vol. 50, December 1983, pp. 1097–1103.

181. Lund, J. and Hansen, P., "An Approximate Analysis of the Temperature Conditions in a Journal Bearing. Part I: Theory," J. Trib. Trans. ASME, Vol. 106, 1984, pp. 228–236.

182. Lund, J. and Tonnesen, J., "An Approximate Analysis of the Temperature Conditions in a Journal Bearing. Part II: Application,: J. Trib. Trans. ASME, Vol. 106, 1984, pp. 237–244.

183. Majumdar, B.C., "The Thermohydrodynamic Solution of Oil Journal Bearings," University of Karlsruhe, Germany, February 1975.

184. Majumdar, B.C. and Saha, A.K., "Temperature Distribution in Oil Journal Bearings," Wear, Vol. 18, May 1974, pp. 259–266.

185. Martin, F. A., "Discussion to Paper by Pinkus and Wilcock," TET, 1980, p. 330.

186. McCallion, H., Yousif, F. and Lloyd, T., "The Analysis of Thermal Effects in a Full Journal Bearing," J. Lub. Tech. Trans. ASME, October 1970, pp. 578–587.

187. Medwell, J.O. and Bunce, J.K., "The Influence of Bearing Inlet Conditions on Bush Temperature Fields," TET, 1980, pp. 56–64.

188. Mitsui, J. and Yamada, T., "A Study of the Lubricant Film Characteristics of Journal Bearings. Part I: A Thermodynamic Analysis with Particular Reference to the Viscosity Variation Within Lubricating Film," Bull. JSME, Vol. 22, No. 172, 1979, pp. 1491–1498.

189. Mitsui, J., "A Study of the Lubricant Film Characteristics of Journal Bearings. Part II: Effects of Various Design Parameters on Thermal Characteristics of Journal Bearing," Bull. JSME, Vol. 15, No. 210, 1982, pp. 2010–2017.

190. Mitsui, J., "A Study of the Lubricant Film Characteristics of Journal Bearings. Part III: Effects of the Film Viscosity Variation on the Dynamic Characteristics of Journal Bearings," Bull. JSME, Vol. 15, No. 210, 1982, pp. 2018–2019.

191. Mitsui, J., Hori, Y. and Tanaka, M., "An Experimental Investigation on the Temperature Distribution in Circular Journal Bearings," J. Lub. Tech. Trans. ASME, Vol. 108, October 1986, pp. 621–627.

192. Mitsui, J., Hori, Y. and Tanaka, M., "Thermohydrodynamic Analysis of Cooling Effect of Supply Oil in Circular Journal Bearing," J. Lub. Tech. Trans. ASME, July 1983, pp. 414–421.

193. Moes, H., Ten Hoeve, P.B.Y. and Van der Helm, J., "Thermal Effects in Dynamically Loaded Flexible Journal Bearings," J. Lub. Tech. Trans. ASME, Vol. 111, January 1989, pp. 49–55.

194. Moore, D.F., "An Energy Balance for Squeeze Films," ASME Paper No. 65-Lubs-7, 1965.

195. Morton, P.G. and Keogh, P.S., "Thermoelastic Influences in Journal Bearing Lubrication," Proc. R. Soc. Lond. A, Vol. 403, 1986, pp. 111–134.

196. Motosh, N., "Cylindrical Journal Bearings Under Constant Load, the Influence of Temperature and Pressure on Viscosity," Proc. Inst. Mech. Eng., Vol. 178, Pt. 3N, 1963–64, pp. 148–160.

197. Motosh, N., "Der Wärmeaustausch Zwischen Ölschicht und einem Gleitlager unter Berücksichtigung der Veränderlichkeit der Ölviskosität," Ingenieur-Archiv, March 1964, pp. 149–161.

198. Muraki, M. and Kimura, Y., "A Simplified Thermal Theory of EHL Traction, JSLE Int'l. Tribology Conference, July 8–10, 1985.

199. Muskat, M. and Morgan, F., "Temperature Behavior of Journal Bearing System," J. Appl. Phys., Vol. 14, 1943, p. 234.

200. Ng, C.W. and Pan, C.H.T., "A Linearized Turbulent Lubrication Theory," J. Basic Engin. Trans. ASME, Series D, Vol. 87, No. 3, Sept 1965, pp. 675–688.

201. Nagaraj, H.S., Sanborn, D.M. and Winer, W.O., "Direct Surface Temperature Measurement by Infrared Radiation in Elastohydrodynamic Contacts and the Correlation with the Blok Flash Temperature Theory," Wear, Vol. 49, 1978, pp. 43–59.

202. Nakashima, K. and Takafuji, K., "Measurement of Oil Film State Using Two Kinds of Electrical Circuit," Proceedings of the JSLE Int'l. Tribology Conference, July 8–10, 1985.

203. Neal, P.B., "Some Factors Influencing the Operating Temperature of Pad Thrust Bearings," TET, 1980, pp. 137–142.

204. Neal, P.B., "Heat Transfer in Pad Thrust Bearings," The Institution of Mechanical Engineers, Vol. 196, No. 20, 1982, pp. 217–228.

205. Neal, P., "Influence of Film Inlet Condition on the Performance of Fluid Film Bearings," J. Mech. Eng. Sci., Vol. 12, No. 2, 1970, p. 153.

206. Nica, A., "Thermal Behaviour and Friction in Journal Bearings," J. Lub. Tech. Trans. ASME, July 1969, pp. 1–8.

207. Niemann, G. and Lechner, G., "The Measurement of Surface Temperatures on Gear Teeth," J. Basic Eng., Trans. ASME, September 1965, pp. 641–654.

208. Nikolajsen, J.L., "The Effect of Variable Viscosity on the Stability of Plain Journal Bearings and Floating-Ring Journal Bearings," J. Lub. Tech. Trans. ASME, October 1973, pp. 447–465.

209. Orlin, A. S. and Zarenbin, V. G., "Calculation of the Temperature Field of Friction in the Ring Seals of Internal Combustion Engines (in Russian)," Izvestiya Vysshikh Uchevnykh Zavedenii, Mashinostroenie, No. 1, 1978, pp. 106–110.

210. Orlin, A.S. and Zarenbin, V.G., "Calculation of Temperature of the Piston Ring by the Method of Finite Integral Transformations (in Russian)," Energomashinostroenie, No. 7, June 1976, pp. 6–8.

211. Orlin, A.S., Zarenbin, V.G. and Litvinenko, N.P., "Determination of Temperatures and Thermal Stresses in Piston Rings of Internal Combustion Engines (in Russian)," Izvestiya Vysshikh Uchevnykh Zavedenii, Mashinostroenie, No. 4, 1974, pp. 97–101.

212. Osterle, F. and Saibel, E., "On the Effect of Lubricant Inertia in Hydrodynamic Lubrication, Z. Angew. Math. Phys., Vol. 6, 1955, p. 334.

213. Osterle, F., Charnes, A. and Saibel, E., "On The Solution of the Reynolds Equation for Slider Bearing Lubrication. Part IV: Effect of Temperature on the Viscosity," ASME Trans., 1953, p. 1117.

214. Ott, H.H. and Paradissiadis, G., "Thermohydrodynamic Analysis of Journal Bearings Considering Cavitation and Reverse Flow," J. Lub. Tech. Trans. ASME, Vol. 110, July 1988, pp. 439–447.

215. Palmberg, J.P., "On Thermo-elasto-hydrodynamic Fluid Film Bearings," Chalmers University of Technology, Sweden, 1975.

216. Pascovici, M.D., "The Circumferential Temperature Distribution of Complete Journal Bearings," Wear, Vol. 16, July–August 1970, pp. 143–147.

217. Pascovici, M.D., "Temperature Distribution in the Lubricant Film of Sliding Bearings Under Intensive Lubricant-Wall Heat Transfer Conditions," Wear, Vol. 29, September 1974, pp. 277–286.

218. Peeken, H., Gleitlagertemperatur bei von 20°C abweichenden Umgebungstemperaturen," Konstruktion, Vol. 18, July 1966, pp. 232–236.

219. Pedroso, R.I., Thermal Analysis of Laminar Fluid Film Under Side Cyclic Motion," Journal of Heat Transfer, Vol. 96, February 1974, pp. 100–106.

220. Pinkus, O. and Sternlicht, B., "Theory of Hydrodynamic Lubrication," McGraw-Hill, 1961.

221. Pinkus, O. and Sternlicht, B., "The Maximum Temperature Profile in Journal Bearings," Trans. ASME, February 1957, pp. 337–341.

222. Pinkus, O. and Wilcock, D.F., "Thermal Effects in Fluid Film Bearings," TET, 1980, pp. 3–23.

223. Pinkus, O. and Bupara, S.S., "Adiabatic Solutions for Finite Journal Bearings," J. Lub. Tech. Trans. ASME, October 1979, pp. 492–496.

224. Pinkus, O., "Transverse Velocity and Pressure Variation in Finite Journal Bearings and Cylinders," Israel Journal of Technology, Vol. 11, Nos. 1-2, 1973, pp. 41–52.

225. Pinkus, O., "Anisothermal Fluid Films in Tribology," Israel Journal of Technology, Vol. 22, 1984–85, pp. 120–141.

226. Pinkus, O. and Wilcock, D.F., "Low Power-Loss Bearings for Electric Utilities," Vol. I, EPRI Document No. CS 4048, June 1985.

227. Pinkus, O., Chapter 6 of *Sawyer's Gas Turbine Engineering Handbook*, 3rd ed., Turbomachinery International Publications, 1985, pp. 22–65.

228. Pinkus, O., "The Reynolds Centennial: A Brief History of the Theory of Hydrodynamic Lubrication," J. Trib. Trans. ASME, Vol. 109, January 1987, pp. 2–20.

229. Podolskiy, M.Y.E., "The Problem of the Temperature Field of a Lubricating Layer in Thrust Slider Bearings," Gos. Nauk. Issl. Institute, IzdvoNauka, Moscow, 1970, pp. 89–104.

230. Pollmann, E., "Use of Similarity Number in Connection with Plain Bearings Under Conditions of Varying Temperature and Viscosity," Proc. Inst. Mech. Eng., Vol. 181, No. 30, 1966–67, pp. 144–155.

231. Poritsky, H., "Lubrication of Gear Teeth Including the Effect of Elastic Displacement," 1st ASLE National Symposium, Chicago, 1952.

232. Prashad, H., "The Effects of Viscosity and Clearance on the Performance of Hydrodynamic Journal Bearings," STLE Trans., Vol. 31, No. 1, pp. 113–119.

233. Purvis, M.B., Meyer, W.E., and Benton, T.C., "Temperature Distribution in the Journal-Bearing Lubricant Film," Trans. ASME, November 1957, pp. 343–350.

234. Quale, E.B. and Wiltshire, F.R., "The Performance of Hydrodynamic Lubricating Films with Viscosity Variations Perpendicular to the Direction of Motion," J. Lub. Tech. Trans. ASME, January 1972, pp. 44–48.

235. Raimondi, A.A., "An Adiabatic Solution for the Finite Slider Bearing (L/B=1)," ASLE Trans., Vol. 9, 1966, pp. 283–298.

236. Raimondi, A.A. and Boyd, J., "A Solution of the Finite Journal Bearing and its Application to Analysis and Design—Part III," Trans. ASLE, Vol. 1, No. 1, 1985, p. 38.

237. Rajalingham, C. and Prabhu, B.S., "Thermohydrodynamic Performance of a Plain Journal Bearing," ASLE Transactions, Vol. 30, No. 3, 1987, pp. 368–372.

238. Rajaswamy, T.G., Rao, T., Muralidhara and Prabhu, B.S., "An Experimental Study of Sector-Pad Thrust Bearings and Evaluation of Their Thermal Characteristics," Proc. 13th Leeds Lyon Symposium on Tribology, Elsevier, 1987.

239. Robinson, C.L. and Cameron, A., "Studies in Hydrodynamic Trhust Bearings Comparison of Calculated and Measured Performance of Tilting Pads by Means of Interferometry," Phil. Trans. R. Soc. Lond. A, Vol. 278, 1975, pp. 367–384.

240. Rodkiewicz, C.M., Hinds, J.C. and Dayson, C., "Inertia Convection and Dissipation Effects in Thermally Boosted Oil Lubricated Sliding Thrust Bearings," J. Lub. Tech. Trans. ASME, January 1975, pp. 121–129.

241. Rodkiewicz, C.M., Hinds, J.C. and Dayson, C., "The Thermally Boosted Oil Lubricated Sliding Thrust Bearing," J. Lub. Tech. Trans.

ASME, July 1974, pp. 322–328.

242. Rodkiewicz, C.M., Jedruch, W. and Skiepko, J., "Thermal Effects in Conical Bearings, Wear, Vol. 42, 1977, pp. 187–196.

243. Rohde, S.M. and Oh, K. P. A., "Thermohydrodynamic Analysis of a Finite Slider Bearing," J. Lub. Tech. Trans. ASME, July 1975, pp. 450–460.

244. Rohde, S.M. and Ezzat, H.A.A., " A Study of Thermohydrodynamic Squeeze Films," J. Lub. Tech. Trans. ASME, October 1973, pp. 1–9.

245. Rohsenow, W.M. and Choi., H., "Heat, Mass and Momentum Transfer," Prentice-Hall, 1961.

246. Rozeanu, L. and Snarsky, L., "Second Order Thermal Effects in Lubrication," TET, 1980, pp. 95–100.

247. Ruddy, B.L., Parsons, B., Dowson, D. and Economou, P.N., "The Influence of Thermal Distortion and Wear of Piston Ring Grooves upon the Lubrication of Piston Rings in Diesel Engines," TET, 1980, pp. 84–94.

248. Rylander, H., "A Theory of Liquid-solid Hydrodynamic Film Lubrication," ASLE Trans. Vol. 9, 1966, pp. 264–271.

249. Safar, Z., "Thermohydrodynamic Analysis for Laminar Flow Journal Bearings," J. Lub. Tech. Trans. ASME, Vol. 100, October 1978, pp. 510–512.

250. Safar, Z., "Thermal and Inertia Effects in Turbulent Flow Thrust Bearings," J. Mech. Eng. Sci., Vol. 22, No. 1, 1980, pp. 43–45.

251. Safar, Z. and Szeri, A.Z., "Thermohydrodynamic Lubrication in Laminary and Turbulent Regimes," J. Lub. Tech. Trans. ASME, Vol. 96, 1974, pp. 48–56.

252. Safar, Z. and Peeken, H.J., "Thermal and Centrifugal Effects in Misaligned Hydrostatic Thrust Bearings," Institut für Maschinenelemente und Maschinegestaltung, Tech. University Aachen.

253. Seireg, A. and Ezzat, H., "Thermohydrodynamic Phenomena in Fluid Film Lubrication," J. Lub. Tech. Trans. ASME, April 1973, pp. 187–194.

254. Seireg, A. and Doshi, R.C., "Temperature Distribution in the Bush of Journal Bearings During Natural Heating and Cooling," Proc. JSLE-ASLE Int. Lub. Conf., Tokyo, 1975, pp. 194–203.

255. Seireg, A., and Ezzat, H., "Thermohydrodynamic Phenomena in Fluid Film Lubrication," J. Lub. Tech. Trans. ASME, 1973, pp. 187–194.

256. Seireg, A. and Dandage, S., "Empirical Design Procedure for the Thermohydrodynamic Behavior of Journal Bearings," ASME Paper 81-Lub-

19, 1981.

257. Smalley, A.J. and McCallion, "The Influence of Viscosity Variation with Temperature on Journal Bearing Performance," Proc. Inst. Mech. Eng., Vol. 181, Pt. 3B, 1966–67, pp. 55–62.

258. Smith, R.N. and Tichy, J.A., "An Analytical Solution for the Thermal Characteristics of Journal Bearings," J. Lub. Tech. Trans. ASME, Vol. 103, July 1981, pp. 443–452.

259. Snidle, F.W., Parsons, B. and Dowson, D., "A Thermal Hydrodynamic Lubrication Theory for Hydrostatic Extrusion of Low Strength Materials," ASME Paper No. 75-Lub 3, 1975.

260. Snyder, W.T., "Temperature Variations Across Lubricant Film in Hydrodynamic Lubrication," Appl. Sci. Res. Sec. A, Vol. 14, 1964–65, pp. 1–12.

261. So, H. and Shieh, J.A., "The Cooling Effects of Supply Oil on Journal Bearings for Varying Inlet Conditions," Tribology International, Vol. 20, No. 2, April 1987.

262. Sokolov, Y.N., "Oil Film Temperature in Multipad Hydrodynamic Bearings," Machines and Tooling, Vol. 35, Dec. 1964, pp. 26–31.

263. Spikes, H.A., "A Thermodynamic Approach to Viscosity," STLE Trans. Vol. 33, 1990, pp. 140–141.

264. Spikes, H.A. and Guangteng, G., "Properties of Ultra-Thin Lubricating Films Using Wedged Spacer Layer Optical Interferometry," 14th Leeds-Lyon Symposium, "Interface Dynamics", September 1987, pp. 275–279.

265. Sternlicht, B., "Energy and Reynolds Considerations in Thrust Bearing Analysis," Conf. on Lubrication and Wear, IME Publ., London, 1957.

266. Sternlicht, B., Reid, J.C., Jr. and Arwas, E.B., "Performance of Elastic, Centrally Pivoted, Sector, Thrust-Bearing Pads," J. Basic Eng., Trans. ASME, 1961, pp. 169–178.

267. Sternlicht, B., Carter, G.K. and Arwas, E.B., "Adiabatic Analysis of Elastic, Centrally Pivoted, Sector, Thrust-Bearing Pads," ASME Paper No. 60-WA-104, 1960.

268. Strömberg, J., "The Plane Pad Bearing of Infinite Width Considering Variable Viscosity Along as Well as Across the Fluid Film. The Sector Thrust-Bearing Considering Variable Viscosity," Royal Swedish Academy of Engineering Sciences, 1971.

269. Strömberg, J., "Sector Thrust-Bearing Considering Variable Viscosity," Mech. Eng. Ser., 1971, pp. 66–122.

270. Strömberg, J., "Plane Pad Bearing of Infinite Width Considering Variable Viscosity Along as Well as Across the Fluid Film," Mech. Eng. Ser., 1971, pp. 3–59.

271. Strömberg, J., "The Sector Thrust Bearing Considering Variable Viscosity," Chalmers University of Technology, Gotenberg, 1970.

272. Suganami, T. and Szeri, A.Z., "A Thermohydrodynamic Analysis of Journal Bearings," J. Lub. Tech. Trans. ASME, Vol. 101, January 1979, pp. 21–27.

273. Summers-Smith, D., "Thermal Failures of Oil-Lubricated Paired or Double-Row Bearings," TET, 1980, pp. 255–258.

274. Szeri, A.Z., "Some Extensions of the Lubrication Theory of Osborne Reynolds," J. Trib., Trans. ASME, Vol. 109, No. 1, Jan. 1987, pp. 21–36.

275. Tahara, H., "Forced Cooling of a Slider Bearing with Wedge Films," ASME Paper No. 67-Lub-19, 1967.

276. Tahara, H., "Forced Cooling of a Slider Bearing of Infinite Length," Lubr. Eng., Vol. 21, No. 5, 1965, pp. 193–200.

277. Tahara, H., "Heat Transfer of Infinite Length Lubrication Film," J. Japan Soc. Lub. Eng., Vol. 9, No. 5, 1964, pp. 269–377.

278. Tanaka, M., Hori, Y. and Ebinuma, R., "Measurement of the Film Thickness and Temperature profiles in a Tilting Pad Thrust Beating," Proc. of the JSLE International Tribology Conference, Tokyo, 1985, pp. 553–558.

279. Taniguchi, S. and Ettles, C.A., "Thermoelastic Analysis of the Parallel Surface Thrust Waster," Trans. ASLE, Vol. 18, No. 4, 1975, pp. 299–305.

280. Tanner, R.I., "Study of Anisothermal Short Journal Bearings with Non-Newtonian Lubricants," J. Appl. Mech., December 1965, pp. 781–787.

281. Tao, L.N., "On Journal Bearings of Finite Length with Variable Viscosity," J. Appl. Mech., December 1958, pp. 1–5.

282. Tichy, J.A. and Smith, R.N., "The Thermal Behavior of Oscillating Squeeze Films," ASME Paper 81-Lub-26, 1981.

283. Tichy, J.A. and Smith, R.N., "An Analytical Solution for Thermal Behavior of the Step Thrust Bearing," J. Lub. Tech. Trans. ASME, January 1980, pp. 34–40.

284. Tieu, A.K., "Oil-Film Temperature Distribution in an Infinitely Wide Slider Bearing: An Application of the Finite Element Method," J. Mech. Eng. Sci., Vol. 15, No. 4, 1973.

285. Tieu, A., "Research Note: A Three-Dimensional Oil Film Temperature Distribution in Tilting Pad Thrust Bearings," J. Mech. Eng. Sci., Vol. 16, No. 2, 1974, pp. 121–124.

286. Ting, L.L., "Development of a Laser Fluorescene Technique for Measuring Piston Ring Oil Film Thickness," J. Lub. Tech. Trans. ASME, Vol. 102, April 1980, pp. 165–171.

287. Tipei, N., "Flow and Pressure Heat at the Inlet of Narrow Passages, Without Upstream Free Surface," ASME Paper 81-Lub-41, 1973.

288. Tipei, N.A., "Solution of the Thermohydrodynamic Problem for Exponential Lubricating Films," General Motors Corporation, November 1972, pp. 1–28.

289. Tipei, N. and Degueurce, B., "A Solution of the Thermohydrodynamic Problem for Exponential Lubricating Films," ASLE Trans., Vol. 17, 1974, pp. 84–91.

290. Tipei, N. and Nica, A., "On the Field of Temperatures in Lubricating Films," J. Basic Eng., October 1966, pp. 1–9.

291. Tipei, N. and Nica, A., "Investigations on the Operating Conditions of Journal Bearings. Part I: Influence of Viscosity Variation," Revue de Mecanique Appliquee, Vol. 4, No. 4, 1959.

292. Tipei, N. and Nica, A., "Investigations on the Operating Conditions of Journal Bearings. Part II: Global Characteristics in the Case of Variable Viscosity," Revue de Mecanique Appliquee, Vol. 5, 1960, pp. 34–45.

293. Tonnesen, J. and Hansen, P.K., "Some Experiments on the Steady State Characteristics of a Cylindrical Fluid Film Bearing Considering Thermal Effects," J. Lub. Tech. Trans. ASME, Vol. 103, 1982, pp. 107–114.

294. Ustinov, A.N. and Chugunov, A.S. "Investigation of the Mechanism of Heat Transfer Through Piston Rings of Internal Combustion Engines (in Russian)," Energomashinostroenie, No. 3, March 1975, pp. 13–16.

295. van Leeuwen, H., Meijer, H. and Schouten, M., "Elastohydrodynamic Film Thickness and Temperature Measurements in Dynamically Loaded Concentrated Contacts: Eccentric Cam-Flat Follower," TET, 1985, pp. 611–625.

296. van Leeuwen, H., Meijer, H. and Schouten, M., "Elastohydrodynamic Film Thickness and Temperature Measurements in Dynamically Loaded Concentrated Contacts: Eccentric Cam-Flat Follower," Proc. 13th Leeds-Lyon Symposium on Tribology, Elsevier, 1987, pp. 611–625.

297. Vohr, J., Prediction of the Operating Temperature of Thrust Bearings," J. Lub. Tech. Trans. ASME, Vol. 103, January 1981, pp. 97–106.

298. Wachmann, C., Malonoski, S.B. and Vohr, J.H., "Thermal Distortion of Spiral-Grooved, Gas Lubricated Thrust Bearings," J. Lub. Tech. Trans. ASME, Vol. 93, January 1971, pp. 102–112.

299. Walowit, J., "The Stability of Couette Flow Between Rotating Cylinders, etc.," A.I.Ch.E. J., January 1966.

300. Walowit, J. and Pinkus, O., "Performance of Oil Pumping Rings," MTI Report 86TR17, 1986.

301. Walowit, J., Tsao, S. and DiPrima, R.C., "Stability of Flow Between Arbitrarily Spaced Concentric Cylindrical Surfaces, etc." ASME Paper No. 64-APM-30, 1964.

302. Wang, C.Y., "Heat Generated by Couette Flow with Porous Walls," J. Lub. Tech. Trans. ASME, Vol. 95, Oct. 1973, pp. 539–541.

303. Watkins, R.C., "Thermal Effects in Cam Follower Scuffing Wear," Proceedings of the JSLE Int'l. Tribology Conf., July 8–10, 1985, Tokyo.

304. Wierzeholski, K., "Estimation of Solutions of the Navier-Stokes and Energy Equs., etc.," Wear, Vol. 58, 1980, p. 15.

305. Wilcock, D.F. and Booser, E.R., "Thermal Behavior in Tilting Pad Journal Bearings," STLE Trans. Vol. 33, 1990, pp. 247–253.

306. Wilcock, D.F. and Pinkus, O., "Effects of Turbulence and Viscosity Variation on the Dynamic Coefficients of Fluid Film Bearings," J. Trib., Trans. ASME, April 1985, pp. 256–262.

307. Wilson, W.R.D. and Mahdavian, S.M., "A Thermal Reynolds Equation and Its Application in the Analysis of Plasto Hydrodynamic Inlet Zones," ASME Paper No. 73-Lub-35, 1973.

308. Wilson, A.R., "An Experimental Thermal Correction for Predicted Oil Film Thickness in Elastohydrodynamic Contacts," TET, 1980, pp. 179–190.

309. Wilson, W.H. and Thomson, A.G.R., "Heat Flow In and Dissipation From Self-Contained Bearings Illustrating the New ESDU Method of Calculation in Comparison with the Measured Temperature from Actual Bearing Assemblies," TET, 1980, pp. 71–83.

310. Winer, W.O. and Kool, E.H., "Simultaneous Temperature Mapping and Traction Measurements in EHD Contacts," TET, 1980, pp. 191–200.

311. Winer, W.O., "Temperature Effects in EHD Lubricated Contacts," Tribology in the 80's, Vol. 11, NASA Conference Publication 2300, 1984, pp. 533–548.

312. Wing, R.D. and Saunder, A.O., "Oil Film Temperature and Thickness Measurements on the Piston Rings of a Diesel Engine," Proc. Inst. Mech. Engs., 186/172, dl–d9 1972, pp. 1–9.

313. Woolcot, R.G. and Cooke, W.L., "Thermal Aspects of Hydrodynamic Journal Bearing Performance at High Speeds," Proc. Inst. Mech. Eng., Vol. 181, Pt. 30, 1966–67, p. 127.

314. Yadav, J.S. and Kapur, V.K., "Variable Viscosity and Density Effects in Porous Hydrostatic Thrust Bearings," Wear, Vol. 69, 1981, pp. 261–275.

315. Yasuna, J.A. and Hughes, W.F., "A Continuous Boiling Model for Face Seals," ASME Paper 89–Trib–44, October 1989.

316. Yoshida, K., "Effects of Sliding Speed and Temperature on Tribological Behavior with Oils Containing a Polymer Additive or Soot," STLE Trans. Vol. 33, 1990, pp. 221–228.

317. Young, J., "Thermal Wedge Effect in Hydrodynamic Lubrication," The Engineering Journal, 1962, pp. 46–54.

318. Yuan, K. and Chern, B.C., "A Thermal Hydrodynamic Lubrication Analysis for Entrained Film Thickness in Cold Strip Rolling," J. Trib. Trans. ASME, Vol. 112, No. 1, pp. 128–134.

319. Zienkiewicz, O.C., "Temperature Distribution Within Lubricating Films Between Parallel Bearing Surfaces and Its Effect on the Pressures Developed," Conf. on Lubrication and Wear, Paper #71, Inst. Mech. Eng., London, October 1957.

AUTHORS INDEX

Abdel-Latif, L.A., 322
Aggarwal, B.B., 92, 93
Andrisano, A.O., 416, 418, 419
Artiles, A., 210
Ausherman, V.K., 422
Barrett, L.E., 225, 229
Basu, P., 331, 334, 336, 340, 341, 344, 345
Beaman, J., 158, 161–164
Bejan, A., 341, 346–350
Berthe, D., 391, 397, 399, 400, 429, 431
Bill, R.C., 433
Blahey, A.O., 266, 272–277
Boncompain, R., 161, 167–171, 201, 221, 224-226
Booser, E.R., 223, 227, 228
Boyd, J., 39
Branagan, L.A., 224–226
Braun, M.J., 351, 353–355
Bruegemann, H., 264, 267–269, 275
Brugier, D., 306, 307, 311–313
Bupara, S.S., 49
Burgdorfer, A., 18
Burr, A.H., 215, 217–218
Cameron, A., 20, 188–190, 195, 294, 295, 301–305, 407, 409–410
Chao, N.H., 329, 333, 335
Cheng, H.S., 279, 280, 285, 286, 288–292
Conry, T.F., 277, 387, 390, 392–394
Conway-Jones, J.M., 315, 316
Cope, W., 22, 35
Craighead, L., 8, 170
Cuenca, R.M., 93, 95–97
Dill, J.F., 435
Di Pasquantonio, F., 243, 249–251
Dong, Z., 272, 279, 281–284
Doshi, R.C., 385, 387
Dowson, D., 108–111, 127, 129, 130, 270
Elrod, H.G., 18, 236–239

Esfahanian, V., 319, 321, 323, 324

Ettles, C.M., 188–190, 195, 301, 308–310, 373, 378, 379, 381–386

Ezzat, H.A., 137, 141–145, 358, 362–367, 370–377

Feng, N.S., 317

Fillon, N., 309, 314

Frene, J., 161, 167–171, 201, 221, 224–226

Guangteng, G., 408, 411

Hamrock, B.J., 270

Hagg, A., 147

Hahn, E.J., 16, 132, 133, 208–210, 226

Hendricks, R.C., 351, 353–355

Heshmat, H., 26, 30, 195, 197, 198, 208–210, 226

Hopf, G., 242, 244–246, 308

Hori, Y., 195

Huebner, K.H., 253, 256–258, 260

Hudson, J.D., 108–111, 127, 129, 130

Hughes, W.F., 329, 331, 333–336, 340, 341, 344, 345

Hunter, W.B., 103, 106

Jain, P.K., 91

Jeng, M.C., 177, 178, 181, 182, 208

Kahlert, W., 233

Kannel, J.W., 420–423

Kettleborough, C.F., 16, 132, 133, 208, 242

Khonsari, M., 158, 161–164, 319, 321, 323, 324

Kim, K., 180, 184–186

Kingsbury, A., 89

Kollmann, F.G., 264, 267–269

Kool, E.H., 422

Lauer, J.L., 422

Launder, B.E., 234, 243, 247

Leopard, A.J., 315, 316

Leschziner, M., 234, 243, 247

Lumley, J.L., 233

Lund, J., 229

Mallory, M.L., 427, 429

Martin, F.A., 23, 311

McCallion, H., 42, 45, 150, 153, 154, 180

McConnel, B.D., 433, 436, 437

Mitsui, J., 193, 195, 205, 416, 417

Motosh, N., 44, 156–160

Nagaraj, H.S., 422

Ng, C.W., 236–239

Oh, K.P.A., 270, 284, 296, 299, 379

Osterle, F., 233

Ott, H.H., 199, 202, 221, 272

Pan, C.H.T., 236
Paradissiadis, G., 199, 202, 221, 222
Pascal, M.T., 306, 307, 311–313
Pinkus, O., 14, 26, 43, 49–51, 73, 74, 82, 195, 197, 198, 208–210, 226
Raimondi, A.A., 39
Raynor, S., 93, 95–97
Robinson, C.L., 294, 295, 301–305, 407, 409, 410
Rodkiewicz, C.M., 109, 115–119
Rohde, S.M., 137, 141–145, 284, 296–270, 299, 358, 362–367, 370–377, 379
Rylander, H., 320, 321
Safar, Z., 249, 253–255
Saibel, E., 233
Sala, P., 243, 249–251
Schneider, G.E., 266, 272–277
Schueler, D., 242, 244–246, 308
Seireg, A., 385, 387
Shieh, J.A., 204, 206
Shi-zhu, W., 272, 279, 281–284
So, H., 204, 206
Spikes, H.A., 408, 411
Stroemberg, J., 59, 62–71, 130, 135, 136
Tahara, H., 141, 146, 147
Tanaka, M., 195
Ting, L.L., 414–416
Van Leeuwen, H., 404
Vohr, J., 26, 190
Walowit, J., 73, 74
Wilcock, D.F., 43, 51, 223, 227, 228
Wilson, W.R.D., 92, 93
Winer, W.O., 422, 425, 427, 428
Zienkiewicz, O.C., 99–103, 106

SUBJECT INDEX

Acceleration effects, 369
Adiabatic constant, 24, 52
Axial temperatures
   - in EHD contacts, 285
   - in journal bearings, 52, 56, 167
   - in sliders, 139
Biharmonic equation, 289
Bimetallic effects, 11
Cavitation
   - constituents, 318
   - effect on temperature, 227
   - extent of, 54, 164
   - form of, 28, 220
   - gaseous and vapor, 351
   - in pumping rings, 76
Centrifugal force, 357
Compressibility, 18, 85, 137, 322
Concentric journal, 13
Concentration, see Solid fraction
Conduction
   - boundary conditions, 14
   - coefficients, 45, 459
   - dimensionless form, 21
   - in energy equation, 35
   - in lubricants, 20
   - order in magnitude, 22
   - and regimes of operation, 13, 20
   - in seals, 342
   - in solids, 457–459
   - in transverse direction, 83
Continuity equation, 82, 87, 94, 245
Convection
   - in air, 217
   - dimensionless form of, 21
   - in energy equation, 35
   - in journal bearings, 168

- order of magnitude, 22
- in seals, 342
- in thrust bearings, 176
Core flow, 361
Cryogenic fluids, 434
Decompression cooling, 289
Density
- expressions for, 19, 273, 280
- or gas-liquid phase, 326
- of hydrogen, 469
- order of magnitude, 22
- of oxygen, 468
- tables, 456–557
- variation with p and T, 362
Density wedge, 11
Diffusivity, 21, 24, 457
Diffusion equation, 379
Dilatation work, 35, 83
Dividing streamline, 31
Dissipation, 83
Dynamic coefficients, 112, 170, 209, 213
Eckert number, 21, 90
Eddy diffusivity, 235
Edge boundary condition, 14, 137, 188, 218
Elasticity equation, 283, 286
Elastic strain, 388
Elasto-plastic model, 398
Emissivity, 426
Energy equation
- adiabatic, 35, 37
- axisymmetric, 92
- in cavitated region, 220
- compressible, 85, 272
- in Couette approximation, 46, 150
- in cylindrical coordinates, 173
- derivation of, 80
- elliptical, 135
- incompressible, 22, 85
- integral form of, 156, 337
- with mean values, 245, 252
- for moving surfaces, 208
- one-dimensional, 66
- in squeeze films, 359
- in terms of enthalpy, 324
- transient state, 83, 368

- with transverse variations, 83
- for two-phase lubricant, 319
- with turbulence, 236, 240

Enthalpy, 85, 330
Equation of state, 18, 351
Euler number, 21
Eyring model, 387
Finite difference methods, 56, 60
Flow of lubricant, 10
Fluorescence, 413
Fourier number, 21, 375, 376
Free convection, 124
Fusion temperature, 324
Gas lubrication, 15
Groove effects, 203
Heat flux, 123, 155
Heat radiation, 212
Heat transfer
  - coefficients, 124, 156, 216, 338, 375
  - as function of Nu no., 252
  - in turbulent flow, 243
Heat transmission, 121, 123, 156
Hydrostatic films, 92
Hysteresis, 369
Index of starvation, 206
Inertia effects, 233
Initial value problem, 197
Internal energy, 19
Intrinsic energy, 81
Isoviscous solutions, 59, 90, 109
Jaeger integral, 127, 208, 266, 272, 279, 395
Knudesen number, 21
Lapalace equation, 123, 137, 148, 175, 283
Law of the wall, 236
Limits of THD solutions, 151, 180
Mach number, 19
Maximum temperature
  - in gas films, 18
  - in isoviscous solutions, 59
  - in journal bearings, 58
  - and temperature rise, 9
  - in thrust bearings, 58
Mean quantities in films, 235
Metalworking, 261
Negative stiffness, 327, 364

Newton's law of cooling, 382
Nusselt number, 21, 252, 335, 339, 375
Parabolic equation, 218, 368
Parallel surfaces, 11, 20, 87, 90, 96, 294
Partial pressures, 353
Peclet number, 21, 84, 129
Photolitography, 406
Piston rings, 414, 434
Poisson's equation, 364
Poiseuille flow, 93
Polytropic relation, 325
Power dissipation, 10, 164
Prandtl number, 21, 90, 240, 252
Pumping ring, 60
Quality, 328
Radiation coefficient, 213
Ratcheting, 4, 5
Recirculating flow, 30, 72
Reichardt's formula, 235
Relaxation time, 204, 395, 397
Reverse flow, 31, 73, 199
Reynolds number, 21, 157
Reynolds equation
  - incompressible, 37
  - one-dimensional, 66
  - polar, 37
  - with squeeze film effects, 359
  - transient, 367
  - with transverse effects, 87, 143, 173
  - for turbulent conditions, 10, 235
Settling period, 367
Side leakage, 38
Sinh law, 368
Solid fraction, 261, 281, 284, 319
Sommerfeld number, 84
Specific heat, 20, 22, 451, 456
Starvation, 203
Stefan-Boltzmann constant, 213
Stiffness coefficients, 12
Stream function, 109
Surface tension
  - in cavitation, 351
  - in two-phase regime, 322
Temperature
  - average or bulk, 38, 52, 55, 60

- axial, 52, 56, 137, 167
- in cavitated region, 29, 221, 225, 308
- effective, 37, 89
- in reverse flow, 30, 197, 199
- runner, 207
- an thermocouples, 27, 416
Temperature distribution
- axial, 52, 56, 137, 219
- in journal bearings, 54
- in thrust bearings, 61
- transverse, 193, 241
Thermal gradients, 6, 11
Thermal compressibility, 83, 137
Thermal diffusivity, 359
Thermal expansion, 298, 309
Thermal gradients, 11
Thermal inertia, 357, 362
Thermal layer, 191
Thermal parameters, 21
Thermal response, 370, 373
Thermal variations
- causes of, 3
- and computational problems, 13
- effects of, 4, 9, 53
- and experimental difficulties, 25
- nature of, 3
- and turbulence, 11
Thermal wedge, 11
Thermocouples, 27, 416, 418
Time constant, 586
Total derivative, 85, 357
Traction regimes, 386
Transition in turbulence, 240
Transmissivity, 426
Transported energy, 80
Transverse convection, 105, 108
Transverse temperature
- in elasto-plastic model, 399
- measurement of, 416
- in melt lubrication, 351
- in squeeze films, 361
- in turbulent flow, 241
Transverse velocities, 87, 107, 150, 257, 307, 398
Turbulence
- coefficients, 10, 235

- effects of, 11
- equations for, 10
- and dissipation, 240
- and inertia, 233
- and Peclet no., 84
- power loss, 10
- and Prandtl no., 239
- velocity profiles, 239
Vaporization effects, 327
Variable velocity, 368
Viscosity
- average, 94
- effect on stability, 170
- effect on performance, 8
- expressions for, 46
- as functions of p, 262, 264, 267, 272, 280
- as function of quality, 332
- of gases, 16
- mixing inlet, 4
- in non-Newtonian fluids, 31
- in two-phase flow, 317, 338